W9-CDH-627

Praise for *The Last Man Who Knew Everything*

"David Schwartz has written a highly readable account of an undervalued figure in the making of the atomic age—one that puts Enrico Fermi in the proper historical context." **—Gregg Herken, author of *Brotherhood of the Bomb***

"In this compelling and well-researched biography, Schwartz reveals both triumph and tragedy in the life and work of Enrico Fermi, one of the greatest and hitherto most enigmatic scientists of the twentieth century." **—Frank Close, professor of physics, Oxford, and author of *Neutrino* and *Half-Life***

"Enrico Fermi was a singular figure of modern science, and Schwartz has written a singular biography. His book is unusually adept and nuanced in its appreciation and explanation of both the scientific and humanistic aspects of its subject. It is also a joy to read, as Schwartz has a beautiful authorial voice that is perfectly appropriate for his subject matter: appreciative and sympathetic, without falling into the hyperbolic or uncritical. It is a rare book that will please both the experts and the novices, but I think this is such a rare book." **—Alex Wellerstein, assistant professor, Stevens Institute of Technology, and author of *Restricted Data: The Nuclear Secrecy Blog***

"Enrico Fermi was part of a great brain drain pre–World War II from Axis nations, when ideology overwhelmed the search for truth and even self-interest. . . . Despite what you might think from the title, *The Last Man Who Knew Everything*, this amazing book by Schwartz is brimming with anecdotes in which Enrico Fermi is not the smartest guy in the room. He is focused on family, colleagues, and meaning. Schwartz really puts us intimately at the table for the historic atomic revolution. This humanization of geniuses and forging public engagement in complex science is crucial today as we become ever more dependent on technological leadership. As fresh and riveting a biography as any you will find." **—George Church, author of *Regenesis***

"A lucid writer who has done his homework, Schwartz . . . delivers a thoroughly enjoyable, impressively researched account. . . . Never a media darling like Einstein or Oppenheimer, Enrico Fermi (1901–1954) is now barely known to the public, but few scientists would deny that he was among the most brilliant physicists of his century. . . . A rewarding, expert biography of a giant of the golden age of physics." ***—Kirkus***

The Last Man Who Knew Everything

ALSO BY DAVID N. SCHWARTZ

NATO's Nuclear Dilemmas

Ballistic Missile Defense
(co-edited with Ashton B. Carter)

The Last Man Who Knew Everything

THE LIFE AND TIMES OF ENRICO FERMI, FATHER OF THE NUCLEAR AGE

DAVID N. SCHWARTZ

BASIC BOOKS
New York

Basic Books
Hachette Book Group
1290 Avenue of the Americas, New York, NY 10104
www.basicbooks.com

Printed in the United States of America

First Edition: November 2017

Published by Basic Books, an imprint of Perseus Books, LLC, a subsidiary of Hachette Book Group, Inc.

Print book interior design by Trish Wilkinson
Set in 11 point Adobe Caslon Pro

Library of Congress Cataloging-in-Publication Data

Names: Schwartz, David N., 1956– author.
Title: The last man who knew everything : the life and times of Enrico Fermi, father of the nuclear age / David N. Schwartz.
Description: New York : Basic Books, an imprint of Perseus Books, LLC, a subsidiary of Hachette Book Group, Inc., [2017] | Includes bibliographical references and index.
Identifiers: LCCN 2017020558| ISBN 9780465072927 (hardcover) | ISBN 0465072925 (hardcover) | ISBN 9780465093120 (ebook) | ISBN 0465093124 (ebook)
Subjects: LCSH: Fermi, Enrico, 1901–1954. | Physicists—Italy—Biography. | Physicists—United States—Biography. | Nuclear physicists—Italy—Biography. | Nuclear physicists—United States—Biography.
Classification: LCC QC16.F46 S39 2017 | DDC 530.092 [B]—dc23
LC record available at https://lccn.loc.gov/2017020558

LSC-C

10 9 8 7 6 5 4 3 2 1

For Susan, with love, affection, and gratitude

CONTENTS

PART THREE: THE MANHATTAN PROJECT

PART FOUR: THE CHICAGO YEARS

PREFACE

My father was a particle physicist. In 1962, he and two of his colleagues conducted an experiment that demonstrated the existence of two distinct types of "neutrinos," ghostly subatomic particles that can pass through hundreds of millions of miles of lead without bumping into a single atom. Hypothesized in a leap of imagination by the acerbic Viennese physicist Wolfgang Pauli, the neutrino's creation in radioactive processes was first explained by Enrico Fermi, who also gave the particle its Italianate name, meaning "little neutral one." The 1962 experiment—a direct legacy of one of Fermi's most famous scientific achievements—made the front page of the *New York Times* and won my father and his collaborators the 1988 Nobel Prize in Physics.

My connection to Fermi might well have been more direct if it were not for my father's stubbornness as an undergraduate. When he was a senior at Columbia in 1953, my father approached his favorite teacher, Jack Steinberger, and said that he wanted to stay at Columbia for his doctorate. He asked Jack to be his thesis adviser. Quite sensibly, Jack explained that it would be a mistake for my father to stay at Columbia and suggested doing a doctorate at the University of Chicago with Fermi. Jack had been one of Fermi's first postwar graduate students, and the experience had changed Jack's life. With all the callowness youth can muster, my father objected that if he went to Chicago he would lose all the graduate credits he had accumulated as an undergraduate at Columbia.

Jack relented and went on to become my father's thesis adviser and then collaborator on the 1962 neutrino experiment.

So I always knew that Enrico Fermi was an important physicist, at least as far as my household was concerned.

My father passed away in 2006, at the relatively young age of seventy-three. Some seven years later, my mother called with the news that she had finally gone through a file cabinet my father kept in the family garage. She found hundreds of papers and documents stashed away and had no idea what to do with them. I suggested that she send them to me. When they arrived, I went through them and pulled out a series of entertaining letters and papers by a physicist named Valentine Telegdi. Telegdi was a young Fermi colleague in the early 1950s and a close friend of my father. One of the things he had sent my father was a paper he wrote about Fermi's years at the University of Chicago after the war. Telegdi's paper on Fermi, subsequently published in a collection of essays on great professors at Chicago edited by Edward Shils, was an eye-opener. I read with fascination about a physicist of astonishing breadth and depth, someone as adept in experiment as in theory, and a world-class teacher, to boot. Finishing the paper, I decided to get a good recent biography of the great man.

To my amazement, the most recent biography in English (as of summer 2013) was the one written by his first graduate student and later close colleague and friend, Emilio Segrè. It was published in 1970. In the forty-odd years since then, an enormous amount of Nobel Prize–caliber research has extended Fermi's legacy in the physics world. In addition, much has been published in the way of memoirs and historical studies to enhance our understanding of Fermi and his place in the world of physics. It seemed unjust to me that Fermi—one of the most dominant and most interesting figures in twentieth-century physics—was not as well known to the general public as Einstein or Feynman or Oppenheimer, about whom the public seems to have an insatiable interest.

By the late fall of 2013, I decided to write a full-length, general reader biography of the man, encompassing his scientific achievements, his personal life, and his legacy in a way that would bring him to life for a new generation of readers. Basic Books was willing to give me the opportunity to do so. It has been quite a journey.

I am not a physicist and this is not a physics book. This is a book about a man who happened to be an extraordinary physicist and who also led an eventful, dramatic life. The physics is important, of course, but you will not find equations, Feynman diagrams, or the like in this book. You will find what I hope to be straightforward descriptions of his science, accessible to the lay person. For anyone interested in a deeper level of understanding of his achievements, the best source by far is Fermi's own writings, lovingly collected by his colleagues in two volumes published in the early 1960s by University of Chicago Press, *Enrico Fermi: The Collected Papers*. This two-volume set stands as his scientific biography and is accessible to anyone who majored in physics in college. The clarity and simplicity of Fermi's style of science writing are hallmarks of his unique approach.

What you will find in this book, I hope, is a narrative that brings the whole person into focus. It is tempting to say, as did many of his colleagues, that he was "all physics, all the time," and there is an element of truth to this. But he was also a husband, a father, a colleague, and a friend. He played a central role in some of the most important events of the twentieth century. The drama of his life can only be appreciated through an examination of all of these aspects.

Unfortunately, the story cannot be told as directly as a biographer would like. Fermi was prolific in his professional publications but revealed very little of himself or his inner life. Diaries simply do not exist, and personal letters are few and far between and provide little, if any, personal insight. His numerous pocket diaries are filled with physics doodles and brief accounts of his expenditures on various trips. One searches in vain for anything intimate. The biographer is left triangulating among the various source materials available: the memoir written by his wife of their life together, published in 1954, the year he died; the 1970 Segrè biography mentioned above; memoirs and reminiscences of those who studied with him and those who worked with him, both in his native Rome and in his adopted country, the United States. Fortunately, the results of such triangulation provide a fairly consistent portrait. Still, some mysteries about what he did, and why, will probably never be resolved. I have tried to illuminate where I can and note carefully where such illumination is impossible.

Strangely, although my father and I discussed the contribution of many physicists over the years, I cannot recall ever discussing Enrico Fermi with him. To my regret, he did not live to see me undertake this project. In retrospect, he might well have been inspired by Fermi. My father was also a fine experimentalist and an outstanding teacher, able to convey complex ideas in simple, compelling ways. Whenever I asked him to explain something to me, his first words were invariably "David, that's trivial!" And he would make it seem so to me. I am sure he would be surprised that I chose to embark on this project and I hope he would be pleased.

INTRODUCTION

He was born in Rome, Italy, on September 29, 1901, and died in Chicago, Illinois, on November 28, 1954. His life spanned two world wars, and though he was too young to participate in the first, his contribution to the outcome of the second was pivotal and made him world famous. His life also spanned the two major intellectual revolutions of the twentieth century—relativity and quantum theory—and though his contributions to the first were notable, his contributions to the second established him as one of the greatest scientists of his day, indeed, one of the greatest scientists of all time.

To no one's surprise he won a Nobel Prize in 1938.

He made friends easily and inspired passionate loyalty. Those who knew him wept openly when they learned of his premature death at the age of fifty-three. Newspapers around the world carried his death on their front pages, befitting his status as one of the most famous scientists of his day.

His name was Enrico Fermi.

The arc of his life is quickly summarized. Born in Rome at the turn of the century, he was a child prodigy, and by the time he arrived at university he had mastered all of classical physics. Because no one in Italy was teaching relativity or quantum theory, Fermi spent his university education teaching himself these subjects; when he graduated he had already published in professional journals. After graduation he studied briefly in Germany and in Holland, returned to the University of Rome for a period as a lecturer, and then won a position at the University of Florence, where he made his first, and some say most important contribution—a method to bring quantum mechanical rules into the field of statistical mechanics. Two years later he won a competition for a professorship in theoretical physics at the University of Rome under his powerful mentor, Orso Mario Corbino. He built one of the major international schools of modern physics and made several extraordinary contributions: a theory that explains a puzzling type of radioactive process called "beta decay"; the discovery that certain elements, when bombarded with neutrons, become radioactive; and the discovery that the intensity of this induced radioactivity increases when the neutrons are slowed down prior to hitting these elements.

He chose the opportunity of his Nobel Prize in 1938 to leave fascist Italy via Stockholm for a faculty position at Columbia University. Shortly thereafter he learned, to his astonishment and embarrassment, that German scientists replicating his 1934 experiments bombarding uranium with neutrons concluded Fermi had been splitting uranium atoms without knowing it. With this knowledge, he and Hungarian émigré Leo Szilard began to explore the possibility of creating a sustained nuclear chain reaction with uranium. After moving the project to the University of Chicago at the request of the US government, Fermi and a large team of fellow physicists and others succeeded in doing so on December 2, 1942, officially ushering in the nuclear age. He was a central figure in the design of plutonium production reactors for the Manhattan Project and in the summer of 1944 moved to Los Alamos, where the first atomic bombs were designed and built. He played a key role in solving the many theoretical and practical problems involved in this final phase of the Manhattan Project. He witnessed the first detonation of an atomic bomb, known as the Trinity test, at Alamogordo, New Mexico, on July 16, 1945.

After the war, Fermi returned to the University of Chicago, where he continued research in nuclear physics and pioneered high-energy physics experiments on Chicago's new particle accelerator. He spent summers at Los Alamos working on the hydrogen bomb, known as "the Super," and pioneering the use of computers for simulating complex physics problems. He also studied cosmic rays and astrophysics and took on a full teaching load at the University of Chicago, eventually advising a string of future Nobel laureates and many others who went on to brilliant, high-profile careers. During this period, he advised the US government on all aspects of nuclear technology policy and came to the defense of his Manhattan Project colleague J. Robert Oppenheimer during 1954 hearings on the latter's security clearance. Fermi died of stomach cancer in November 1954 at the age of fifty-three, leaving behind an indelible mark on virtually every aspect of physics.

THESE ARE THE FACTS ON WHICH ALL AGREE. SEARCHING FOR A richer portrait, one comes across the inevitable outliers. One author portrays him as a "puerile" prankster consumed by jealousy of his more brilliant student Ettore Majorana. Another paints the picture of the greatest scientist in Western history. The consensus not surprisingly lies somewhere in between.

In his youth he was fond of the occasional juvenile prank, but he matured out of this well before he left Rome for the United States. Far from impeding Majorana's career, Fermi strongly promoted the brilliant introvert's groundbreaking work.

Fermi was certainly an extraordinary physicist, one of his generation's greatest, but to argue that he was history's greatest reflects more the passion he could inspire in those who worked with him than it does his actual place in history.

He had a formidable power of physical intuition and a disciplined, methodical technique that allowed him to crush physics problems in ways that amazed and awed his colleagues. He had a charisma that defies easy analysis—modest, yet fully aware of his superiority over most of the physicists with whom he worked, personally reticent and yet highly gregarious, able to discern the objective and lead others relentlessly toward it, blunt but never nasty, and capable of a self-deprecating wit that immediately put people at ease. No other physicist has ever received such affectionate postdeath tributes. One looks in vain for tributes to other physicists that compare to *To Fermi with Love,* a two-record set of reminiscences by those who worked with him at Argonne Labs outside Chicago, or *The World of Enrico Fermi,* the Canadian Broadcasting Company's lovingly produced documentary of his life and times. Those who worked with him often jostled with each other to secure the mantle of Fermi's legacy.

He could be collegial but was also highly competitive. Reminiscences from his students paint an inconsistent picture. Those from the early days in Rome speak of a man insensitive to the career difficulties of those around him and completely disinterested in their personal travails. Those who studied with him in Chicago universally comment on his generosity of spirit and ability to connect with those around him and attribute their future successes to his fortuitous interventions.

In other words, the picture is a complex one, hardly surprising given that he was a complex individual in a highly complex world.

Some would argue, as did his old friend and colleague Isidor Isaac (I. I.) Rabi during an interview for the CBC documentary, that the only thing interesting about Fermi was his science, that beyond the science the details of his life are trivial and not worth exploring. Although this is a common view among scientists, who tend to view scientific achievement as distinct from the individual who achieved it, it misses the point. The circumstances of Fermi's life determined much of what he achieved, and had the chips fallen some other way, his career—and our world—would be different. If Laura had agreed to move to the United States when Fermi first wanted to in 1930, how different would have been the trajectory of his science? We can imagine his coming to the same conclusions regarding beta decay, but would his work on slow neutrons have proceeded the same way with a different, American team? Would that team have discovered fission in 1934, with the benefit of better (or luckier) radiochemists? One can imagine the arc of Fermi's research altering considerably with an earlier immigration to the States, with unpredictable results. Perhaps he would have delved into high-energy particle physics earlier, although the accelerators available in the 1930s did not have the energy to explore the subatomic world that became the focus of Fermi's postwar research in Chicago. Even as late as 1939, much of his research agenda seems accidental, particularly the odd set of circumstances that threw Fermi and Szilard together in an historic partnership beginning in early 1939. If in January 1939 Fermi had shown up in Ann Arbor instead of Manhattan, would Szilard have sought him out? Would Fermi have been a central player in the experiments leading to the first chain reaction?

All these questions are essentially imponderables. Still, the very fact that they are imponderable leads to the conclusion that, in common with all scientists of great stature, the specific circumstances of Fermi's life had an enormous impact on his scientific career. So though one doesn't need to know much about Fermi's personal life to study his specific scientific achievements—it is possible, for example, to read the beta decay paper without having any insight whatsoever into the circumstances of its creation—it is incorrect to conclude that an understanding of Fermi's life is irrelevant to our understanding of his life as a scientist. It is, indeed, essential to grasp the relationship of circumstance to scientific creativity and achievement, to comprehend how history, personality, and circumstance combine to shape the development of any particular scientific achievement. In another context, the British historian of science Charles Percy (C. P.) Snow put it succinctly when he wrote, "If Fermi had been born a few years earlier, one could well imagine him discovering Rutherford's atomic nucleus, and then developing Bohr's theory of the hydrogen atom. If this sounds like hyperbole, anything about Fermi is likely to sound like hyperbole." Snow was alluding to Fermi's brilliance, but this assessment underscores the importance of circumstance in the development of a scientific career. Yes, Fermi may have discovered the atomic nucleus, he may have thought of the Bohr model of the atom, but he never had the opportunity to do so because he was born two decades too late. We are all are prisoners of the era into which we are born, scientists being no exception.

THE RECEIVED NARRATIVE OF HIS LIFE RINGS TRUE, BUT IT OBSCURES as well as illuminates. Why is there such a difference between the memories of his Italian students and those of his American students, particularly in regard to Fermi's willingness to encourage them and to promote their careers? Why did he remain so long in Italy, working under a thuggish, essentially evil regime? Was there a part of him that grudgingly supported the fascist dictatorship? Did he really decide to come to the United States only when Mussolini promulgated the anti-Semitic laws that would have targeted his wife? Was he an enthusiastic participant in the Manhattan Project, as so many of the histories of the period suggest? Or was he perhaps somewhat reluctantly pulled along by events out of his control? In October 1949, he was an outspoken opponent of the development of the hydrogen bomb, but by the summer of 1950 he was working intensively on the project. Why? Any biographer must grapple with these questions, even if there are, in the end, no clear-cut answers.

HIS FAME DIMINISHES, EVEN AS HIS LEGACY GROWS.

Perhaps the most enduring of his discoveries as far as the general public is concerned are those relating to his work on the atomic bomb, for which he earned the sobriquet "father of the nuclear age." Even this, though, raises some important questions. The history of the Manhattan Project is the history of many thousands of scientists, engineers, soldiers, and others who had a hand in bringing about the development of nuclear weapons. Where the epithet most clearly fits is in his role in the development of the first nuclear reactors, devices that demonstrated the possibility of nuclear fission chain reactions that form the basis for nuclear explosions and that served as the production engines for plutonium, the element that formed the core of one of the two atomic bombs that ended World War II. These reactors were built in great haste, under enormous pressure, largely without well-formed engineering plans. Indeed, the first one, at the University of Chicago, emerged more or less fully formed directly from Fermi's brain. In retrospect, the amazing fact is that they worked as anticipated and that the effort was scalable to a degree that astonishes engineers even today.

His role in the development of the atomic bomb itself is more difficult to assess, being that of a highly valued adviser rather than an architect or designer. In the traditional narratives he is overshadowed by the scientific director of the project, J. Robert Oppenheimer, a physicist as different from Fermi as it is possible to be. But when in the summer of 1944 the work at Los Alamos came to a grinding halt owing to unforeseen technical problems, it was to Fermi that Oppenheimer appealed, asking him to come to the secret city on a New Mexican mesa to help inspire and lead. Fermi did so effectively and without complaint.

For physicists, several of his other achievements rank far higher than those of the Manhattan Project, if not existentially then certainly scientifically. His success in integrating quantum rules into statistical mechanics, in what we now call Fermi-Dirac statistics, is the basis for virtually all condensed matter physics and much else besides. Fermi-Dirac statistics are, if anything, even more useful today than they were when they were proposed in 1926. His 1933 theory of beta radiation, though not considered precisely accurate today, gave rise to an enormous amount of fascinating research in particle physics, resulting in more than a dozen Nobel Prizes to date. After World War II his experimental work in high-energy particle physics helped to lay the groundwork for the quark theory of matter and the Standard Model of particle physics, producing another string of Nobel Prizes. Alone among his true peers, his expertise extended across both theory and experiment, a significant anomaly among world-class physicists. And though he may have had a handful of peers in either theory (Paul Dirac, Werner Heisenberg, Wolfgang Pauli) or experiment (Arthur Compton, James Franck, I. I. Rabi), in the art of teaching, he had none. Some five of his graduate students went on to win Nobel Prizes and several other future Nobel Prize winners thought of him as their primary graduate or postgraduate mentor. In terms of influence as a teacher and mentor, he was truly unique.

IT IS THIS COMBINATION OF LASTING SCIENTIFIC ACHIEVEMENT AND profound influence on several generations of physicists, in the United States and in Italy, that make his story directly relevant to us today. Underlying these achievements was a foundation built on enormous talent, but equally important, on a disciplined, almost terrifyingly comprehensive effort as a young man to teach himself all of known physics. During the period in which he laid down this awe-inspiring foundation, he also developed a unique way of thinking about problems that allowed him to achieve what he did and to inspire those around him. He learned at an early age how to strip a problem to its essentials and structure the solution in a straightforward manner, invariably starting at the right place and avoiding complications that might bedevil others. He used this technique in a wide variety of settings, notably in solving problems that now bear his name. "Fermi problems" can be simplified into a finite set of variables whose values can be estimated to within an order of magnitude. Linking those variables together not only provides a quick, rough-and-ready solution but also forces one to think about the elements of the problem that are essential and those that can be safely discarded.

Fermi problems often have at their core estimates of the probability of one event or another occurring. This was the type of problem that Fermi excelled in solving, in part because during that formative period of his intellectual development he mastered probability and statistics as a central part of his scientific repertoire. Calculations of probabilities run like a bright thread throughout his work and at several crucial points in his career provided a focus for his most important breakthroughs—the Fermi-Dirac statistics, for example, or his later fascination with Monte Carlo simulations. This way of thinking, which he passed along to colleagues and students, is one of his greatest legacies. Recruiters at firms like McKinsey and Goldman Sachs pepper potential hires with Fermi problems to see how they think and probably never realize the debt they owe to this giant of modern physics.

Fermi's ability to grind out solutions to difficult problems using a well-developed toolkit of techniques was paired with an extraordinary sense of what problems were important and an affinity for the quick-and-dirty solution when appropriate to his needs. The former set him apart from most of his contemporaries and ensured that he would be at the forefront of his field throughout his career. The latter was sometimes misinterpreted as laziness or, worse yet, a fear of complex mathematics. He was neither lazy nor afraid—he had enormous reserves of energy that drove him to work longer and harder than many younger colleagues, and he was a fine mathematician, able to hold his own with geniuses like John von Neumann—but he valued his time and chose to work only hard enough to get a practical solution. He once quipped to his daughter, Nella, "Never make something more accurate than necessary." Offered in the context of some unattractive but functional carpentry for the living room of the family home, it was a philosophy that also guided him in physics.

HE MAY HAVE BEEN A WORLD-CLASS PHYSICIST, BUT HE WAS NOT A world-class family man.

As a husband he could be a frustrating and sometimes infuriating person to live with, as his wife Laura makes clear in her largely affectionate but sometimes arch account of their marriage, *Atoms in the Family*, published in 1954, just prior to his death. The incessant teasing, the long periods—sometimes months—spent away from her, his unwillingness to take her into his confidence during his work on the Manhattan Project all took their toll. Physics was the most important thing in his life and everything else took second place. Laura knew this when she married him, of course, and if she had any illusions to the contrary, they were shattered the afternoon during their honeymoon when he insisted on teaching her Maxwell's equations. There is no doubt, however, that they loved each other and eventually accommodated themselves to each other's idiosyncrasies, as most successful couples do.

He was also not the best of fathers. He helped Laura little in domestic matters relating to raising their children, and Laura seems not to have expected anything different. His daughter, Nella, had great affection for him, although even she admits he could be distant. His son, Giulio, chafed at living in his father's shadow and ultimately put as much distance as possible between himself and his family legacy. Whether Fermi was any worse a father than other successful, driven men of the time is an open question. Parenting in the 1940s and 1950s was not the art it is today, and the profile we have of Fermi as a father is not substantially different from the profile of many others at that time. In his final years he took his parenting role a bit more seriously, but by then much of the damage had already been done. It was, to say the least, difficult being a child of Enrico Fermi.

THAT HE IS NOT BETTER KNOWN, THAT PHYSICISTS LIKE RICHARD Feynman and Stephen Hawking are more well known to the general public, may simply reflect the circumstances of his death in 1954 at an early age and before the widespread advent of television. Few films of his lectures survive, and his television appearances were rare. In later years he may have resented the adulation of Einstein, but in response he did little in the way of self-promotion. Not that he had any sense of false modesty. One of Fermi's University of Chicago colleagues reported being told of the following conversation between Fermi and his brilliant but troubled graduate student Majorana:

> MAJORANA: There are scientists who "happen" only once in every five hundred years, like Archimedes or Newton. And there are scientists who happen only once or twice in a century, like Einstein or Bohr.

FERMI: But where do I come in, Majorana?
MAJORANA: Be reasonable, Enrico! I am not talking about you or me. I am talking about Einstein and Bohr.

He had an enormous confidence in his own abilities, confidence that was well placed. Perhaps he felt that self-promotion was a bit unseemly. Though he sometimes loved to show off to his colleagues, deriving complex theorems from scratch when he could just as easily have looked them up, he did not have the compulsion, notable in some less secure geniuses, to always make sure that everyone knew he was the smartest guy in the room. His lack of self-promotion, particularly after World War II, may also have resulted from a belief that his proper focus was on his research and teaching and not on his public profile. To the extent that he was a public figure, it was as a private adviser to public bodies such as the US Atomic Energy Commission. He was one of the first major scientists involved in public affairs, but his advice was highly classified and remained so for years after his death. His involvement in these matters was fraught, involving him in controversies that pitted friend against friend. When Oppenheimer asked Fermi to continue advising the Atomic Energy Commission in 1951, Fermi demurred, convinced that he was better suited to a world in which truth was clear and where opinion mattered little.

Another reason for his relative lack of profile may have been his resistance to pronouncements on broader political or philosophical matters. After World War II many physicists raised their voices in public on issues related to the development of nuclear weapons and the accelerating arms race between the United States and the Soviet Union. Fermi rarely spoke publicly about such matters, reserving most of his commentary, such as it was, for private councils of state. Nor was he deeply philosophical, never pondering the underlying reality behind quantum theory, never engaging in the kind of metaphysical debate that characterized the schools of quantum theory that arose in Copenhagen and Göttingen. He confined his intellect to matters of physical reality and to physical problems that could be solved using the physics he so loved. Perhaps for this reason quantum theorist Wolfgang Pauli once quipped, in his typical, acid-tongued way, that Fermi was a mere "quantum engineer." The general public is fascinated by the philosophical conundrums of quantum theory, but Fermi chose to ignore these in favor of focusing on real physics problems with solutions that could be developed clearly through theory and experiment. He was at core an empiricist, driven by the empirical observations of the world around him. Yet Fermi's antiphilosophical orientation was shared by some, like Feynman, whose fame has only grown over time. Feynman's wit, colorful story-telling, and involvement in the *Challenger* shuttle investigation have contributed to the mystique that still surrounds him. Biographers continue to find him a fascinating subject, as they do Einstein and Oppenheimer. In the case of Fermi, some forty-six years have passed between English-language biographies.

Fermi's reputation with the public has faded for all these reasons and perhaps others as well. As others have recently realized, a correction is long overdue.

No one really knows how genius works, why it manifests in some instances and not in others. Even more perplexing, why do some scientists make great discoveries and others do not? Why do some, like Einstein, Wolfgang Pauli, Marie Curie, John Bardeen, or Fermi himself, make several great discoveries, while other scientists with apparently equal talent, like Oppenheimer, have no such discoveries to their name? Fermi used a well-defined set of techniques to attack problems in physics, and if a problem came up that did not seem to fit any of these techniques, he bent the problem to succumb to them. There were, of course, hundreds of physicists who knew these techniques, but almost none of them were able to apply them with Fermi's effectiveness. At some point early in his career, working almost entirely on his own, Fermi learned how to look at a problem carefully, find the right starting point to address it, and cut through it steadily with an enormously powerful intellect, avoiding all the false starts and dead ends that might trap less talented thinkers, to get at its solution. How he did this, how Fermi became Fermi, is in some sense perhaps an irreducible mystery. Nevertheless, it is one worth trying to solve, because understanding how a young boy from Rome became, in the words of one of his former students, "the last man who knew everything" can help us to appreciate the full potential of the human mind and spirit.

PART ONE

BECOMING FERMI

CHAPTER ONE

PRODIGY

On the streets of Rome, not far from the central train station, a middle-aged engineer meets an adolescent boy, the son of a colleague. The boy wants to talk about mathematics and science. The engineer quickly realizes that he is dealing with someone who has a profound gift for the subject matter, a sponge who absorbs complex ideas faster and more thoroughly than seems possible. The engineer decides to take the lad under his wing and give him a thorough education in mathematics and physics, one that goes far beyond what is available in the boy's high school.

It is impossible to know what might have happened if thirteen-year-old Enrico Fermi had not met his father's friend and colleague Adolfo Amidei during the summer of 1914 and if Amidei had not taken a deep and sustained interest in the adolescent and his scientific education. How many promising intellects have withered on the vine because no one was nearby to cultivate them? What we do know is that Amidei decided to give young Enrico Fermi an undergraduate education in mathematics and physics, thus beginning the transformation of a teenage Roman boy into a master physicist.

Enrico Fermi's father, Alberto, arrived in Rome in the 1880s in pursuit of his career at the Italian Ministry of Railroads. He was born in north-central Italy near the town of Piacenza, nestled in the fertile

Po Valley some forty miles southeast of Milan and twenty-five miles due west of Cremona, the famous home to the great violin makers of the seventeenth century Antonio Stradivari and Andrea Guarneri. The Fermis had worked the land in the region of Piacenza for centuries, but Alberto's father, Stefano, was ambitious and found himself an administrative job with a local nobleman, the Duke of Parma. Alberto, the second of Stefano's children, did well in the local high school, but the family did not have the financial wherewithal to send him to university. He was, however, quite bright and, like his father, ambitious. The combination enabled him to land a job with the Italian railroads.

The late nineteenth century found Italy in the midst of a belated but intense period of industrialization. Unification of the country in 1870 set off a national effort to catch up with the northern European industrial powerhouses. One result was rapid urbanization. Rome, a sleepy midsized city of about 150,000 in 1849, when the founder of modern Italy, Giuseppe Garibaldi, first attempted to bring it into a unified Italian state, grew to slightly over 225,000 people at the time of unification in 1870. By 1901 it had doubled in size to some 460,000 residents. Any such dramatic increase in an urban population places extraordinary demands on a city's infrastructure. In the case of Rome, entire new residential sections were thrown up virtually overnight, fundamentally changing the look and feel of the city, particularly in the elevated eastern section, perched above the Quirinal, Viminal, and Esquiline Hills.

Another result was the rapid development of a national railroad system that, though predominantly focused on the more prosperous northern communities of Milan, Genoa, and Turin, provided both employment and a certain cultural cachet for Roman workers, as well. Italians were, and remain today, proud of their train system. Because it was developed somewhat later than the English and French railway systems, Italy's railroad benefited from technological advances unavailable during the earlier efforts of its industrial neighbors to the north. It provided employment for laborers and for the better educated technological elite and quickly moved through a series of regional mergers to centralized national ownership. The Ministry of Railroads was a prestigious place to work, and Alberto could be legitimately proud of his career.

In the early years his work had moved him from town to town, and he ended up in Rome around 1888. Quickly recognizing his abilities, his employers promoted him first to accountant, then to inspector, and eventually to *Capo Divisione,* a fairly high level in the Italian civil service roughly equivalent to a brigadier general in the Italian military. By the time Enrico was born, in 1901, Alberto had even been named a *Cavaliere,* or knight. Though not quite as prestigious as its British counterpart, the title reflected the value his employers placed on his skills and performance. More titles were to follow.

In surviving photographs, Alberto appears reserved and even a bit aloof. Attractive but intense eyes peer out from behind wire-rimmed glasses, set above hollow cheeks and a mouth obscured by a generous, and presumably fashionable, handlebar moustache. Setting aside the complete lack of humor in his eyes, one could imagine him singing Verdi arias while shaving, a penchant commented upon by Emilio Segrè, the Italian physicist. He clearly had ability and ambition and climbed the bureaucratic ladder at the ministry with little trouble.

Along the way, he met a woman named Ida de Gattis. Ida came from Bari, just above the "heel" of the Italian boot in the southeast of the country. She was a schoolteacher and younger than Alberto by some thirteen years. Trim and attractive, with delicate features and soulful, gentle, timid eyes, she caught Alberto's attention, and they married in 1898. The newlyweds moved to an apartment not far from Rome's central train station, Termini, at Via Gaeta 19.

The neighborhood just northwest of Termini was designed to accommodate the influx of workers to the new Italian capital in the most efficient and direct way possible. Streets formed a grid pattern and were lined with small and relatively undistinguished apartment buildings painted in a variety of Mediterranean pastel colors. Via Gaeta 19 stands today, a five-story ochre edifice with two apartments on each floor and a plaque indicating its illustrious pedigree as Enrico Fermi's birthplace. Do Not Disturb signs hang above the doorbells of the two apartments on the third floor, presumably to ward off pilgrims to Enrico Fermi's first home.

Here, Ida gave birth to three children: Maria in 1899, Giulio in 1900, and Enrico in 1901. Perhaps overwhelmed by the arrival of three infants in quick succession, or perhaps because as the wife of a secure and increasingly prosperous civil servant she felt she had the resources and social position to do so, Ida Fermi packed Giulio and Enrico off

FIGURE 1.1. Giulio, Enrico, and Maria Fermi, 1904.
Courtesy of Rachel Fermi.

with nannies to the countryside almost as soon as each was born, a practice common among wealthier Romans. Maria stayed home under Ida's direct care. We do not know whether the brothers were kept together, nor do we know when Giulio arrived home, but it seems to have been before Enrico. Enrico's wife, Laura, later wrote that, owing to "delicate health," Enrico was kept in the countryside for two and a half years. At that point he was returned to Rome and, confronting his real family for the first time, he burst into tears of fright. Ida delivered a stern scolding. Crying was forbidden in her household. He quieted down, and from that time on, Laura speculates, he was responsive to strict authority.

Giulio and Enrico were inseparable. Enrico was somewhat shy and awkward, and Giulio evidently decided that he would be Enrico's constant companion and protector. The relationship is apparent in one of the first portraits of all three children, taken in 1904. Maria stands at the right side of the frame, strong and proud, while the brothers hold hands at the left side of the frame, Giulio offering protection to his timid younger brother.

It was not long before the boys became interested in science. Perhaps they were inspired by Guglielmo Marconi, the inventor of radio, who, in the first decades of the twentieth century, was grabbing headlines around the world with his experiments in electricity. Although much of his work was done in Britain, Marconi had become a hero in his native Italy. In the process, he served as an inspiration to countless Italian children, and it would have been strange indeed if the brothers did not come to worship him. As they grew older the two boys became obsessed with electricity, engineering, and everything related. They would come home from school and spend their free time designing and making electric motors, mechanical gadgets, and other devices. Like many boys of their age, they were also inspired by the dawn of the age of powered flight and reportedly designed an aircraft engine that impressed experts. They ate, drank, and slept science and technology.

The boys were apparently equally bright, but there the similarity ended. As a child Enrico was quiet and withdrawn, always most comfortable in the presence of his outgoing, more socially adept older brother. When very young he had a bit of a temper, which earned him the family nickname "little match," but he eventually learned to control that. Later in life Enrico developed a robust, sturdy, even gregarious presence, but as a child he was distinctly delicate and introverted, perhaps taking after his mother. Giulio had greater physical presence, not only because he was a year older but because he cut a more social, more verbal, outgoing figure, and initially he was a better student. He was also, according to the accepted narrative, his mother's favorite. Enrico was a slow starter, physically awkward, not particularly articulate, and incapable of making a good impression during his first years of school. This began to change, though, and somewhere along the way Enrico developed an affinity for mathematics. Although the main fare of Italian elementary education at the time was classical studies, Enrico began to impress teachers and became one of the top students in his class. The brothers spent time after school supplementing their somewhat mediocre technical education at a nondescript *liceo* (middle and high school) by reading science books and

FIGURE 1.2. Drawing of Giulio by Enrico, 1914. *Photo by Susan Schwartz. Courtesy of Department of Physics library, University of Pisa.*

magazines, keeping themselves abreast of a rapidly scientific world. They may not have understood all that was going on around them, but they must have been aware of developments and excited by them.

In the archives of the University of Pisa's physics department resides a drawing that Fermi made of his brother, dated June 20, 1914. Enrico was no artist, but he enjoyed doodling and sketching, and the profile he drafted of his brother, with obvious care, affection, and attention to detail, demonstrates better than perhaps anything else his feelings toward Giulio.

Less clear is the relationship that Enrico had during early childhood with his older sister, Maria. It may be that the two brothers created a hermetically sealed bubble around themselves, with little room for anyone else. It may also be that, from a cultural point of view, girls were neither expected nor encouraged to have scientific or technical interests, so Maria would probably have not been responsive even if the brothers had tried to engage her. We know that sometime later Enrico tried to talk to Maria about a physics book he picked up at a used bookstall, but she paid little attention. At this point in his life, his relationship with his sister seems to have been distant, particularly in contrast with the relationship with Giulio.

In 1908 the family moved from Via Gaeta to a larger apartment building a few blocks away in Via Principe Umberto. They had more space, but still no hot water. It was home for Enrico until he left for university in 1918.

Many photographs of the Fermi children have come down to us from this period, suggesting the Fermi family's relatively comfortable financial circumstances. Alberto's income was such that photographic portraits and even some candid photography were regular features of family life. In many of these photographs, Giulio and Maria have bright, engaged expressions, while Enrico usually looks dreamy and distracted, reserved, and somewhat uncomfortable. Ida has a slightly nervous, anxious look about her. Alberto is generally absent.

On January 12, 1915, Giulio died.

The boy developed an abscess in his throat, and Ida and Maria accompanied Giulio to the hospital to have it removed surgically. Doctors assured Ida that the operation was routine, an outpatient procedure, and there was nothing to be worried about. As Laura Fermi describes it:

> On the appointed morning Mrs. Fermi and Maria accompanied him to the hospital and set themselves to wait quietly in the hall. Suddenly there was a great commotion. Nurses rushed into the hall, saying aimlessly, "Don't worry; you should not worry." Their tone was strained. The surgeon came. He asked the women to keep calm. He could not explain, he could not quite understand himself what had happened. The boy had died before the anesthesia was completed.

She concludes by observing, "The blow could not have been heavier, nor the family less prepared to receive it."

Ever since returning home from his first two years in the countryside, Fermi and his brother had been inseparable. Enrico had relied heavily on his more outgoing sibling. Their bond was probably closer than any Enrico established later in life. He was devastated, but characteristically would not show it. It was the first of many incidents in which he forced himself to hide his emotions from the outside world, even from those closest to him. He did, however, resolve after several weeks to walk by the hospital where Giulio died, to confront the tragedy head-on to prove that he could deal with his grief.

Many years later he named his son after the brother he lost.

His mother's grief was more visible, more sustained, and more debilitating. For years afterward she would break into tears suddenly, without provocation, and she spent long periods in a state of depression. She gradually withdrew from the family and died in the spring of 1924 at the age of fifty-three, an end certainly hastened by Giulio's passing.

What might have happened had Giulio lived into adulthood? Knowing what Enrico would eventually accomplish, it is almost inconceivable that Enrico was the less talented of the two. Perhaps, incredibly, Giulio would have continued to outshine his little brother. Or maybe little Enrico would have eventually eclipsed the more socially adept Giulio. Maybe their interests would have diverged at a certain point, with one going into a field other than physics. Even more satisfying to contemplate, perhaps they might have worked together in brilliant harmony throughout their lives. In the end, however, we are left with unsatisfying, unanswerable questions.

AT ABOUT THIS TIME, TWO IMPORTANT PEOPLE ENTERED FERMI'S LIFE.

One was a classmate of Giulio's named Enrico Persico. Persico had observed the brothers from afar for some time and had concluded that there was no room for a third wheel in the relationship. With Giulio gone, Persico reached out to the younger brother. His efforts were greeted enthusiastically. The two Enricos shared a love of science and technology, and soon they picked up where the two brothers had left off, absorbing science and math books, concocting experiments, and hanging out together whenever possible.

Persico was not much taller than Fermi, but he had the face of a tall man, dramatically elongated with a prominent aquiline nose, an almost nonexistent forehead, and eyes that sparkled with intelligence and good humor. He became something of a fixture around the Fermi household.

Persico soon realized that his new friend was exceptional. Years later he wrote about these early days of their friendship:

> We formed the habit of taking long walks together, crossing the city of Rome from one side to the other, discussing all kinds of subjects with the brashness of youth. But in these adolescent talks Enrico brought a precision of ideas, a self-assurance, and an originality which continually surprised me. Furthermore, in mathematics and in physics he showed a knowledge of many subjects well beyond what was taught at school. He knew these topics not in a scholastic fashion, but in such a way that he could use them with extreme facility and familiarity. For him, even at this time, to know a theorem or a law meant chiefly to know how to use it.

On these walks, the two Enricos would sometimes go hunting for books to satisfy their hunger for science, a quest that brought them from the grid streets of new Rome to the ancient, meandering alleys and passageways of the historic center of the city, off the Corso Vittorio Emanuele II. A left turn down Via del Paradiso, a particularly narrow alley, led them eventually to a large, ancient square, the Campo de' Fiori, where every Wednesday a used book market attracted visitors in search of the odd special volume unavailable at regular bookstores. The boys scoured the stalls for items of interest. One particular edition they found was a nineteenth-century physics text by a Jesuit priest named Andrea Caraffa. A two-volume summary of all that was known about classical physics in 1840, the year of its publication, the boys quickly snapped it up, and Fermi absorbed it enthusiastically. As reported by Laura Fermi, he couldn't stop talking about it to his sister, Maria, as he read it. Indeed, so enthralled was he that he barely noticed it had been written in Latin.

The two boys conducted a variety of experiments, some of them fairly sophisticated given the boys' ages and education. They measured the density of Roman tap water. They calculated the strength of the gravitational and magnetic fields in Rome. They even tried to explain the behavior of a spinning top, even though neither knew the mathematics and physics that would have made this difficult problem a bit easier.

Particularly impressive, in retrospect, was Fermi's knowledge of and enthusiasm for the theory of relativity. In later years Persico recalled extended conversations in which his young friend would enthuse about Einstein's radical theory of gravity. Persico served as Fermi's first "student," if one discounts Enrico's failed earlier attempts to interest Maria in the Caraffa book. Fermi enjoyed explaining ideas to his new friend, and Persico was a ready pupil, eagerly absorbing Fermi's explanations of fairly complex concepts in physics. This was the first time that Enrico discovered his knack for conveying physics in ways that others less gifted could clearly understand, a knack that was to figure prominently in future years.

With Persico, Fermi was educating himself with the Caraffa volumes and other scientific books, and at *liceo,* he was exposed to Latin, Greek, history, and Italian literature. Fermi particularly enjoyed the traditional Italian epic poem "Orlando Furioso" (the Italian version of the "Song of Roland") and in later years would impress friends with recitations of Dante from memory. Languages came easily to him, which may explain why he was hardly aware that the Caraffa volumes were written in Latin. In these subjects, he was a good student, but not extraordinary. It is clear that any advanced mathematics and physics he absorbed had nothing at all to do with what he was learning in school.

Fermi and Persico separated in 1918 when Fermi headed to Pisa for university and Persico remained in Rome. Over the next four years, they stayed in close touch by letter and saw each other often during school holidays. When Fermi eventually returned to Rome for good, they resumed their deep and lasting friendship.

Enrico had fun with other youngsters, as well, in the process developing a love of outdoor physical activity that would continue unabated throughout his life. He loved to play soccer and the perennial Italian schoolboy favorite French War, described by Laura Fermi as an Italian form of the Cops and Robbers game so popular in America, with just a hint of youthful nationalism. Yet he never drew personally close to these playmates, saving his friendship and affection for the other Enrico.

While the two Enricos' bond strengthened as they explored the world of physics, another person entered Fermi's life, an older man who knew a great deal more than either of the boys about mathematics and physics. This man would become central to Fermi's early intellectual development.

Adolfo Amidei was an engineer for the railway company, with a rank of chief inspector—one rung above Alberto Fermi. In spite of the difference in rank, the two men became friends, and during the summer of 1914 they began to walk home together after work, which suggests that they lived close to one another.

Amidei was from Volterra, about fifty miles south of Pisa. Seven years younger than Alberto Fermi, he showed early technical proficiency and was admitted to the pure mathematics program at the University of Pisa, where he eventually broadened his studies to include physics. He joined the regional railway system as an engineer and junior inspector, and when various regional companies merged to create a national railway company, he made the move and was soon promoted to full inspector. By the time Amidei met young Enrico, he was a principal inspector. He was promoted repeatedly throughout his long and successful career and retired in 1940 with the title of *Capo Compagnia 1ª Classe*. Along the way, like Alberto, he was named a *Cavaliere* by the Italian government.

Sometime during the summer of 1914, as Europe edged toward catastrophic war, Enrico began to meet his father after work and accompanied him on the forty-minute walk from the ministry to their apartment on Via Principe Umberto. Amidei occasionally joined the father and son, and soon Enrico learned that his father's colleague was an engineer with a strong mathematics and physics background. He summoned up the courage to ask Amidei a particularly abstruse question: "Is it true that there is a branch of geometry in which important geometric properties are found without making use of the notion of measure?"

Amidei explained to the youth that this branch of mathematics was known as projective geometry. The idea puzzled Fermi, and he asked, "But how can such properties be used in practice—for example, by surveyors or engineers?" Amidei recalled that in Pisa he had studied a book on projective geometry by a German mathematician named Theodor Reye that had an excellent introduction explaining the practical uses of the discipline. He lent Enrico the book. Two months later Enrico revealed he had mastered the material, having worked through all of the theorems and solved all the problems at the back of the book. Amidei was understandably skeptical because the book had been difficult for him as a university student and, as a result, he had never completed the proofs himself. When Fermi gave Amidei the proofs for the theorems, the older man's doubts vanished.

Enrico Fermi was thirteen years old.

This anecdote about Fermi's grasp of projective geometry is the first example we have of what would become Fermi's typical style of learning. He studied Reyes's book by himself (or if it was with someone, it was with Giulio, who had not yet died) and, not satisfied with a cursory reading, worked through the proof of every theorem in the volume until he mastered the entire text. At this young age he displayed independence, thoroughness, and a willingness to grind through difficult material to ensure he mastered it. In this, as in all future work, he was never satisfied with a superficial grasp of a subject.

Amidei was suitably impressed and later wrote, "I became convinced that Enrico was truly a prodigy, at least with respect to geometry. I expressed this opinion to Enrico's father, and his reply was: Yes, at school his son was a good student, but none of his professors had realized that the boy was a prodigy."

This may not strictly be true. Fermi scholar Roberto Vergara Caffarelli cites a letter to Laura Fermi from Ida's sister Olga, dated August 27, 1951, in which Olga recounts a chance encounter during this early period of Ida, Olga, the young Enrico, and Enrico's teacher from his middle school. The teacher shook hands enthusiastically with mother and aunt, proclaiming the young Enrico a "second Galileo." He may have been the first, but would not be the last, to do so. Of course memories shift and distort through the lens of time, and by 1951 Olga de Gattis's memory of the event may well have been colored by Enrico's subsequent development into a master physicist, but somewhere during this period a towering intellect started to make itself known. Amidei may not have been the first to notice it, but he certainly was the first to do anything about it.

Confronted with this young phenomenon, Amidei took a fateful step. Enrico had mentioned his forays into the bookstalls at Campo de' Fiori and his efforts to learn physics and mathematics from odd books he picked up. It was just at the time when Fermi and his brother (and later Persico) would try to explain the physics of spinning tops. Responding to this, Amidei decided to impose a certain intellectual structure on Fermi's education. He would guide the youth through a carefully arranged sequence of textbooks to educate him in undergraduate physics. He explained that a fundamental understanding of the behavior of a top would require a thorough grounding in classical mechanics, which first required a foundation in trigonometry, analytical geometry, algebra, and calculus, including differential equations. Once he had this under his belt, he assured his young friend, Enrico would find the equations of motion for a spinning top easier to understand.

This disciplined approach appealed to Fermi, although in retrospect it is apparent that he also continued his forays into physics textbooks independently of the informal but highly accelerated plan Amidei outlined. The curriculum does not look that much different from what a young entry-level physics undergraduate would undertake today. Amidei first intended to give Fermi a sufficiently strong foundation in mathematics to begin a serious study of classical mechanics.

Amidei sequenced the mathematics carefully over the next three years, beginning with trigonometry and moving through analytical geometry to calculus. It helped that Amidei was a proficient mathematician and that Italian mathematicians at that moment stood at the forefront of the field internationally. It also helped that Amidei was teaching a unique genius.

By 1917—Fermi was fifteen or sixteen—Amidei believed that Fermi was ready for a thorough course in classical mechanics and lent him the classic treatise on the subject by French mathematician Siméon-Denis Poisson, originally published in two volumes, in 1811 and 1833.

For the study of classical mechanics and the mathematics required to understand it, no better curriculum could be provided, not even today. Not only is the subject matter virtually unchanged since these books were written but also the volumes represent the very best thinking on the subjects they cover.

Years later Amidei remained astonished by Fermi's ability to absorb material. Some forty years after the fact, he recalled that by the time Fermi graduated from *liceo* (in 1918, a year early):

> I had already ascertained that when he read a book, even once, he knew it perfectly and didn't forget it. For instance, I remember that when he returned the calculus book by [Ulisse] Dini I told him that he could keep it for another year or so in case he needed to refer to it

> again. I received this surprising reply: "Thank you, but that won't be necessary because I'm certain to remember it. As a matter of fact, after a few years I'll see the concepts in it even more clearly than now, and if I need a formula I'll know how to derive it easily enough."

Amidei's efforts notwithstanding, Fermi could not keep himself from some side reading in physics. At the central public library of Rome, he tackled the magisterial five-volume, four-thousand-page *Treatise on Physics* by Russian physicist Orest Chwolson. Not for the faint of heart, it covers in depth every aspect of classical physics—mechanics, thermodynamics, statistical mechanics, optics, electromagnetic theory, even acoustics. After discovering that he had already mastered some thousand pages of the text, presumably as a result of his study of Poisson, he spent time every morning in the library poring over the balance of the book, assimilating its material, working out problems. It took him a little over a year, from August 1917 until September 1918, but master it he did.

According to Persico, he and Fermi also studied a relatively new book by British physicist Owen W. Richardson, *The Electron Theory of Matter*. Published in 1914, it was a graduate-level text, incorporating British physicist Joseph John (J. J.) Thomson's 1897 discovery of the electron into the broader framework of electromagnetic theory. By the time they tackled it, the two youths had significant physics and mathematics under their respective belts. (A few years later, Fermi also recommended the book to his university friend Franco Rasetti.)

For Amidei, the experience of mentoring Fermi was sufficiently profound that he decided to write down statements made by the boy as a record for posterity—hence the reliability of his recollections four decades later.

For Fermi, as well, the experience was extraordinary. Here was an adult he could talk to about the subjects that mattered most to him, someone who cared enough to tutor him on all things related to mathematics and physics. It may also have been the first time Fermi fully realized his great gifts in these areas. Time and again he discovered that he understood the material Amidei threw at him faster and more comprehensively than Amidei ever could. An experience he would have throughout his subsequent education, it gave him a sense of confidence in his abilities that was only to grow in later years.

During this last year, Amidei inquired as to whether the young man, now sixteen years old, wanted to pursue mathematics or physics at university. Fermi's reply was straightforward: "I studied mathematics with passion because I considered it necessary for the study of physics, to which I want to dedicate myself exclusively." Mathematics was the means to an end. He would always be proud of his mathematical ability, sometimes to the point of boastfulness, but physics would always be his one true love.

Amidei understood that Fermi's decision at such a young age to become a practicing physicist meant that Fermi needed to learn German. The world's leading physics journals—most notably *Annalen der Physik* and *Zeitschrift für Physik*—were published in German. Amidei insisted, and Fermi complied, once again demonstrating his ability to quickly master a new language. When he arrived at university in the fall of 1918, he was, according to friends, reading German as he read Italian.

All of this was taking place against the backdrop of World War I.

When Italy entered the conflict in mid-1915, on the side of the Allies, neither Amidei nor Fermi's father was drafted. They were too old and, also, to the extent that the railways were a strategic asset in the war, they both held important positions right where they were.

It may have been a relief for Amidei to have the distraction of training young Enrico in the intricacies of mathematics and physics. The war effort cost Italy dearly, although not as much as other European combatants: between 460,000 and 610,000 men were lost, fewer than for England, France, or Germany, but an enormous loss nevertheless. Fortunately, the war did not directly affect the Fermi family, but it was a major concern for them. And it did have a direct effect on the intake of students at the Scuola Normale Superiore in Pisa, where Fermi landed in the fall of 1918, just as the war was ending. Nine of his twelve classmates were admitted in 1915 but were deferred for military service until the war ended.

That said, Enrico's distance, both psychologically and physically, from the travails of the war is notable. It seems not to have affected him at all. Perhaps the loss of his brother was enough. Perhaps compared to Giulio's tragic and untimely death, the impersonal statistics of Italian war dead simply did not register. We know that Fermi threw himself into his friendship with Persico and his studies with Amidei with an energy and passion that bespeaks a gritty determination to overcome that loss, a far greater personal loss than anything the war raging around him could deliver.

BY THE FINAL YEAR OF *LICEO*, AMIDEI HAD WORKED WITH ENRICO for almost four years and understandably took a proprietary interest in the young prodigy's future. In Amidei's view, that future lay in one direction: Pisa's Scuola Normale Superiore.

The school was founded by Napoleon in 1810 as an Italian equivalent to the École normale supérieure in Paris. By 1918 it was the most prestigious institution of higher education in Italy. Only a dozen or so students were admitted every year, half of whom would focus on the humanities, half on the sciences. In Fermi's year it was even more competitive than in previous years, given the deferrals granted because of the war. The competition for the three open spots was intense.

The University of Rome was another obvious option, but Amidei's advice to Alberto and Ida was to send Enrico to Pisa. The Scuola Normale had produced towering figures in mathematics and the humanities, and its mathematics faculty was world class. That there were no notable physicists (except perhaps for Vito Volterra, more of a mathematician than a physicist) among its graduates was more a commentary on the state of Italian physics at the time than on the school. Indeed, physics students would be expected to take a parallel set of courses in physics at the University of Pisa, just a short walk away.

Fermi's parents objected. They were proud of Enrico and wanted the best for him, but they also wanted him to remain at home. In an age before widespread access to telephones, sending Enrico off some two hundred miles north would mean that they would have no contact with him except for the occasional letter. The devastating loss of Giulio made Enrico's parents, especially Ida, doubly reluctant to part with their son. The University of Rome was strong and the physics department was close enough to walk to, in an area just to the west of the grand Santa Maria Maggiore church on the Esquiline Hill, on a street called Via Panisperna. A famous physicist, Orso Mario Corbino, had just taken over the physics department, with the intention of making it an elite center for teaching and research. He was also a rising figure in the Italian government, having been named chairman of the Public Works Council (he would soon become a member of the senate, too). Why couldn't Enrico study with him? That Amidei succeeded in persuading them to allow Enrico to apply to the Scuola Normale and, once accepted, to attend is a testament to Amidei's persistence and commitment to Enrico's advancement. It is also clear that Enrico wanted to go, even though doing so would separate him from his parents and from Persico, who had decided to attend the University of Rome.

The entrance exam paper Fermi submitted for physics remains legendary to this day. The subject was a close analysis of the vibrations of a rod fixed at one end. He brought to bear all that he had learned from Poisson and Chwolson about harmonic waves and their behavior, and the analysis he presented demonstrated a graduate level of sophistication. The examiners were more than impressed. They may have suspected fraud and in any case wanted to meet the youngster who submitted the essay. The examiner, a professor of geometry at the University of Rome named Giuseppe Pittarelli, called Enrico in for an interview, something that rarely if ever occurred. In the course of the interview, the young prodigy satisfied Pittarelli that the work was his own. Pittarelli considered the exam to be at the level of a doctoral thesis and told him so. He also told Fermi that he was destined to become an important scientist. Fermi placed first among those who took the exam and was admitted without reservation. He would, in time, become the fabled institution's most famous graduate.

With trepidation and heavy hearts, Alberto and Ida bade farewell to Enrico in October 1918. He was off to Pisa, where the next phase of his life as a physicist would begin.

CHAPTER TWO

PISA

PISA IS A LITTLE OVER TWO HUNDRED MILES NORTH OF ROME, but in 1918 it was, and remains, centuries apart in time. The Rome Fermi left behind was a turbulent, noisy, crowded mess—a beautiful mess, an inspiring mess, but a mess just the same; a nineteenth-century city in full-throated transition to the twentieth century. In sharp contrast, Pisa was a quiet, medieval university town, frozen somewhere in the early to mid-fourteenth century, when it lost out to its great commercial and cultural Tuscan rival, Florence. The university officially dates itself from 1343, but like its counterparts throughout Europe, Pisa was probably home to scholars at least as early as the eleventh century. When Fermi arrived in late 1918, the main academic departments, including physics, were housed in old buildings scattered around the medieval heart of the city; now the only departments that remain there are those relating to humanities, philology, history, and law. The technical and scientific departments, as well as the medical school, have migrated to modern, functional facilities on the city's outskirts. In Fermi's time the physics and mathematics departments were in an ochre nineteenth-century edifice in Piazza Torricelli, a short walk from the Scuola Normale Superiore in Piazza dei Cavalieri. They are now about half a mile east, just inside the old city walls. The older building was not architecturally distinguished, but neither is the new one. It is, in fact, a 1974 renovation of a textile factory built in 1939 by the famous Pisan Jewish family, the Pontecorvos.

However, the building that houses the Scuola Normale Superiore is indeed distinguished and must have seemed magical to the young Fermi. Largely the work of Renaissance architect Giorgio Vasari, it is one of the most beautiful buildings in a city known for the magnificence of its architectural heritage.

Students who are admitted to the Scuola are also admitted to the university and are required to take classes at the university in addition to the seminars offered at the Scuola Normale. They receive degrees from both institutions, a *licenza* from the Scuola Normale and a *laurea* from the university. The *laurea* entitles the holder to be called *dottore,* although it is not the equivalent of a PhD, which was officially adopted in Italian education only in 1980. In later years a formal graduate program, called a *perfezionamento,* was added to the Scuola Normale's offerings. Today only about sixty undergraduates out of a pool of one thousand applicants are admitted to this most prestigious and competitive academic institution in Italy.

Students, called *normalisti,* were housed in small rooms within the palazzo, each equipped with a simple desk and bed. There was no central heat, and the students relied on small coal burners to provide what little comfort they could get during the raw Tuscan winters. The Scuola Normale provided free room and board as well as a small stipend. In this, *normalisti* had it easier than other less exalted university students, who were on their own when it came to finances and to housing. Unlike Oxford or Cambridge, the University of Pisa was not organized along a residential college system, and to this day students are expected to find their own accommodations within the city, although the education is heavily subsidized.

And so Enrico Fermi arrived in this quaint ancient college town, intent on mastering his chosen field.

Fermi's reputation surely preceded him, and professors like famed mathematician Luigi Bianchi and Luigi Puccianti, the professor of physics at the university, were probably already aware of the young man and his brilliant entrance exam in physics. Fermi must have sought them out, too, given his training with Amidei and his independent reading. It was clear to everyone that he had already mastered the physics coursework on offer, with the exception of experimental physics. Much of his time would be spent on independent reading and study.

Oddly enough, given what he told Amidei regarding his desire to study only physics, he started off as a mathematics student, perhaps reflecting the Scuola Normale's worldwide reputation for its mathematics faculty. Faculty members had written many of the textbooks Amidei had provided during Fermi's informal studies in Rome. The mathematicians of the Scuola Normale were not merely textbook writers, they were at the cutting edge of the field and quite powerful at the school, a fact that could not be said for the physicists either at the Scuola Normale or at the university. This may account for Fermi's decision to enroll in the mathematics degree program. Within a year, however, he had shifted his formal focus to physics, taking courses at the university under the guidance of Puccianti, who soon realized that the best way to educate the young prodigy in physics was to let him loose in the university library.

Fermi attended lectures and seminars, both at the Scuola Normale and at the university, but in letters to Persico, he indicated that the lectures took relatively little time and effort and he spent much of his time in independent study. The Scuola Normale expected its science students to take a full load of physics, mathematics, and chemistry. Fermi earned perfect grades across the board. He took courses in German and received perfect grades in that subject, too. At the university, his transcript shows perfect performance in every science course, with honors in most. The one course in which he did not receive a perfect grade was in Freehand Drawing (*Disegno a mano libera*), in which he received a "24" out of "30." He had covered almost all of the required material in mathematics and physics during his time with Amidei, so the class work was simple and he had plenty of free time to spend in the library, keeping up with research articles on quantum physics and relativity in the major physics journals. Enrico's letters to Persico suggest that the only courses that involved some work for him were those in chemistry—not surprising, because he had not studied chemistry with Amidei and had little interest in the subject.

THE ENTERING CLASS AT THE SCUOLA NORMALE IN 1918 WAS UNUSUAL, the bulk of the class having been admitted in 1915 but only now arriving in Pisa with the war's end. The Scuola Normale had long been a training ground for Italy's intellectual and cultural elite, and Fermi's class was to prove no exception. Among his classmates were individuals destined for great achievement in mathematics, astronomy, and physics. One of those he quickly befriended was a young man named Nello Carrara, who had arrived a year earlier. Skinny and athletic, he and Fermi became fast friends.

There was, however, another student, not at the Scuola Normale but enrolled as a physics student at the university, with whom Fermi would develop a legendary, lengthy professional and personal relationship. His name was Franco Rasetti.

Born in a small village in Umbria near the town of Castiglione del Lago, Rasetti was about six weeks older than Fermi, but in appearance he could not have been more different. Tall and beanpole thin, with a scrawny neck and bony face, he gave the impression of a human ostrich. He was close to his mother, and when Rasetti was admitted to university the whole family moved to Pisa, where they lived together in an apartment in town.

Rasetti was a true eccentric. He initially chose physics as a subject for university study because, he later explained, it was not a science that involved memorization and classification and would thus be a challenge for him. But in fact he had an extraordinary memory and focused it on classification projects in marine biology and botany; he was able to identify thousands of different species of mollusks and flowers. In later years he would turn from physics to biology as his primary field of study.

Rasetti met Fermi in science classes at the university, and the two immediately hit it off. Fermi became a frequent visitor to the Rasetti household and often dined with them despite the fact that *normalisti* received free room and board. The young men shared an impish sense of humor and soon formed an "Anti-Neighbor Society" dedicated primarily to irritating members of the public with pranks and public displays of disrespect. One of their favorite pastimes was tossing bits of sodium into Pisa's public urinals, enjoying the reaction of those at the urinals when the sodium exploded upon contact with water. They would walk up to strangers and put padlocks through buttonholes on their jackets, locking the jackets closed. Fermi even played this prank on Rasetti, padlocking the front door of the Rasetti household closed while the family was inside. They loved to place a pan of water atop a door that was slightly ajar and watch as an unsuspecting victim brought the pan down on himself. They fought mock sword fights on the rooftops of Pisa. They concocted a stink bomb and exploded it in a classroom during a lecture, for which they were almost expelled. They were saved only through the intervention of Professor Puccianti, who prevailed on them to limit pranks to those who would not get them into further trouble with university authorities.

When they were not wreaking havoc in town, they were hiking in the mountains north of Pisa, often joined by Nello Carrara. The three of them loved strenuous physical activity and spent most free weekends, weather permitting, hiking in the Apuan Alps north of the town, home to the famous Carrara marble quarries. For Fermi this was the beginning of a love affair with mountain hiking, an activity he pursued to the end of his life. The Anti-Neighbor Society eventually grew into a larger circle of friends, both male and female, with slightly more congenial ambitions. The members would spend weekends exploring the hills and mountains above Pisa, Fermi typically out in front, charting the path for the day's walk, leading the way, and determining when they would return. He was the group's natural leader, and in a pattern that was to be repeated throughout his life, people around him were happy to follow wherever he decided to go.

Fermi's exposure to the opposite sex was not limited to weekend jaunts with the Anti-Neighbor Society. Both the Scuola Normale and the university admitted women. In a letter he wrote to Persico, Fermi revealed a distinctly unattractive attitude toward his fellow female students, composing a cruel skit in which he ridiculed them—"barring one or two exceptions ugly enough to scare anybody"—by portraying them as incapable of reducing a simple fraction. Yet he must have been attractive to women. They enjoyed hiking with him and found him intellectually impressive and self-confident. They might well have also found him to be slightly immature. The combination of Fermi and Rasetti, with their constant teasing and posturing, would have been enough to try anyone's patience, as it clearly did with young Laura Capon a few years later. Fermi's eagerness to ridicule the intellectual abilities of the women around him would not have been particularly unusual in his day, or decades later, for that matter. Perhaps it reflected merely the widely shared prejudices of his time and culture. More than likely, it was also a

FIGURE 2.1. Fermi, Rasetti, Carrara in the Apuan Alps north of Pisa. *Courtesy of Amaldi Archives, Department of Physics, University of Rome, La Sapienza.*

bravado with which he could mask his awkwardness around women, an ineptness he would eventually outgrow.

AT THE UNIVERSITY, FERMI AND RASETTI TOOK MANY COURSES together. One such course was analytical chemistry. In one memorable lab session, they were asked to identify the components of a chemical mixture using the analytical chemistry techniques they were supposed to master. Looking at the mixture, Fermi decided that it would be easier and more straightforward to examine it under a microscope and visually identify the components. They scored a perfect grade for the lab. No one was the wiser.

University life was not all fun and games. Fermi was a serious, passionate student of physics and his independent reading stayed with him throughout his working life. He mastered Poincaré's classic work on the hydrodynamics of whirlpools and absorbed two other classic books, Appel's *Mechanics* and Planck's *Thermodynamics,* so well that he could recall proofs from them years later. His knowledge of relativity and quantum theory was soon greater than that of his teachers, and he frequently gave lectures on relativity. Puccianti, a generous, very old-fashioned physicist, would ask Fermi to lecture him on theoretical subjects that puzzled the older man. In these sessions Fermi honed his pedagogical skills, skills that would play a central role in his subsequent career. Puccianti appreciated these mini-seminars so much that he eventually dubbed the younger scientist his very own expert on relativity.

Fermi spent the summer of 1919 between his home in Rome and his grandfather's home just outside Piacenza and began a process that would continue throughout his life and distinguish him from virtually every other physicist of his generation. He dutifully recorded the entire sum of his knowledge of physics in a carefully organized notebook, written in miniscule handwriting. Subsequent notebooks—twenty-five of which are in the Fermi archives at Pisa's Domus Galilaeana and several dozen more in the archives at the University of Chicago—reflect the astonishing, continuous intensity of his physics activities. He was, in a real sense, the Voltaire of physics, always scribbling away in his notebooks, solving problems.

The 1919 notebook is 102 pages long and contains a wide range of material on advanced physics. One senses that this notebook, and all subsequent ones, served as his way of internalizing a deep foundational knowledge of physics that he carried with him for the rest of his life. Given his appreciation for his own abilities, he may even have written the notebook with a view toward future historians interested in how he learned his discipline.

It soon became apparent to those around him, students and faculty alike, that they were in the presence of a rare mind. As would happen often during the Manhattan Project some two decades later, professors would call him to solve equations that proved resistant to straightforward solution. In response, he would work his way steadily to the solution the professor had been struggling toward. He also developed some of the personal tics that would stay with him. When lost in thought he would grab some nearby object—chalk, pencil, or, in one case, a pen knife—and fiddle with it absently. With the pen knife, he tapped it in the open position against his forehead near his right temple and sliced himself by accident, leaving a small scar that he carried for the rest of his days.

After his second year in Pisa, he was allowed to focus full time on a university physics program that lasted another two years. Rasetti and Nello Carrara were admitted to the program with him, and Puccianti gave them free rein over the department's laboratory facilities, primitive though they were. Experimental physics was, as Rasetti later explained, the only subject in which one could receive a degree. Theoretical physics was not considered a separate, legitimate academic field. Puccianti's labs were, unfortunately, poorly set up for advanced experimental work. Much of the apparatus in the lab was fit only for demonstrations during lectures. The three students spent a few weeks exploring the equipment they found in cabinets and drawers, and Fermi decided—for all three of them, apparently—that the most promising research lay in the direction of X-ray research. They soon discovered that the X-ray tubes at their disposal were not suitable for the kind of experiments they wanted to do, so they built their own, with the help of a local glassblower.

In later years, Rasetti noted that Fermi's disinterest in any distinction between theory and experiment was evident even this early. Fermi was "from the first a complete physicist." The temperaments of theorists and experimentalists are often thought to be quite different, the former more comfortable in an isolated setting solving theoretical problems, and the latter, more collaborative, enjoying the design and execution of experimental projects. From an early age Fermi effectively and eagerly bridged this gap. He combined a willingness to noodle on a theoretical problem for a sustained period with an enjoyment of working with his hands on an experimental design. Perhaps this was because the Italian physics establishment did not take theory particularly seriously and encouraged this gifted theorist early on to get his hands dirty in the lab. Perhaps if theory had been held in higher regard in the corridors of the University of Pisa, he would not have developed into the balanced physicist he became. However, it is clear from his childhood efforts with Giulio and later with Persico that he loved to tinker with objects, to make machines, to do physical experiments. Naturally gifted in both, he saw no need to choose between them. Fermi saw physics as an integrated whole. By the time he was ready to present his experimental dissertation to the examiners at the university, Fermi had already published several theoretical papers in scholarly journals.

His first was a paper on electrodynamics of a rigid, charged body. The second and third papers focused on his first love, relativity theory; the third presented an important theorem about how the theory works within very small distances and proposed a system of coordinates to make the analysis of these small distances easier to compute. The fourth was a highly successful effort to reconcile the different ways that the electromagnetic mass of a rigid spherical charged body—that is, the mass measured by application of force in an electromagnetic field—is measured in classical electrodynamic theory and in relativity.

A fifth paper, apparently commissioned for a German publication while he was still at Pisa but published after graduation, was an appreciation of relativity. This essay was one of the very few published by an Italian physicist to evince any enthusiasm for Einstein and his outlandish theories of space, time, and gravity. Fermi's main purpose was, characteristically, to call attention not to the puzzling philosophical and metaphysical consequences of the theory but rather to one of the theory's most compelling physical predictions:

> If we could liberate the energy contained in one gram of matter we would get more energy than exerted by a thousand horses working continuously over three years. (Comments seem superfluous!) It will be said, with good reason, that in the near future at least that it does not appear possible to find a way to liberate this awesome amount of energy. This is indeed as one can only hope; an explosion of such an awesome amount of energy would blow to pieces the physicist who had the misfortune of finding a way to produce it.

He may not have been the first person to notice this consequence of Einstein's work, but he was certainly one of the very few at the time who emphasized its importance. In light of what life had in store for him, his words are particularly prophetic. Years later, as he witnessed the first test of an atomic bomb and considered his role in making that test possible, he certainly thought back to these words, written when he was just twenty-one years old.

By the time he graduated in July 1922, Fermi was an expert on relativity and its strongest—perhaps *only*—proponent within the Italian physics community. Although the subject was not yet on the radar screen of most Italian physicists, it was of great interest to Italian mathematicians, who pioneered the mathematics necessary for the theory, in particular the brilliant University of Rome mathematician Tullio Levi-Civita. Einstein had consulted Levi-Civita as he struggled with the theory. The Italian had made some essential suggestions and kept abreast of Einstein's revolutionary work. So impressed was Levi-Civita with Fermi's work on relativity that he used Fermi's coordinate system in his own treatise in 1925. After Fermi's graduation, it was the mathematicians who first appreciated his potential for transforming physics in Italy.

FERMI WENT ON TO COMPLETE HIS UNIVERSITY PHYSICS DISSERTATION on X-ray diffraction with little trouble. The results were a series of images made with X-rays passed through crystals and Fermi's analysis of these images. He graduated with honors. His dissertation at the Scuola Normale was more theoretical, involving the solution of a particular theorem in probability and its application to the orbit of a comet. Fermi was more nervous about this dissertation for the Scuola Normale than he was about the university degree. He was concerned that someone else had already proven a theorem he was working on and that someone had applied these theorems to the orbit of asteroids. So fretful was he, in fact, that he wrote to Persico several times to see if his friend in Rome could do some independent research to determine whether his worries were founded.

As it turned out, the exam was not smooth sailing, but not because of the issue of prior work. During his oral exams, several of the examiners were mathematicians who grilled him hard on the equations he used. In the end he received a *licenza,* but the examiners did not offer him the customary handshake, nor was the thesis published, as was the tradition at the Scuola Normale. Years later his wife would claim that the thesis went over the heads of the examiners. This seems unlikely. It is more likely that they were simply exacting revenge on a renegade mathematics student who offended them by abandoning the rarified field of mathematics for the less cerebral, earthier pursuit of physics.

In the summer of 1922, degrees in hand, Fermi, already a legend in Pisa, headed home to Rome. Because he had come to the notice of major figures in the mathematics community in Italy, Fermi was a frequent visitor to the Saturday evening salons held at the home of Levi-Civita's friend and colleague Guido Castelnuovo in Rome. Word of Fermi's brilliance had spread to the dean of Italian physics, Orso Mario Corbino. A man of extraordinary power and influence, Corbino was the director of the Institute of Physics at the University of Rome. He soon understood exactly what Fermi could do for Italian physics and he was determined to make sure the young man succeeded.

CHAPTER THREE

GERMANY AND HOLLAND

FERMI'S FOUR YEARS IN PISA WERE SCHOLARLY AND CONTEMPLATIVE, but they were years of turmoil and chaos for the rest of the country. Though Italy was one of the victors in November 1918, the war left a power vacuum in its wake, and the following four years saw one faction after another trying to grab the political high ground. Strikes and labor unrest led to regular breakdowns in production and transportation. In the streets of the northern industrial cities of Milan, Turin, and Genoa, communists and right-wing agitators fought each other with increasing violence and lawlessness. Virulent Italian nationalists sought a voice in parliament. Italy seemed increasingly ungovernable.

In the midst of this turmoil, a brash journalist named Benito Mussolini began to promote a nationalist movement with increasing appeal to the majority of the Italian people. Bright but unsophisticated, he was drawn to socialist causes as a young man and fought briefly in World War I. Leaving service, he wrote for socialist papers opposed to Italian participation in the war, but in a turn that remains somewhat mysterious to this day, he shifted away from opposition to more of a nationalist, pro-war stance. When hostilities ended, he started a newspaper, *Il Popolo d'Italia* (*The People of Italy*), which gave voice to a new and coherent philosophy of statist nationalism that eventually became known as fascism.

As Fermi's summer holiday of 1922 came to an end, he began to consider what to do next. Mussolini and his confederates were planning an audacious move: a national coup d'état, culminating in a fascist march on Rome to take control of the government. The march took place on October 28, 1922, the very day that Fermi and Corbino were scheduled to meet and discuss the young man's future.

Corbino was a central figure in early-twentieth-century Italian physics and was to become a central figure in Fermi's life as well. Unlike most senior scientists of his era, he played an important role in public service as well as in the development of his academic field. Born in Sicily in 1876, Corbino was just slightly younger than Fermi's father, Alberto, a year younger than Fermi's mentor Amidei, and he grew up in the years following Italy's unification in 1870. He proved a promising physicist and did important work in the burgeoning field of spectroscopy that garnered him the attention of senior Italian physicists. They plucked him from his position as a teacher at a *liceo* in Palermo and brought him to the university in Messina on the eastern coast of Sicily. Corbino taught there from 1904 till 1908, when he survived a disastrous earthquake that destroyed much of the town. Pietro Blaserna, the director of the University of Rome's Institute of Physics and holder of the physics chair at the university, invited Corbino to take a position at the institute. When Blaserna passed away in 1918, Corbino ascended to the directorship of the institute, a position he held until his own sudden and untimely death from pneumonia in 1937.

Corbino was even shorter than Fermi and rounder, bald with a bushy moustache and bright eyes. He was a fine teacher. Laura Fermi enjoyed a course she took with him, pronouncing him "impressive."

Corbino's early graduate work in Palermo was critical in shaping his internationalist view of the field. His professor had studied with the great Dutch astrophysicist Hendrik Lorentz and, unlike many of his Italian contemporaries, was eager to keep abreast of developments outside Italy. He imparted this enthusiasm to Corbino, who devoted much of his career to keeping track of external developments and, at the same time, to bringing Italian physics to the attention of the rest of the world.

Corbino was also an administrative genius. Successive Italian governments recognized this and placed him in positions of increasing importance. After the war he served as head of the committee overseeing water resources, and in 1921 he was appointed Minister of Public Instruction. In 1923, Mussolini named him Minister of National Economics, in spite of the fact that he was not, and never would be, a member of the Fascist Party. Along the way, Corbino also became a senator in the Italian parliament. He was close to Italian industry and served on the boards of several Italian electricity companies.

His academic prestige grew in tandem. He was a member of the *Accademia dei Lincei,* at the time the most prestigious scientific society in Italy, and president of the *Società Italiana di Fisica* (Italian Physical Society) from 1914 to 1919.

With each new role he further cemented his reputation as a brilliant technocrat with a sound sense of judgment and an ability to move serenely and efficiently through the bureaucratic labyrinths of Italian power. He openly regretted the various governmental responsibilities that prevented him from continuing what had been a distinguished research and teaching career. In young Fermi he saw his opportunity to put Italy where he believed it deserved to be, at the forefront of world physics. For his part, Fermi could have found no one better placed to guide his career. It was to be an historic partnership.

On that day in October 1922, when the two of them were supposed to be discussing Fermi's future, Mussolini's supporters marched on Rome and the prime minister petitioned the king to declare a national state of emergency, something that, under Italy's constitutional monarchy, only the king could do. Corbino and Fermi could not help but fixate on the situation. As reported by Laura Fermi years later, they speculated as to whether the king would sign the declaration. Corbino expressed distaste over the fascists' embrace of violence for political ends but viewed a signature as the start of a prolonged, bloody civil war. Fermi noted that the king rarely countered his cabinet's recommendation. "Do you think he may go against his cabinet?" Fermi asked the older and wiser Corbino. "He has never been known to take the lead but has always followed his ministers."

Corbino paused before replying. "I think there is a chance that the king may not sign the decree. He is a man of courage."

"Then there is still a hope," his younger colleague suggested. Fermi had clearly misunderstood Corbino.

"A hope?" he replied. "Of what? Not of salvation. If the king doesn't sign, we are certainly going to have a Fascist dictatorship under Mussolini."

Corbino was right. The king refused to sign, sparing Italy a long civil war, and within a week, Mussolini was prime minister of Italy, well on his way to creating the twenty-one-year-long dictatorship foreseen by Corbino.

Corbino had advised Fermi to apply for a scholarship at the University of Göttingen, in the central German state of Lower Saxony. The competition for this particular scholarship, sponsored by the Italian Ministry of Education, was intense, but Fermi won it without apparent effort. On October 30, 1922, two days after the Fascist march on Rome, the fellowship committee met and awarded him the fellowship. Thus began one of the most mysterious and inexplicable periods of Fermi's entire career.

The University of Göttingen was one of the world centers of physics. The eminent theorist Max Born and his close colleague experimentalist James Franck had already started to build a major center, recruiting brilliant students, including young Werner Heisenberg, who would shortly become world famous. Another theorist, Pascual Jordan, had recently arrived from Hanover and would in time make his own indelible mark on quantum theory. Fermi just missed overlapping with Wolfgang Pauli, an Austrian who had already written a seminal treatment of relativity theory. The two of them would meet only five years later, at a conference at Lake Como in Italy in 1927.

Born was a slightly shy, somewhat formal gentleman in his early forties, who had studied at Göttingen early in the century with three of the world's greatest mathematicians, David Hilbert, Felix Klein, and Hermann Minkowski. His original work on relativity won lavish praise from Einstein himself. After stints at the universities in Berlin and Frankfurt, he was recruited to Göttingen and brought Franck, an old friend and colleague, along with him. Born was at Göttingen barely two years when Fermi arrived, a newly minted graduate with some five published papers to his name. It should have been a wonderful moment for Fermi.

However, quantum theory, for which Göttingen would soon become world famous, was at a momentary impasse. Niels Bohr had developed a theory of the atom that incorporated the basic insights of quantum theory, but a number of important experimental observations remained unexplained. In early 1925, all that would begin to change, but when Fermi arrived, late in 1922, these breakthroughs were in the future.

Not only was quantum physics at a standstill but also Germany itself was an exceedingly unpleasant place. Reeling from four years of unsustainable reparation payments as a consequence of the Treaty of Versailles, the German economy entered a period of hyperinflation that destroyed the economic and financial fabric of the country and destabilized the precarious Weimar Republic. Laura Fermi writes that for the first time Enrico enjoyed a feeling of relative wealth. Because his stipend was paid in Italian lire, he was able to maintain and even improve his standard of living as the financial crisis progressed. He splurged and bought himself a bicycle. A sleepy university town with little in the way of industry, Göttingen might not have suffered from some of the more unpleasant aspects of the crisis—industrial unrest, strikes, and riots—but it could not have been a happy place.

Most importantly, Fermi felt ignored. Born was not a demonstrative man and apparently paid little attention to his new young Italian visitor. It seems that Heisenberg, Jordan, and the other scholars at Göttingen neglected him, as well. Though his letters home to his father during his stay at Göttingen show no signs of unhappiness—rather, they focus on money, family, food, vacation, his new bicycle—Laura Fermi writes that this was a moment in Fermi's career when he needed validation and did not get it. Yet Fermi knew he was special and throughout the course of his career never needed anyone else to tell him so.

Segrè puts some of the blame on Fermi, suggesting that the young physicist was "proud, shy, and accustomed to solitude." Years later Fermi's future Manhattan Project colleague Leona Libby went further, observing that he had "a stored-up, never forgotten bitterness against the physicists he encountered on his first visit to Germany as a very young man. He most of all resented the fact that Marie Curie and Werner Heisenberg, in particular, had completely ignored him, to the point of exceeding rudeness," adding, "This was the winter [1923] that made

Fermi miserable because, as he told us later, he was completely unappreciated and ignored, almost ostracized from the in-group." The intensity of the resentment is surprising at such a remove in time and place, yet it is hard to see why Libby would invent such an anecdote.

Finally, and not inconsistent with the other explanations, this was Fermi's first time living and working outside of Italy for any substantial length of time. In spite of his comfort with the language, he may have felt alien and lonely simply by virtue of his being so far away from home.

If Fermi was deliberately shunned by the Göttingen "in-group," this may also have reflected snobbery among German physicists toward their Italian colleague. Fermi never carried himself as a particularly cultivated or cultured individual. He was obsessed with physics and aside from his outdoor activities had little interest in anything else. His regular habits involved getting up early, working all day long, with a lunchtime break for several hours in the early afternoon. When he came back from lunch, he would work until late afternoon, have dinner, and go to sleep. Each day was very much like the last. He had little time for nightlife, café life, or cultural life. Though Born was no night owl, the younger students loved to hang out in cafés and talk about culture and philosophy well into the night.

Furthermore, the group that Born and Franck recruited was certainly more inclined toward philosophy than Fermi was. Heisenberg was obsessed by what quantum theory really meant about the nature of reality. Pauli, although not quite so obsessed, thought deeply about these issues, as well. In Fermi they did not find a fellow traveler. Fermi preferred to stick with questions that had definite answers, answers that he could find through careful, systematic work. As quantum theory developed, the philosophical issues became more complex, and yet throughout his life Fermi resisted any speculation of a philosophical nature. His aversion to the deeper questions could well have influenced his German colleagues to ignore him.

One wonders why Curie would have, in Libby's words, "ignored him, to the point of exceeding rudeness." She may have been a bit prickly, but by the time she visited Göttingen in 1923, she had nothing to gain from being rude to a young Italian physicist. On the other hand, she was known to be strongly skeptical of theory and theorists. To the extent that Fermi presented as a theorist—which is quite likely, given that he did no experimental work while he was in Göttingen—it is possible that she lumped him together with other theorists and thus had no time for him.

The work he did was indicative of the problems that interested him at the time and in many ways throughout the rest of his career. His Scuola Normale dissertation focused on probability theory, and continuing the theme, the papers he published while he was at Göttingen focused on various aspects of the field in which probability and statistics are central—statistical mechanics. Fermi loved the mathematics of probability and statistics and enjoyed statistical mechanics for exactly this reason.

One of his Göttingen papers involved a probabilistic concept called the "ergodic theorem." The ergodic theorem is central to the study of statistical mechanics. In statistical mechanics, a system is said to be ergodic if, in principle, it can start in a given state and, over time, pass through every possible state, eventually returning arbitrarily close to its original state. Fermi became interested in ergodic processes during this period. In particular he wondered whether certain well-specified systems were actually ergodic in their behavior or only seemed to be so. He developed a proof of the ergodic theorem that, though not mathematically rigorous, was sufficiently useful and penetrating to attract the attention of physicists outside Italy.

Fermi returned to statistical mechanics in late 1925 and early 1926 with historic results. Indeed, probability and statistics remained a central preoccupation thereafter. Why did Fermi find probability and statistics so compelling? Not all physicists do. Whereas knowledge of probability is essential to an understanding of most of modern physics, in and of itself it is not a field on which many physicists focus. Fermi loved to gamble, particularly on the outcome of his own athletic competitions. He also loved to pose riddles that required an understanding of probability. His way of solving problems, now known as Fermi problems, often consisted of evaluating the probabilities of particular events, to at least an order of magnitude, and linking those probabilities together to arrive at a solution that seemed reasonable. The so-called Fermi paradox—Fermi's back-of-the-envelope observation that if intelligent life exists elsewhere in the universe we should have been visited by it long ago, given the size and age of the universe—is in itself a study in probability.

One possible reason probability and statistics captivated him may stem from the trauma of Giulio's death. Here was something unexpected, totally improbable, and yet it occurred, with devastating results. Perhaps the experience awoke within Fermi a desire to understand, delimit, and prepare to the extent possible for events that were out of one's control. The odds were low that Giulio would die when he did and yet it happened. Fermi may have taken away from this trauma the need to understand the likelihood of any particular event and a feeling that in understanding that probability he was in a better position to anticipate it, prepare for it, and perhaps even shape its outcome. We will, of course, never know. What we do know is that the study of probability ran like a thread through his career and was at the root of many of his most important contributions.

FERMI ARRIVED BACK IN ROME FOR THE SUMMER OF 1923, HAPPY TO be home, and Corbino soon found him a one-year appointment teaching physics to engineering students at the University of Rome, alongside his old friend Persico, who had stayed in Rome and received a physics degree at the university under Corbino's watchful eye. During the academic year, one spring weekend Fermi met another young person, someone destined to have perhaps the greatest impact on his life.

Her name was Laura Capon.

Laura Capon was a bright, vivacious young woman, just shy of seventeen years old, the daughter of a prominent Roman Jewish family. She grew up in relative splendor, in a home in the elegant district just north of Viale Policlinico, at Via dei Villini 33. Her father, Augusto Capon, was an officer in the Italian navy. By the time Laura met Enrico, her father was an admiral.

The admiral valued intellectual ability and achievement, which explains in part why he decided that, unlike most young women of her socioeconomic status, Laura should eventually earn a degree in general science from the University of Rome. There, she would take a course in physics taught by a young man named Enrico Persico.

The admiral's intellectual and social ambitions brought him into contact with major Roman intellectuals, and he entered the circle of the two great mathematicians of Rome, Tullia Levi-Civita and Guido Castelnuovo, both of whom also happened to be Jewish. The two

FIGURE 3.1. Young Laura Capon with friend, date uncertain. *Courtesy of AIP Emilio Segrè Visual Archives, Uhlenbeck and Crane-Randall Collections.*

mathematicians were friends, and the latter held a Saturday evening salon at his home to which Levi-Civita and a variety of other prominent mathematicians were invited. Augusto Capon was also a regular at the salon. The adults often brought their children along, and the children became close friends. Laura became particularly fond of Castelnuovo's daughter Gina. It was with this group of friends that Fermi found himself one Sunday in the spring of 1924. He was older than the rest of them but had gotten to know them because he, too, was a regular visitor to these salons and enjoyed the company of the younger group as much as he enjoyed being with the older intellectuals.

Early photos of Laura reveal a slim woman with a round, cherubic, happy face. It is easy to see why Fermi found her appealing. Years later, when a colleague complimented him on how attractive his daughter Nella was, he replied, "You should have seen Laura when she was young." Their first meeting was hardly auspicious. Laura was unimpressed with this man, introduced as a brilliant young scientist who was already teaching at the university at the age of twenty-two:

> He shook hands and gave me a friendly grin. You could call it nothing but a grin, for his lips were exceedingly thin and fleshless and among his upper teeth a baby tooth lingered on, conspicuous in its incongruity. But his eyes were cheerful and amused: very close together, they left room only for a narrow nose, and were gray-blue, despite his dark complexion.

The group of friends went to play in a park on the northern outskirts of the city just on the other side of what is now the Via Salaria, near a bend in the Tiber River. Fermi, who was a bit older than the others, took the role of leader and decided the group should play soccer. Laura explained she had never played the game before. Fermi assured her it would be fun and made her the goal keeper for his team. Laura describes what happened next:

> There was an easy self-reliance in him, spontaneous and without conceit. Luck, however, was against him: at the height of the game the sole of one of his shoes came loose and dangled from the heel. It hampered his running, made him stumble and fall on the grass. The ball zoomed above his fallen body and sped toward the goal. It was up to me to save the day: while I was observing our leader's predicament with more amusement than pity, the ball hit me in the chest. Stunned, I wavered, almost fell, recovered my balance. The ball bounced back into the field and victory was ours.
>
> Our leader pulled out of his pocket a large handkerchief, wiped off the perspiration that streamed down profusely from the roots of his hair over his face, then sat down and tied his loose sole with a piece of string.

It was, she concluded, "the first afternoon I spent with Enrico Fermi, and the only instance in which I did better than he." She would not see him again for some two years.

The year went by quickly, but for Fermi the academic year was cut short by the death of his mother on May 8, 1924.

The family had been caring for her through a series of respiratory illnesses, so her passing was no surprise. Yet we can be sure it had an effect, though he would never speak directly about it. Giulio may have been her early favorite, but there is no reason to think that Ida and Enrico were not also close. Ida found some pride and a bit of respite from her grief in Enrico's increasing and gratifying success. Segrè notes that Fermi in later years expressed admiration for his mother's ability to make things for herself when she had to—a makeshift pressure cooker being an example. Fermi's ability with his hands may have been inherited from this fragile, disciplined former schoolteacher. Her death was a blow, but it was a blow that he dealt with in typical fashion, that is, privately.

His father, Alberto, had been planning to move the family out of the city and to a modern but modest suburban area, Città Giardino Aniene, about five miles north of central Rome. He had bought a plot at what became Via Monginevra 17 and was in the process of having the home built when Ida passed away. The family moved there in 1925. It was an improvement over Via Principe Umberto 133, but by no means grand. Alberto and his surviving children, Enrico and Maria, called it home for the next few years. Alberto died in 1927 and Fermi married the following year, moving to his own apartment and leaving Maria installed at the family home.

Together Corbino and Fermi plotted Fermi's next step. Corbino was aware of an International Rockefeller Foundation fellowship that might be available to Fermi for another overseas scholarship. In the interim, Fermi had made the acquaintance of a young Dutchman, less than a year his senior. The two events came together in a happy coincidence.

The Dutchman was named George Uhlenbeck. A graduate student in physics at the University of Leiden, in southern Holland about five miles outside of The Hague, Uhlenbeck had taken a year off and was working with the household of the Dutch ambassador to Rome, serving as a tutor in math and science to the ambassador's two sons. Shortly after he arrived, Uhlenbeck received a note from his thesis adviser, Paul Ehrenfest, that a young physicist in Rome named Fermi had written an insightful paper on the ergodic theorem. This was exactly the kind of paper that would have appealed to Ehrenfest, a keen student of statistical mechanics who had been looking for ways to integrate it with the emerging quantum theory. Ehrenfest had an instinct that Fermi was a potential intellectual soul mate and urged Uhlenbeck to make Fermi's acquaintance and report back.

The two young men got on famously. In 1962, Uhlenbeck recalled those first meetings with Fermi: "He was younger than I was, but he was in a sense a wonder child, like Pauli. Anyway, we had then a little seminar, with Pontremoli, Persico, and me, in this old building. . . . He was so much ahead of all three of us, that he was the one who talked all the time."

The Dutchman and the Italian became lifelong friends.

Enthused at the prospect of working with Uhlenbeck's colleagues in Holland, Fermi spent the rest of 1924 in Leiden on his Rockefeller fellowship and enjoyed his time immensely. Ehrenfest had enormous respect for his new young colleague. After the disappointment of Göttingen, Fermi was delighted to find his abilities appreciated. No doubt this had to do with the outgoing character of Ehrenfest and the team he had around him, including Uhlenbeck and another graduate student, Samuel Goudsmit. Uhlenbeck and Goudsmit were the Dutch equivalent of Fermi and Rasetti—extremely close friends who shared an impish sense of humor and a passion for their chosen profession. They even physically resembled their Italian counterparts: Uhlenbeck, tall and gaunt, towered over the shorter, sturdier Goudsmit. The two Dutch physicists were inseparable. Fermi found them engaging companions.

Much of Fermi's success in Leiden was certainly due to Ehrenfest, and Fermi wrote to his sister Maria of his delight in getting to know the Dutch physicist, describing him as a "*persona molto simpatica*"—a very kind person. Ehrenfest, a close friend of Einstein and one of the participants in the philosophical dialogue between Bohr and Einstein over the meaning of quantum physics, was, in Einstein's eyes, the best teacher of his generation. Indeed, Einstein visited Ehrenfest while Fermi was there, and Fermi met the great physicist for the first and only time. Fermi kept a group photo of Einstein, Ehrenfest, and a gathering of other colleagues; Fermi is not pictured, leading at least one scholar to presume that Fermi took the photo himself. Einstein clearly impressed the young man, who wrote an enthusiastic letter to Maria describing his encounter with the great scientist, in which the two spoke German since Fermi's Dutch was not up to par. However, in later years, Fermi expressed annoyance at the level of adulation given Einstein by an adoring world. The creator of the theory of relativity was the only physicist about whom Fermi ever overtly expressed a sense of envy.

Ehrenfest shared Fermi's deep interest in matters relating to statistical mechanics and probability. This shared intellectual passion formed the basis of a productive and influential relationship for both men. Ehrenfest also clearly appreciated just how special Fermi was, how much the young man had mastered, and how clearly he thought about complex physics problems. He was impressed and let Fermi know it.

Fermi got to work quickly and by November he presented a theoretical paper analyzing problems associated with the intensity of multiple spectral lines that had arisen during experiments at the University of Utrecht. In it he expressed his appreciation for an idea suggested by Ehrenfest. In contrast, none of his earlier Göttingen papers made any reference to ideas suggested by his German colleagues. The rest of his time at Leiden went quickly and by the beginning of January 1925, Corbino's efforts resulted in a lectureship in Italy for his gifted protégé. This time, it was Florence. Fermi took on a teaching position at the university and rejoined his old friend Franco Rasetti, who had himself found a position teaching experimental physics.

He may not have known it, but in Florence Fermi was destined to make scientific history.

CHAPTER FOUR

QUANTUM BREAKTHROUGHS

THE BREAKTHROUGHS IN QUANTUM THEORY DURING 1925, including the one made by Fermi toward the end of the year, were brilliant. The characters involved were fascinating and the source of endless, often amusing anecdotes. To appreciate these achievements and how they affected Fermi's own work, it is important to understand the state of theoretical and experimental physics at the time and the contributions of two quantum theorists in particular—Wolfgang Pauli and Paul Dirac.

IT WAS MAX PLANCK IN THE 1890S WHO FIRST PROPOSED THAT ENERGY was not "continuous" but came in discrete, tiny "packets," or "quanta." ("Quantum" is the singular of *quanta* and gives the theory its name.) This was the only way Planck could explain the odd problems encountered by German engineers who were trying to build a more energy-efficient light bulb to compete with the American engineers at General Electric and Westinghouse. Building on Planck's solution, Einstein later proposed that light was composed of such packets, which he dubbed "photons." Further experimentation showed, rather confusingly, that photons appeared to display the characteristics of both particles and waves—a very strange duality that remains at the heart of quantum theory to this day.

The concept that energy seemed to come in discrete quanta was great news for scientists who were working with spectroscopes. For over a century it was known that when elements were isolated, heated, and the light produced put through a spectroscope (essentially a high-precision prism), the spectrum that resulted was not a continuous rainbow of light at all but consisted of distinct lines of color. Each element had its own unique pattern of lines, corresponding to different frequencies of color. That energy came in discrete packets and that spectra were not actually continuous suggested a connection between the two.

The early twentieth century had more than its share of geniuses. Planck and Einstein were two. A third, the Danish physicist Niels Bohr, had an idea based on the Planck-Einstein quantum work and on the experimental discoveries of a fourth genius, Ernest Rutherford, which suggested the underlying mechanism that produced these lines. Rutherford and his colleagues in Manchester, England, conducted elegant experiments to show that gold atoms seemed to have a very specific structure: they seemed to consist of a "cloud" of light, negatively charged matter surrounding a core of heavy, positively charged matter buried deep within that cloud. In 1897, Rutherford's colleague J. J. Thomson isolated a negative particle, the electron, which led to the conclusion that the negative cloud consisted of electrons. Physicists began to call the heavy central core of the atom the "nucleus." (Rutherford discovered the positive particle in the nucleus, the proton, in 1919. Neutral particles in the nucleus, called neutrons, were only discovered in 1932. They played an important part in Fermi's story in due course, but when Fermi arrived in Florence, no one yet knew that they existed.)

With Rutherford's discoveries in mind, Bohr began playing with a compelling idea. Perhaps electrons were confined to specific "orbits" around a nucleus and could not exist in the spaces between these orbits. All electrons had a minimum orbit, called a ground state, below which they could not go. Otherwise, an electron with negative charge would be attracted to a nucleus with positive charge and all matter would collapse into itself. Electrons could, though, "leap" from one orbit to another, stimulated by the absorption or emission of a particle of light, that is, Einstein's photon. The leap from one orbit to another would correspond to the emission or absorption of a specific frequency of light, depending on the frequency of the electron's orbit around the nucleus. Bohr imagined this movement to be similar to the way in which planets orbit the sun.

Bohr's model was the work of a true genius and was perhaps *the* critical breakthrough (although in most important details it was wrong). Good as it was, though, it left some important questions unanswered. It seemed clear that when an electron was stimulated by a photon, it would absorb that photon and jump to a higher energy orbit, but it was not at all clear that it would always jump the same way. Of equal interest: What would determine whether it jumped back down to a lower energy level, thus emitting a photon? What determined the actual energy level of each orbit? These puzzles defied explanation, but they were reflected and magnified by some spectroscopic phenomena.

One such phenomenon was called the Zeeman effect, named after the Dutch physicist Pieter Zeeman who first observed it. In a magnetic field, the spectral lines produced by an element split. Another phenomenon was also puzzling: even without a magnetic field, some lines in a given spectrum are notably more intense—that is, they shine brighter—than other lines. Were there laws that determined why some lines would be brighter and others dimmer?

For physicists these puzzles were far from trivial. They were deeply disturbing. A full theory would be expected to explain all observed phenomena and to predict with precision any particular observation. The classical mechanics theory of Isaac Newton and the classical electromagnetic theory developed by James Clerk Maxwell each produced specific and precise predictions about the phenomena they addressed and were the standards against which physicists judged a new theory's success. From this vantage point, quantum theorists were proving distinct failures.

Fermi was certainly aware of these problems. The nine months he spent in Göttingen exposed him firsthand to the frustrations of those struggling with these issues, including Born, Heisenberg, and Jordan. That he continued to keep abreast of their work is apparent in his work during his subsequent stint in Leiden, where he published his paper on the problems of predicting the intensity of spectral lines in November 1924. Fermi cited work by Sommerfeld, Heisenberg, and Born, so he certainly was aware of the progress being made during 1925. His own contribution came toward the end of the year, based largely on the work of a young theoretical physicist from Vienna named Wolfgang Pauli.

Pauli's enormous gifts in mathematics and physics were recognized early and cultivated by those around him. He studied in Munich with the great theorist Arnold Sommerfeld, who famously said that by the time Pauli arrived in Munich there was little that Sommerfeld could teach him. Like Fermi, Pauli made his first contributions in relativity theory, writing a treatise on the subject at the age of sixteen. A subsequent treatment of the subject, written for a German encyclopedia on mathematics, established his international reputation while he was still at university.

Like Fermi, he was a child prodigy, and like Fermi he was short, about five-foot-five. In practically every other way he was Fermi's antithesis. Physically, Fermi was solid but fit and unremarkable in appearance. Pauli was round, tending toward obesity, but he was darkly attractive, with soulful eyes and sensuous lips. Fermi was not a drinker; Pauli drank heavily and struggled with alcoholism throughout his life. Fermi habitually retired early; Pauli was a night owl, enjoying the sybaritic life at local cafés and cabarets wherever he happened to be. While at university in Munich, Pauli mixed with artists, writers, musicians, and other bohemians in the Schwabing district of the city. The most bohemian activity Fermi would indulge in was a hike in the mountains.

Fermi was gifted as both a theorist and an experimentalist. As a theorist Pauli was perhaps even more gifted than Fermi, but as an experimentalist he was a disaster. Physicists joked that if a piece of equipment wasn't working properly, Pauli must be in the vicinity. Pauli also had a legendary mean streak, something Fermi completely lacked. Pauli gave new meaning to the phrase "acid-tongued." He is said to have described a somewhat undistinguished colleague of his with the incredibly dismissive "So young and already so unknown." He once derisively called Fermi a "quantum engineer." Of a particularly murky and speculative theoretical paper, he famously exclaimed, "It's so bad it's not even wrong." He delighted in calling his close friend Werner Heisenberg a "fool." As a young man he informed the eminent British astrophysicist Arthur Eddington that work the older man was pursuing in general relativity theory was "meaningless for physics." When Einstein came to lecture at the University of Munich, Pauli was the first to speak after the great man's talk. "What Professor Einstein has just said is not really as stupid as it may have sounded," he explained helpfully.

As noted previously, Fermi was not a philosopher and rarely showed any interest in intellectual or cultural matters beyond physics. He had little time for religion or spirituality. Pauli was born Catholic and took his religion seriously, rejecting the opportunity to join his irreligious colleagues in disparaging religious beliefs. He had a deep mystical streak, which he indulged later in life in conversations and in a decade-long correspondence with the great psychoanalyst Carl Jung.

He was also a man of obsessions, and by the time he was doing a post-doc in Copenhagen under Niels Bohr, over the winter of 1922–1923, Pauli became obsessed with the "anomalous" Zeeman effect, a problem with which he would struggle over the next several years. At this time a colleague ran into him walking the streets of Copenhagen in a state of despair. "You look very unhappy," his friend commented. "How can one look happy when he is thinking about the anomalous Zeeman effect?" came Pauli's reply.

As mentioned above, the Zeeman effect involved the splitting of spectral lines in the presence of a magnetic field. Bohr's model explained the normal tripling of spectral lines, but often more than three lines would appear—sometimes four, sometimes six. By early 1925, as a professor in Hamburg, Pauli developed a solution of sorts but wasn't completely satisfied.

In the Bohr model, electron orbitals required three numbers to specify their location, their frequency, and their orientation. Location corresponded to their distance from the nucleus. Electrons could only be in certain specific orbitals and could not exist in the spaces between orbitals. The frequency related to the speed of their orbits. The orientation number referred to how the specific orbital oriented to the axis of the nucleus. Each one of these three numbers was a quantum "number" that, taken together, identified the quantum "state" of an electron.

Pauli realized that if a new quantum number was added to the mix, a number that could have only one of two opposite values, the anomalous Zeeman effect could be explained. He also posited—because the mathematics explained the phenomenon so beautifully—that no two electrons could share all four quantum numbers. This may not seem terribly subtle, but the implication is astonishing: two electrons can be in exactly the same place at the same time, moving at the same speed and in the same orientation, as long as their fourth quantum number is different. Thus was born Pauli's "exclusion" principle.

Pauli tried to find physical interpretations of this fourth number, but eventually gave up. One idea, first suggested by a young colleague of his, Ralph de Laer Kronig, was that electrons actually "spin," and that the spin can either be "right-handed" or "left-handed," producing intrinsic angular momentum in opposite directions. Pauli ridiculed the idea, because it implied, at least to him, that if you measured the speed of the electron at its "equator" (think of a spinning globe) that speed would have to be faster than light, which was impossible. Suitably chastened, Kronig dropped the idea. It was picked up again at the end of the year by Fermi's good friends Uhlenbeck and Goudsmit, who did not know of Pauli's objection. If they had, they might not have published their idea, but as it turned out, the two of them did not know of Kronig's spin proposal and are thus often given credit for the idea of electron spin. Pauli still resisted the idea. It was Dirac who, in 1926, pointed out that a relativistic interpretation permitted—even required—electron spin.

Pauli's exclusion principle was audacious. To understand his achievement and his particular way of doing physics, one must appreciate that he proposed a mathematical solution that had no physical interpretation at all—it was simply the only way he could imagine to solve the problem he was confronting. It was, in a sense, like Planck's invention of the quantum, a mathematical solution without an underlying physical interpretation. Like Planck's idea, it took Pauli enormous courage to propose it and even more to resist the efforts of colleagues to propose the most obvious physical interpretation. That he was wrong in resisting the idea of electron spin takes nothing away from the achievement itself, a beautiful one which bespeaks his own particular genius.

THE EXCLUSION PRINCIPLE ALONE WAS A GREAT STEP FORWARD. It provided an explanation of the anomalous Zeeman effect and gave a more complete understanding of how electrons "fill up" the orbits surrounding the nucleus. More breakthrough work was done that year. A young German named Werner Heisenberg, working in Göttingen with Max Born and a young theorist named Pascual Jordan, formulated a method to analyze the way electrons move between orbits, using a mathematical technique called matrix multiplication that successfully explained the varying intensity of spectral lines. Late in the year, a Viennese physicist named Erwin Schrödinger, closeted away in a ski chalet with one of his numerous lovers, came up with a differential "wave" equation that did much the same thing, using techniques more familiar to the average physicist. Physicists now had two highly effective ways of delving into previously incomprehensible physical phenomena.

Why did both approaches, so different in form and function, provide the same answer to the thorny questions raised by quantum theory? It took yet another genius to show that matrix mechanics and wave mechanics were two sides of the same coin.

He was, in the words of his biographer Graham Farmelo, "the strangest man."

Paul Adrien Maurice Dirac was the baby of that generation of quantum pioneers, a year younger than Fermi and Heisenberg, two years younger than Pauli. Born in Bristol, England, he was the son of a strict father whose severe discipline had a lasting effect on the young man. Tall, thin, and extremely reserved, he rarely spoke in sentences longer than two or three words. His teachers at the Merchant Venturer's Technical College considered him brilliant, and he easily obtained a place at St. John's College, Cambridge, but his family finances prevented him from attending. Instead, he attended the University of Bristol for his undergraduate education and St. John's for his graduate degree, the first to be given in the field of quantum theory. In his thesis he provided an independent derivation of Heisenberg's quantum mechanics based on his observation that the underlying mathematical structure of the matrix algebra, in particular its noncommutative property, was analogous with a particular form of the mathematics physicists used to express classical mechanics.

Dirac was, in some sense, the anti-Pauli. We might think of him today as suffering from Asperger's syndrome. He was socially awkward in the extreme and literal-minded to an exasperating degree. He once asked Heisenberg why people danced. Heisenberg replied, "When there are nice girls, it is a pleasure." Dirac considered this for a moment and blurted out, "But how do you know beforehand that the girls are nice?" Once during a lecture a student raised his hand and said that he did not understand an equation that Dirac had written on the blackboard. Dirac remained silent because, as he explained later, the person in the audience had not asked a question. His colleagues at Cambridge reportedly defined a "dirac" as a unit of one word per hour. He was also aggressively irreligious. In one famous exchange at the 1927 Solvay conference, talk among the younger physicists ventured into the area of philosophy and religion. Heisenberg recounts that Dirac made what was for him an impassioned plea to the effect that religion had no place in the world of a physicist. Pauli, who was silent for much of the discussion, reportedly ventured, "Well, our friend Dirac has got a religion and its guiding principle is 'There is no God and Dirac is His prophet.'" Heisenberg reported that everyone had a good laugh, none more so than Dirac himself.

Among the brilliant young theorists of his generation, Dirac may well have been the most brilliant. His PhD thesis was impressive and attracted the attention of the physics world at large. Born was stunned that a doctoral student could master the field so completely, especially since the final Born-Heisenberg-Jordan paper had not yet been completed. Others shared Born's surprise. Almost immediately, Dirac was propelled into the highest ranks of theoretical physicists. His was the first PhD degree ever granted in the new field of quantum theory and heralded the arrival of a superstar. Later in the year, Dirac followed it up during a post-doc in Copenhagen with an even more impressive paper, which demonstrated mathematically the underlying unity between the Heisenberg and Schrödinger approaches, seeing both as special cases of something called transformation theory. It was a significant finding, but Dirac's best was yet to come. In 1927, he produced the first paper to develop the concept of a quantum *field* using the electromagnetic field as his focus. The paper had an historic impact on physics and laid the groundwork for Fermi's second great contribution.

In 1925, however, all of Dirac's major contributions were still in the future. As the year came to an end, while Schrödinger was holed up in the Austrian Alps with his girlfriend during Christmas wrestling with his equation and putting it in the proper form and while young Dirac was grinding out his thesis, Fermi was also thinking deeply about quantum problems. But the problems he considered were slightly different. He had been contemplating them for several years and now the work of Pauli, in particular, showed him the way toward a solution.

CHAPTER FIVE

OF GECKOS AND MEN

WHEN HE ARRIVED IN FLORENCE IN 1925, FERMI ASSUMED HIS position as a lecturer at the Institute of Physics, which served as the University of Florence's physics department. Perched on a low hill in the Arcetri area just south of the old town, the institute was far away from the rest of the university, which was closer to the city center. The setting may have been inconvenient, but it was quite beautiful and Fermi loved it.

Andrea Garbasso, the head of the institute, sought to staff the institute with the brightest physicists of the younger generation. The previous year he brought Rasetti on board to teach experimental physics. Now he was delighted to take on Fermi to teach theory. Fermi spent the next two years lecturing on physics to engineering students at the university. He taught two courses, one in mathematical physics and the other in classical mechanics (called "rational mechanics" at the time). The notes for his mechanics lectures were compiled into a beautiful handwritten manuscript, which could serve as a good introduction to the subject even today.

When he was not teaching and when he was not in the library reading about the pathbreaking work being done in Göttingen, he and Rasetti were hiking in the hills around the institute or playing pranks on members of the institute staff or, occasionally, both at the same time. The two pranksters noted the somewhat timid nature of the local girls who served meals at the institute cafeteria. Together they hatched a plan: they would collect bagfuls of small lizards, catching the tiny but plentiful creatures using six-foot-long rods of Pyrex glass fitted with silk loops at the end. Then they would release them in the cafeteria and watch the horrified reaction of these unsuspecting and unsophisticated women. It sounds like a Pisa prank, with all the attendant childishness and unspoken misogyny.

Fermi had little interest in geckos or their habits. Lying in wait for the geckos, he thought about Pauli's work and its implications for a problem he had been grappling with since 1924—how to describe a "perfect gas" in quantum theoretical terms. A paper he wrote in January 1924, while he was still in Göttingen, is an important piece of evidence. He was struggling with the application of quantum theory to statistical mechanics, particularly with respect to the concept of entropy. He needed a hook on which to hang an analysis that incorporated the work of the Göttingen group into one of his favorite fields.

Pauli's work provided just such a hook.

Given his voracious reading, Fermi almost certainly was aware of Pauli's original paper when it was published in March 1925. He may, as Segrè speculates, have spent more time discussing it with Pauli's colleague Kronig during the summer of 1925, when Kronig accompanied Fermi and others—including a young Edoardo Amaldi—on a hiking trip in the Dolomite Mountains. However it happened, he developed a working knowledge of Pauli's exclusion principle and began to consider how the puzzle over statistical mechanics might be resolved in light of it. The leap of insight he had, waiting quietly for a gecko to ensnare itself in one of his lassos, was that Pauli's exclusion principle could be extended from electrons surrounding an atomic nucleus to single atoms in a gas. With this insight, he could explain the strange behavior of gases under very high pressure or at very low temperatures, when they seem to lose their thermal capacity, or what physicists call specific heat. They become degenerate, filling all the lowest energy states possible within Pauli exclusion constraints.

As Fermi's close friend and colleague Hans Bethe later explained:

> The quantum states of an atom in a gas differ by their velocities; there is one and only one state in which the atom is completely at rest, a

> second state in which it moves extremely slowly, a third with somewhat higher velocity, and so on. Since no two atoms can be in the same state, only one single atom can be completely at rest, all others must be in motion, even at absolute zero temperature.

This, in a nutshell, is the difference between traditional statistical mechanics and Fermi's new interpretation that takes into account the exclusion principle. In the traditional version, the absence of heat implies the absence of motion. In Fermi's version, even in the absence of heat, atoms jostle for position because no two atoms can share the same state (velocity).

In his famous paper, "On the Quantization of a Perfect Monatomic Gas," Fermi laid out the statistical approach required to account for the energy level of such gases, incorporating Pauli's exclusion principle. The idea must have gelled in early September 1925, before the Arcetri geckos went into hibernation. It took him the rest of the year to get it all down on paper in a form that satisfied him. He presented a short version of the paper at the institute on February 7, 1926, and published it in the journal of the Accademia dei Lincei soon thereafter. Six weeks later he expanded it considerably and sent it off to *Zeitschrift für Physik,* where it was received on March 26, 1926.

The paper had an immediate impact, particularly because it gave a mathematical explanation of degeneracy. It was known that electrons in metals exhibit degeneracy. Now people like Pauli, Sommerfeld, and a young graduate student of Sommerfeld's, Hans Bethe, were able to calculate the behavior of electrons in metals using Fermi's approach and discovered that it predicted the observable results. Word spread as far as England, where Paul Dirac read the paper—presumably the German version—and promptly forgot about it because it solved a problem that held no interest for Dirac at that particular moment, focused as he was on finishing his PhD thesis. Later in the year, Dirac developed an interest in the problem, approached it from scratch, and came up with a slightly different method that led to the same analytic conclusions. Dirac's approach was broader, extending to particles that do not obey Pauli's exclusion principle but rather do obey what was then known as Bose-Einstein statistics, now called "bosons." He presented his paper to London's Royal Society on August 26, 1926. The editors of the Royal

Society's *Proceedings* may not have been aware of Fermi's prior work on the subject.

When he read Dirac's paper, Fermi was surprised and perhaps a bit annoyed that Dirac made no mention of Fermi's previous paper. He sent a letter in somewhat stilted English to Dirac calling the latter's attention to his prior work:

> In your interesting paper "On the Theory of Quantum Mechanics" . . . you have put forward a theory of the Ideal Gas based on Pauli's exclusion Principle. Now a theory on the ideal gas that is practical [sic] identical to yours was published by me at the beginning of 1926. . . . Since I suppose you have not seen my paper, I beg to attract your attention to it.

Contrary to Fermi's presumption, Dirac had indeed read Fermi's paper. He had simply forgotten it. As he later explained:

> When I looked through Fermi's paper, I remembered that I had seen it previously, but I had completely forgotten it. I am afraid it is a failing of mine that my memory is not very good and something is likely to slip out of my mind completely, if at the time I do not see its importance. At the time I read Fermi's paper, I did not see how it could be important for any of the basic problems of quantum theory; it was so much a detached piece of work. It had completely slipped out of my mind, and when I wrote up my work on the antisymmetric wave functions, I had no recollection of it at all.

He immediately sent an apology to Fermi and from then on always referred to the statistics as Fermi-Dirac, generously giving Fermi primary credit for it. That we now refer to particles that obey the exclusion principle as fermions and those that do not as bosons is a direct result of Dirac's acceptance of Fermi's claim of priority.

As is the case with many important developments, other talented theorists were thinking about these matters at the exact same time, including Born's young protégé Pascual Jordan. In December 1925, Jordan gave Born a manuscript to read while Born traveled to the United States, where he was to present lectures at MIT. Born put it in his suitcase and quickly forgot it. Six months later, Born was rummaging through his suitcase and discovered the unread manuscript. As he browsed through it, he realized that he had inadvertently cheated Jordan of credit for the discovery because the Jordan paper proposed using the Pauli exclusion principle in exactly the same way as had Fermi and Dirac, but even earlier. For his part, Fermi would never have allowed six months to elapse before hearing back from Born. Indeed, so great was Fermi's self-confidence, he probably would not have bothered to ask Born to review it in the first place.

Fermi had no intention of remaining in Florence as a lecturer for the rest of his career. He was ambitious and wanted a chair at a major university. In this he had an ally in Senator Corbino. Fermi's work on quantum statistics had earned him a global profile among the elite of the physics world and would, under normal circumstances, have led to offers at any number of prestigious Italian institutions. That it had not yet done so reflected the highly complex, political nature of Italian academic life. Fermi needed the influence and support of someone like Corbino to make his ambition a reality.

There was a formal system in place in Italy for the recruitment of professors at Italian universities. When an opening was available, the Italian Ministry of Education would run a competition, or *concorso,* in which it would assign a small committee of senior professors to make a recommendation of a candidate to fill the opening. Each of the committee members would nominate potential candidates, and candidates' respective qualifications, including publications and any prior teaching assignments, would be evaluated and compared. During 1925, a position opened at the University of Cagliari in Sardinia. In the *concorso* Fermi found himself in competition with Giovanni Giorgi, a man old enough to be his father who had been teaching at the University of Rome since 1913. He was a solid physicist of the old school, with no fundamental contributions to his name. Fermi's candidacy fell victim to the reluctance of powerful professors to embrace the new physics. The five members of the committee split 3–2 in favor of Giorgi. Levi-Civita and Volterra voted for Fermi; three others put their support behind Giorgi. The three who voted for Giorgi were eminent Italian physicists of the old school; they only reluctantly accepted relativity theory and were relatively immune to the new quantum theory. Levi-Civita, of course, was one of the main advocates of relativity theory, and Volterra was also a strong advocate of the new physics. Segrè notes that Fermi viewed the outcome as "unjust" and never forgave those who voted against him.

The *concorso* for the University of Rome was a much bigger prize and Fermi was clearly eager to fill the opening once it was established. Corbino worked hard to get approval for a position in theoretical physics, which would be the first chair of this type in Italy. The other physics chairs were in either experimental physics or mathematical physics. There was some thought that Corbino might get official approval for the position as early as 1925; however, the matter was postponed until the fall of 1926. That Fermi emerged as the *concorso*'s unanimous first choice was hardly surprising, because Corbino, who chaired the committee, had stacked it with Fermi supporters. Persico and another member of the Rome Institute, Aldo Pontremoli, came in second and third, respectively. Persico went north to Florence, and Pontremoli, to Milan.

Segrè notes that Fermi paid a great deal of attention to his career during this period and seemed obsessed, at least in his letters to Persico, with the mechanics of the *concorso* process and his ability to prevail and beat the competition. He also notes, with barely disguised bitterness, that Fermi did not recall his own struggles to achieve his career objectives:

> That Fermi had been intensely interested in his formal career may come as a surprise to those who knew him in his later years. I remember that as early as 1930, or shortly thereafter, he showed little interest in his friends' and collaborators' problems of academic advancement. He was of course right in giving overriding priority to scientific achievement and to favorable working conditions over career questions; nevertheless, he seems to have forgotten very early the way he himself felt in his youth.

Perhaps Segrè was correct in his interpretation of Fermi's reluctance to intervene on behalf of the careers of more junior colleagues. Perhaps it reflects, as Amaldi once suggested, that Fermi believed the world of physics was a tough one, and one needed to be tough to survive in it. His behavior may have been his way of preparing his younger colleagues. It is also possible, however, that as Fermi became acutely aware of his growing power and influence within Italian academic circles he felt that any intervention he might make would give the recipient an unfair advantage. Fairness, rather than insensitivity, may have been his top priority. In any case, as Segrè acknowledges, this changed noticeably in his later years.

In the report of the Rome *concorso* committee, published in November 1926, Fermi receives fulsome praise, but one particular sentence, crafted by Corbino and expressing his hopes for his young colleague, stands out: "he moves with complete assurance in the most difficult questions of modern theoretical physics, *in such a way that he is the best-prepared and most worthy person to represent our country in the field of intense scientific activity that ranges the entire world.*" This was Corbino's dream and he had used his considerable influence and political savvy to make it possible. He happened to be right. Fermi was indeed a unique figure, someone who had demonstrated that he could fulfill the Corbino plan and probably the only one who could do so. Now, with all the expectations of Corbino and his colleagues resting on his shoulders, Fermi had to deliver.

PART TWO

THE ROME YEARS

CHAPTER SIX

FAMILY LIFE

THE YEARS FERMI SPENT IN ROME—1926 THROUGH 1938—WERE some of the most important years, personally and professionally, of his entire life. He married and had two children. He developed, with the help of his mentor Corbino, a major international center for physics education and research, attracting not only the best minds in Italy but also major young physicists from other European centers of excellence. He organized two major international physics conferences, which put Italy on the physics map. As a theorist, he championed Dirac's quantum field theories to the broader physics world and used those theories to develop the first theoretical explanation of beta decay. Finally, during these years he discovered induced radioactivity through slow-neutron bombardment. In the process he became—reluctantly, but inexorably—a celebrity of the fascist regime.

FERMI ARRIVED BACK IN ROME IN EARLY 1926 AND TOOK UP RESIDENCE with his father and sister in their new home in Città Giardino Aniene at Via Monginevra 12. Alberto and Maria had moved in just a few months before to a house built to Alberto's specifications, more comfortable than the cramped apartment on Via Principe Umberto where the family had lived since 1908. Città Giardino Aniene was what the British would call a "garden suburb," a development within the Roman metropolis where each house had its own plot of land with a private backyard.

It was, in its time, suburban living at its finest, a little over four miles from Fermi's new offices at Via Panisperna. Fermi would walk or bike to work, or sometimes take public transport part of the way. He had not yet bought his first car.

On Saturdays, Fermi returned to the vibrant salon of the great Italian mathematician Guido Castelnuovo. He had attended briefly in the year prior to his assignment to Florence. Now he was far better known to the eminent mathematicians who joined Castelnuovo every Saturday, including Vito Volterra, Tullio Levi-Civita, Federigo Enriques, and Ugo Amaldi, who brought along his son Edoardo, just eighteen years old. These were some of the greatest mathematicians of their generation, and they understood the potential Fermi had to revolutionize Italian physics. They welcomed him—all the more so because of his new work on quantum statistical mechanics.

At these salons, Fermi would spend part of his time with the older gentlemen discussing developments in math and physics. There was much to discuss, especially now that the logjam in quantum physics had given way to major breakthroughs. He would also pull away from the adults and enjoy the company of the youngsters who amused themselves in a separate room. There he joined a small circle of social friends, including Edoardo Amaldi and Castelnuovo's attractive daughter, Gina. Persico, for the time being lecturing at the University of Rome, also attended the salons and soon was part of the group of young people. Eventually, Emilio Segrè joined them, as well. One other friend admitted later to the ring was a fabulously beautiful young woman named Ginestra Giovene, a Roman who started taking classes in physics at Via Panisperna. In short order she would become Edoardo Amaldi's wife.

On Saturdays they would enjoy themselves at the salon, while the older people spoke of more serious things. Fermi would invent silly games, one of which, "fleas," consisted of making small coins bounce across a felt-covered table. Fermi would also play "director" and pretend to direct the others in an imaginary movie. They would often meet up the next day as a group, with Fermi in the lead, setting out on hikes and climbs in the parks and hills that surrounded Rome. Among the group was a young woman Fermi had met a few years before, the daughter of prominent Italian admiral Augusto Capon. Her name was Laura.

Laura Capon had grown into a young woman since the day Fermi had met her just prior to his Florence assignment. In 1924 she was only sixteen years old, and they met once more briefly during the summer of 1925. By the time Fermi began to notice her, at these salons and on numerous group hikes and outdoor adventures, she had become a lovely woman of eighteen years. She was slim and attractive, gifted with a sharp mind and an equally sharp tongue. In Laura Capon, Enrico Fermi had met his match.

During her time as a science student at the University of Rome, she had taken courses in math and physics with Persico. Indeed, she writes that it was the thought of seeing the handsome, blond Persico in a social setting that made her persuade her father to accept the open invitation to attend the Castelnuovos' Saturday salons. Laura would sometimes bring her sister Anna to these gatherings. Anna, an artist with little patience or interest in her sister's science-obsessed friends, dismissively dubbed them "the logarithms," a name that stuck. The friendships Laura developed with Persico, Amaldi, Ginestra, Gina Castelnuovo, and the others were completely independent of the connection to Fermi, and they lasted her entire life, well beyond Fermi's death.

The reticence Fermi showed in every aspect of his personal life prevents us from following their courtship from his point of view. He wrote neither letters nor diaries that narrate the development of their relationship. The only written accounts are Laura's, and they do not provide a detailed picture either. What we do know is that the two drew closer during these outings. Fermi must have found her personality attractive, and she was quite pretty. She also came from a well-to-do family, with greater resources than Fermi's own family. For her part, she clearly understood how gifted he was and also admired his charisma and outgoing personality.

What she didn't appreciate, however, was his tendency to tease her, especially in the presence of his old friend Rasetti, who took up a position at Via Panisperna as a professor of experimental physics not long after Fermi arrived. On outings and road trips, the two of them would gang up on her mercilessly, testing her on all sorts of trivia and ridiculing her when she got an answer wrong. On one occasion they grilled her on the name of a shell they found on the beach at Ostia, a favorite summer destination of the group. Another time they quizzed her on geography: what is the capital of Afghanistan? When she could not answer correctly, they collapsed in laughter. They would also gang up on the others, including the young Segrè. The two of them would feed off each other, but it appears that Rasetti was usually the instigator.

Laura recounts one particular question Fermi posed and then solved himself. It is perhaps the earliest account of what has come to be known as a Fermi problem—a problem of seemingly insurmountable complexity that can be answered to within an acceptable degree of accuracy by making some simple, reasonable assumptions. On one excursion the group came upon an anthill:

> We could see nothing of interest, only a common anthill.
>
> "How many cerebral cells work at building this mound? Would you say that ant brains yield more or less work than human brains per unit of cerebral matter?" Enrico would pull out of his pocket the small slide rule that never left him. "Let's see . . . in a cubic centimeter of neurons . . . " In a short while he would raise his triumphant eyes on us. "I have figured the answers. And you?"

Fermi had the analytical skills to figure out almost anything by himself. Rasetti had a seemingly inexhaustible base of factual knowledge on a vast range of subjects. Together, they were more than intimidating and could give anyone an inferiority complex. Laura was up to the challenge. At one point, after a few years of humiliation, she and Ginestra, by now a close friend, resolved to gain the upper hand in these inquisitions and mastered the entry for the old Egyptian city of Alexandria in the *Enciclopedia Italiana*. That Sunday they grilled Fermi and Rasetti on the topic and silenced them, for the first and only time.

It was during the summer of 1925, when Fermi vacationed with other members of the Castelnuovo salon and spent a brief time with Laura, that he confided to her, in a matter-of-fact way, that he grouped people into four categories of intelligence: lower than average, average, intelligent, and exceptional. Laura describes how she gave back as good as she got, teasing Fermi:

> "You mean to say," I commented, assuming the most serious expression I could manage, "that in class four there is one person only, Enrico Fermi."

> "You are being mean to me, Miss Capon. You know very well that I place many people in class four," Fermi retorted with apparent resentment; then he added on second thought: "I couldn't place myself in class three. It wouldn't be fair. . . . Class four is not so exclusive as you make it. You also belong in it."
>
> He might have been sincere at the time, but later he must have demoted me to class three. Be that as it may, I have always liked to have the last word in any argument and so I said with some finality: "If I am in class four, then there must be a class five in which you and you alone belong." To everyone except Fermi, my definition became a dogma.

This exchange, so early in their courtship, reveals what they found so attractive in each other. She knew he was uniquely brilliant and admired him for it, even as she teased him about it. He understood that she, too, was quite intelligent and difficult to intimidate. Only over time did his constant teasing wear her down. Nevertheless, they must have found a deep mutual compatibility because by 1928 they were married.

Sadly, Alberto did not live to see the couple wed. He had been ailing since Ida's death in 1924, and he died in May 1927, his children at his side. For the next twelve months, the two siblings shared the home in Città Giardino.

Enrico appears not to have expressed any outward grief. None of his early biographers mention it and no letters or diaries exist that allow us to pull back the curtain on what must have been a traumatic moment in Fermi's life. It is completely consistent, of course, with the universal view of those who knew him that he rarely if ever expressed personal feelings. He was still in the grip of the moment when, at the age of two, he broke down in tears at the first sight of his family and his mother scolded him for it.

THAT SUMMER FERMI BOUGHT A CAR.

Generally frugal, he chose the cheapest model Peugeot, a "bébé" Peugeot, a yellow two-seat convertible that looked as silly as it was cheap. He had grandly announced to his friends that soon he would be either getting married or buying a car. Alarmed when she heard the news about the Peugeot, Laura's sister Cornelia wrote to Laura, who was spending the last weeks of summer at her uncle's Tuscan country villa. Everyone in their circle must have known that Laura and Enrico were interested in each other, and Cornelia wanted Laura to hear directly from her as soon as possible the presumably bad news, that Fermi had chosen car over wife, before she found out from someone else. Laura claims to have been pleased by the news and assured her other sister Anna, who was with her at the villa, that she had decided to become a professional woman and had no interest in marriage. Besides, Fermi had described his ideal wife to her and she bore no resemblance to that ideal: "He wanted a tall, strong girl of athletic type, and blonde if possible; she must come from sturdy country stock, be nonreligious, and her four grandparents must be alive." Laura put this down to Fermi's belief in eugenics, common at the time, as well as his love of sports and his general skepticism when it came to matters metaphysical. It appears to have crossed neither her mind nor anyone else's that his description of the ideal wife might have been meant to distract Laura from discovering just how interested he was in her or that he bought the car specifically to help woo her. Soon he was taking her, along with other "logarithms," out for drives in the countryside, often accompanied by Rasetti, who had his own car and could be relied upon to give Fermi a hand when his car broke down, as it often did. For Fermi, driving became a lifelong passion.

Over the next few years, well after their marriage, the car became a character in the Fermi story. Its egg-yolk-yellow color, its dense trailing cloud of black exhaust fumes, and its faintly ridiculous shape made it a sort of minor celebrity on the streets of Rome. Often after a movie or dinner, Laura and Enrico would return to the parked car only to find an amusing (or insulting) note left behind on the windshield or seat.

WE DO NOT KNOW WHEN FERMI PROPOSED—IT WAS PROBABLY IN late 1927—but propose he did. Laura was quick to accept, and the civil ceremony was set for July 28, 1928. In Rome, civil weddings were held in the city hall atop the ancient Campidoglio, or Capitoline Hill, the legendary site of Romulus's founding of the city. According to Laura, it was a very hot day in Rome—104 degrees in the shade, in the days before air conditioning. The wedding party was supposed to meet at the spacious Capon residence at Via dei Villini 33 and head downtown from there by car. At the appointed hour, all except the groom had arrived.

FIGURE 6.1. Laura and Enrico in their "Bébé Peugeot." *Courtesy of Rachel Fermi.*

It wasn't a case of nerves that delayed Fermi. The suit he had ordered arrived with about three inches to spare in the sleeves and trousers and, always one to do his own handiwork, Fermi told his sister to go on while he adjusted them to the proper length. Maria brought word to the group that the groom was suffering not from cold feet but from short arms. Everyone was relieved. Soon the groom himself arrived and the group made their way up to the Campidoglio.

The ceremony was short, and soon afterward the wedding party spilled out into the plaza for a group photo. It is a wonderful photo—so wonderful, in fact, that Fermi's Argonne lab colleagues chose it as the cover of the tribute record album *To Fermi with Love,* produced after his death. The newlyweds are at the center, Fermi grinning from ear to ear, Laura smiling more demurely under a round-topped hat. Sharing center stage with them are Laura's father, in full ceremonial uniform; Senator Corbino, the closest thing to a father figure for Fermi since his own father passed away a year earlier; Maria, Laura's sisters, and various other relatives; and, at the back, his head rising ostrich-like above the rest of the group, Fermi's old friend Franco Rasetti. The joy of the occasion is obvious, despite the sweltering midday heat.

Figure 6.2. Enrico and Laura's wedding party, 1928. Laura's father, Admiral Capon, is at center, next to Enrico. Laura is to Enrico's right. Corbino is seen just behind the admiral's left shoulder, and at the back, peeking over Corbino's left shoulder, is Rasetti. *Courtesy of Rachel Fermi.*

The newlyweds took their honeymoon at a mountain inn nestled in the village of Champoluc in the Italian Alps, ten miles due south of the Matterhorn. The flight from Rome to Genoa, in a Dornier sea plane operated by Italy's fledgling commercial air transport system, was Laura's first. She was apparently terrified but was proud to have hidden it from her new husband. From Genoa they took a train into the mountains.

Settled in their hotel, the couple spent part of their time hiking and exploring the magnificent landscape of the Val D'Aosta, one of Italy's most beautiful alpine valleys. Naturally, Enrico decided that this was the right time to teach Laura electromagnetic theory. A student of general science at the University of Rome, she had a good understanding of basic physics. Incapable of *not* teaching—it was perhaps his principal mode of communication—Fermi resolved to use the honeymoon to teach his new wife Maxwell's famous equations on the electromagnetic field. She was patient and, to her credit, a willing and able student. She was also unafraid to poke holes in her husband's explanations when she saw them. She knew he was one of the most brilliant men in his field, but that didn't stop her from challenging him when she thought he was wrong:

> Patiently I learned the mathematical instruments needed to follow each passage. Faithfully I went over Enrico's explanations, trying to keep my eyes from the window and the inviting meadow that I saw through it, until I had digested my lesson and made it material of my own brain. Thus we arrived at the end of the long demonstration: the velocity of light and of electromagnetic waves were expressed by the same number.
>
> "Therefore," Enrico said, "light is nothing else but electromagnetic waves."
>
> "How can you say so?"
>
> "We have just demonstrated it."
>
> "I don't think so. You proved only that through some mathematical abstractions you can obtain two equal numbers. But now you talk about the equality of two things. You can't do that. Besides, two equal things need not be the same thing."

"I would not be persuaded," she concludes, "and that was the end of my training in physics."

Back in Rome, the material fruits of Fermi's marriage to Laura became quickly apparent. Laura's parents settled a dowry with which the couple bought a relatively spacious apartment in Rome at Via Belluno 28, a short walk from her childhood home. At the penthouse level of a mustard-colored six-story building, the apartment was airy and gave the newlyweds an expansive view of the attractive neighborhood.

Laura also could afford domestic help, something her parents had enjoyed and to which she had become accustomed. By the time of their move to the United States, she managed a staff of three maids, who helped take care of her two children and maintained the household to her standards. By all accounts she was a good cook but left the rest of the household chores to her staff.

Fermi, in the midst of building up one of the most important centers for physics research and education in the world, had but one domestic responsibility when they started out. He was, by Italian tradition, the person to furnish the family home. He provided the money, but showed no interest in actually purchasing the furnishings, a task he left to Laura and her mother. His only instructions were that the tables and chairs should have straight legs, in keeping with his penchant for simplicity. Laura reports that she and her mother decided on pieces with some curves, but these were sufficiently gentle not to bother her demanding husband, who furnished his own study in the apartment with extreme simplicity—a chair, a large table that served as a desk, and a small bookcase. Laura was not the only one to comment with surprise that her husband had few books by his side as he worked. Fermi never amassed much of a collection of physics texts, shunning all but the most essential reference books in favor of working out material on his own.

He settled in to a routine that rarely if ever varied. He would rise at five thirty in the morning, don a blue flannel bathrobe, disappear into his study for two hours, work at whatever physics problem he was trying to solve, emerge at exactly seven thirty—he never used an alarm clock, but had a precise sense of time embedded in his brain—and prepare for the day. After a quick breakfast, he would leave the apartment at exactly eight o'clock. He would return from Via Panisperna at one in the afternoon for what the Italians call "dinner" and spend the next two hours reading or playing tennis until three, when he would promptly return to Via Panisperna to pick up where he had left off, and then return home at eight in the evening for "supper." By nine thirty, he would be dozing off and never went to bed later than ten o'clock. At five thirty the next morning, the routine started all over again. It would take some major development at Via Panisperna to make him vary this rigid routine. In at least two cases we know of—the discovery of slow neutrons in October 1934 and the operation of the first atomic reactor in Chicago in December 1942—he refused to work through lunch and broke off experiments so he and his colleagues could enjoy a civilized meal.

All who knew him observed this habitual regularity, leading Segrè, in an unkind moment, to describe Fermi as someone with the personality of an Italian bureaucrat. Segrè may have been right, but organizing his daily life as he did merely reflected the tidy layout of Fermi's thinking. Though such orderly habits might have been mistaken for a lack of creativity or imagination by some colleagues, it is difficult to picture Fermi living any other way, or achieving what he did in any other fashion.

Laura Fermi was not quite telling the truth when she stated, in her description of her honeymoon encounter with Maxwell's equations, that this was the end of her learning about physics. Throughout her life she generally gave her husband a wide berth when it came to work and usually did not interest herself in the various projects he was working on. She was fond of describing some of his more important contributions as "obscure." Though she must have understood the importance of the work her husband was involved with during the Manhattan Project, she never asked questions about the work and only discovered at the end of the war just how central was her husband's role in it. However, there was one scientific project on which the two of them truly collaborated.

When they were married, Fermi was determined to supplement his somewhat meager academic income. Eventually, he would find a variety of sources that would keep him and his family quite comfortable by any standard, but in 1928 he decided that the most effective way to earn more income was to write a physics textbook for *liceo* students. He invited Laura to join him as a collaborator. He would dictate and she would transcribe and prepare clean copy. Given her level of intelligence and her refusal to be intimidated by her illustrious husband, however, it should surprise no one that she ventured criticism when she felt it was appropriate. Points that seemed obvious to Fermi did not always seem obvious to her, and she told him so. A back-and-forth would ensue, the result of which was, presumably, of great benefit to the students who would use the text.

It took several years for the two-volume set to make it from Fermi's mind through Laura's pen and typewriter to published volumes. Much of the work took place during weekends and summer holidays at her uncle's Tuscan villa, where they took advantage of the quiet solitude to make progress. When *Fisica* came out, it was adopted nationwide as the main physics textbook for high school students. After the war, Edoardo Amaldi assumed responsibility for preparing revisions.

Laura found that she loved writing so much that she and her close friend Ginestra Amaldi collaborated on another book. Published in 1936 and aimed at a general audience, *Alchimia del Tempo Nostro* (*Alchemy of Our Time*) told the story of modern physics, including the discovery of radioactivity and the work of their respective husbands in furthering the understanding of the atomic nucleus. Its success was helped by the fame of their husbands and the fulsome praise it received in the preface by none other than Orso Mario Corbino.

THE WORLD OF PHYSICS IS INTERNATIONAL, AND ANY PHYSICIST WHO wishes to keep abreast of the field needs to travel the globe to conferences, colloquia, summer schools, laboratories, and universities—indeed, to any place where physicists discuss and analyze ongoing developments. It is true even today, but in Fermi's time before the advent of the Internet and instantaneous electronic communication, it was not only important—it was vital.

Prior to moving permanently to the United States, Fermi traveled to the United States at least five times, for summer sessions at Ann Arbor, Stanford, Berkeley, Columbia, and elsewhere. These were opportunities for him to develop a strong network of like-minded scientists eager to share new discoveries and new ideas. He also grew to love the United States, its wide-open spaces, its gregarious people, its optimism, and its openness, all in contrast to the tradition-bound culture of his native Italy. Only on the first trip did he bring Laura along with him. In following years, Laura usually stayed home during his summer travels, partly because she had young children to care for but also because she did not like the United States, at least not at first.

As the summer of 1930 approached, Fermi was invited to attend the international summer symposium for theoretical physics at the University of Michigan. His Dutch friends Sam Goudsmit and George Uhlenbeck came to Ann Arbor in the late 1920s and took over the coordination of the symposium, bringing the first European speakers to the campus and using summers to establish a major center in theoretical physics. The effort worked and by 1941, when the war ended the project, many of the world's greatest physicists had lectured there, including Heisenberg, Pauli, Bohr, Lawrence, and Wigner. The two Dutchmen invited Fermi to the 1930 session, the first of five he would attend. Fermi chose to present a major paper interpreting Dirac's quantum electrodynamics in a strikingly clear, accessible fashion.

At that time, Fermi's command of English was tenuous. Over the years he had learned just enough to be able to read papers published by American and English scientists, but he realized he would need some extra education, so he did what to him seemed the natural thing: he borrowed ten Jack London novels from the library and read them with an Italian-English dictionary by his side. For her part, Laura had learned English in her formal education and decided she knew it well enough not to review it before the trip.

Not surprisingly, they both found out just how little English they really knew the moment they arrived in New York, on the first leg of their voyage to Ann Arbor.

The language impediment was quite upsetting to Laura, but more upsetting was the impression that New York City made on her. The overwhelming size of the city; its verticality; its noise; the hodgepodge of nationalities, races, languages, cultures; the grime and filth on the streets; the summer thunder-and-lightning storms; even the difficulty of getting wine (the Fermis were never big drinkers, but America was in the grip of Prohibition, and the only way they could get a little wine to drink was at speakeasies)—all of these provoked a profoundly negative impression. She was born and raised in genteel circumstances in one of the world's oldest, most beautiful cities. She may have been an urban person, but she was certainly not yet cosmopolitan and was blind to New York's gregarious culture and noisy charm.

Ann Arbor was an entirely different experience. The classic American campus of the university, already one of the finest research institutions in the country, was set in the middle of a charming, quiet, Midwestern town, adjacent to extensive parks and forests. At the summer school she found Europeans she enjoyed being around—both Goudsmit and Uhlenbeck had young wives, and though they were in the process of being "Americanized," as Laura put it, they were still European enough for Laura to relate to them easily and comfortably. The two Dutchmen, for their part, couldn't help but have some fun at Fermi's expense. They promised to sit in the back of the lecture hall when he spoke, take careful notes, and correct his grammar and pronunciation as the summer progressed. What they didn't tell him, though, was that he mispronounced a few words in such a wonderful way they decided they wouldn't correct him.

Fermi loved his time there and returned several times before the outbreak of the war. Laura, though, still found herself out of place. She brooded on what she considered defects of American culture, its "primeval" instincts, exemplified for her in the American tradition of the bounty hunter. She continued:

> There seemed to be a total incomprehension of some instinctive human feelings in the Americans' insistence on separation of the sexes, asking husbands to stag dinner parties, leaving poor young wives to mope at home, or planning wives lunches, where the same poor wives were to find their way among strangers speaking an idiom strange in words and meanings, without the much-needed support of those pillars of strength, their husbands.

It wasn't an antipathy to travel or to foreign cultures. In 1934 the couple spent the summer in Argentina and Brazil, where Fermi gave lectures to standing-room-only crowds. She loved it there. There was something specific about American culture that did not appeal to her. She would change, of course. The decision to stay in the United States after the war may have been her husband's, but the decision to remain in Chicago after his death in 1954, until her own death in 1978, was hers alone and shows that over time she felt at home.

In the meantime, however, her attitude about America may have been a source of strain within the marriage. Fermi was an enthusiast of American culture. He wasn't imbued with high European culture in the way so many of his northern European colleagues were. He had read classical literature as a *liceo* student, but his liberal education ended there. From the time he entered university, it was all physics, all the time. Many of his American colleagues had absorbed high culture, most famously J. Robert Oppenheimer, but it was possible to gain the respect of the American intellectual elite without ostentatious erudition. He enjoyed the openness, the relative lack of hierarchy and respect for tradition. In science he may not have been particularly inclined toward democracy—as his future colleague Luis Alvarez once put it, "There is no democracy in physics. We can't say that some second-rate guy has as much right to opinion as Fermi"—but the lack of deference to seniority appealed to him.

In addition, as emphasized by Italian scholar Giovanni Battimelli, the United States was rich, far richer than Italy, and funds were available for projects that seemed impossible to get off the ground in Italy. Fermi would visit Berkeley, with its magnificent cyclotron accelerator, and feel more than a pang of envy. He always had confidence in his ability to lead the pack, even in the underfunded Italian scientific community. Yet he also knew that his research would be easier to pursue at any of a half dozen US universities.

Finally, he loved the outdoors and, though Italy had much to offer, he was deeply attracted to the wide open spaces of the United States.

It is clear that he wanted to move to the United States and any number of universities would have welcomed him onto their faculties. He broached the topic with Laura on numerous occasions—presumably every time he returned from a trip there without her—but she resisted. She found little to attract her in the United States, and she loved her native Rome. She knew it intimately, and her friends were there. Why should she move? True, the fascist regime was repugnant, but the regime never meddled with her husband's work; indeed, it celebrated that work. Nor did it meddle with her, at least not until 1938. Her father was a distinguished military officer and was not an opponent of the regime. The anti-Semitic laws that Mussolini would pass under pressure from Hitler were well in the future, and cultural or political anti-Semitism was virtually unknown in Italy, although Mussolini's 1929 Lateran Accords, regularizing the regime's relationship with the Vatican, may have gradually served to cool the government's relationship with Italy's Jewish population. Certainly, what anti-Semitism there was did not yet extend to Laura's set or to the Jewish members of the distinguished intellectual circles in which her husband traveled. Every time Enrico would bring up the subject of leaving Rome for the United States, Laura resisted. No, their place was in Rome.

A DAUGHTER WAS BORN ON JANUARY 31, 1931. THEY NAMED HER Nella. Laura had plenty of help around the house to ease the burdens of new motherhood, but her husband was of little use. Laura may have envisioned the life of a professional woman for herself without the obligations of family, but she readily assumed the more traditional model of mother and housekeeper. Her husband was, however, awkward around the newborn. Laura confirms that he had no idea how to relate to this new person in his life, someone with whom he could not speak or engage in any of his favorite pursuits. At times he referred to her as the *bestiolina,* the little beast. Nella physically resembled Enrico more than she did Laura, and Laura, with a misplaced faith in genetics, assumed that Nella would inherit Enrico's intellectual abilities, as well. She could not hide her disappointment that Nella seemed to be a normal, curious, happy child, with little inclination toward mathematics and science.

Little brother Giulio was born just five years later, on February 16, 1936, at the height of Mussolini's ill-fated adventure into Ethiopia. Laura was perhaps a bit kinder to him, having had half a decade of experience raising Nella. She describes him as a strong, loud baby, but there is little evidence that Enrico paid any more attention to him than he had to Nella, even though the boy's name was chosen to honor the memory of Enrico's dead brother.

Neither Enrico nor Laura was an ideal parent. In her memoir Laura virtually concedes this in the self-deprecating title of the section describing the children's arrival, "How Not to Raise Children." There is, however, nothing to suggest that they were particularly bad parents, given the standards of the time. Mother and father both lived a life of the mind, and neither was naturally inclined toward the art of parenting. They were lucky in that they had the resources to hire help as the babies grew into young children. Much suggests they provided well for the children during these years and beyond, although their emotional and psychological distance, and Enrico's fame after the war, eventually took a toll on the siblings.

FERMI'S FAMILY LIFE WAS GOVERNED BY ROUTINE: HIS OWN DAILY schedule, the annual summer holidays, the regular trips to the Tuscan country home that Laura's aunt and uncle owned, the winter sojourns in the mountains for skiing, and the academic cycle around which Fermi's professional life revolved. As Fermi became more of a celebrity, as he won recognition from the physics community and the fascist regime under which he reluctantly labored, there were undoubtedly more distractions, but his was mainly a well-ordered, bourgeois lifestyle, with a marked absence of domestic or psychological drama. Any perturbations during this period were confined to Fermi's professional life, to his efforts to build a world-class physics research institute, and to his major discoveries.

FIGURE 6.3. Ski lodge lunch. *From left:* Antonio Rostagni, Gleb Wataghin, Persico, Fermi, and Maria Rostagni. Gressoney La Trinité, 1932. *Courtesy of Amaldi Archives, Department of Physics, University of Rome, La Sapienza.*

CHAPTER SEVEN

THE ROME SCHOOL

UNDER THE SPONSORSHIP OF CORBINO, BETWEEN 1927 AND 1934 Fermi brought together a young, dynamic team of professors and students who eventually established Rome as a major center for research into nuclear physics. This "Rome School" of physics has been the subject of intensive interest and study. Its hallmarks were a balance of theory and experiment; systematic development of expertise; cautious conservatism in terms of results claimed, married to a keen eye for publicity; disinterest in metaphysical or philosophical inquiry; focus on an empirical approach to all issues under study; and division of labor that enabled the team, consisting of widely differing personalities and temperaments, to work sufficiently well together to achieve major results.

Beginning in 1935, however, the group began to fall apart. The increasingly poisonous political environment was partly to blame, as were financial restrictions that made it attractive to find work abroad. Nevertheless, during the crucial years prior to 1935 the Rome group made an indelible mark on the world of physics.

Throughout the period, Fermi was the clear leader of the group, a fact acknowledged by his "disciples" when they began referring to him—much to his delight—as *il Papa,* "the Pope." It wasn't simply that he was almost invariably right when solving a thorny physics problem. At least one of his Rome students, Ettore Majorana, was nearly as brilliant as Fermi, if not more so. Yet no one would have considered referring to

FIGURE 7.1. Orso Mario Corbino. Photograph by Studio D'Arte Luxardo. *Courtesy of Amaldi Archives, Department of Physics, University of Rome, La Sapienza; AIP Emilio Segrè Visual Archives.*

FIGURE 7.2. Enrico Fermi as the young professor of theoretical physics in Rome. *Courtesy of Amaldi Archives, Department of Physics, University of Rome, La Sapienza.*

Majorana as *il Papa* simply because, outstanding though he was, he was not a leader. Fermi had a charisma and a magnetism that were apparent to all who met him. He was as close as anyone in the physics world has ever come to being a natural leader. The same magnetism that the "logarithms" felt when Fermi organized outings, hikes, afternoons at the beach drew to him the students and researchers who made up the Rome School.

When Fermi first returned to Rome, however, there was no one for him to lead. Corbino and Fermi understood that in order to have a school they needed students, but students were not lining up for entry into physics. Engineering was a popular and prestigious subject within the Italian university system. It was technical, modern, and led to good, stable career prospects. Physics was a bit different. It was more technical, more challenging, perhaps a bit less practical, and the career prospects were generally confined to teaching. Few students of high caliber chose it, and those who did ended up working for somewhat stodgier old school physicists. With Fermi's arrival, Corbino hoped it would be possible to draw a few students from the engineering program to physics.

The first to make the jump was a young engineering student named Emilio Segrè. In Segrè's connection with Fermi we see the influence of the tightly knit community of intellectual Italian Jews into whose salon Fermi had been welcomed. Segrè was just a few years younger than Fermi, born in 1905 into a prosperous Sephardic Jewish family in a suburb of Rome. His father owned a successful paper mill and the family lived a comfortable, bourgeois life. Segrè had always been interested in physics, but a practical bent and a lack of inspirational physics guided him to the engineering program at Rome in 1924. He had, however, taken one of Fermi's lecture classes on mathematical physics prior to Fermi's stint in Florence and knew Fermi to be a brilliant teacher and an original thinker. In Fermi's absence he simply continued his engineering studies.

In the spring of 1927 Rasetti, who had met Segrè through a mutual friend, mentioned that Fermi was returning to Rome. Segrè asked for a personal introduction. Soon Fermi and Segrè were spending time during the summer hiking in the country, informally getting to know one another. Fermi gently probed Segrè to assess his abilities and Segrè tried his best to answer Fermi's questions. Though not deeply prepared, Segrè passed Fermi's informal entrance exam. It probably didn't hurt his chances that he too was an avid outdoorsman.

Fermi invited Segrè to attend the 1927 Como conference, and Segrè was inspired by the giants in the field—Bohr, Heisenberg, Pauli, Planck, and Sommerfeld—discussing new developments in physics. By the time he returned to Rome, Segrè had decided to take the plunge and transferred from engineering to physics. Fermi had his first student.

One student does not a school make. Corbino and Fermi both understood this, and Corbino used the opening of the fall 1927 term at Rome to recruit a few other students into the physics department. Word went out to the engineering classes that there was room for one or two more physics students and that Corbino would facilitate the transfer process for any capable applicant. Only one student applied. He happened to be the young, cherubic Edoardo Amaldi, that same Edoardo Amaldi whom Fermi met at the Castelnuovo salons, a charter member of the "logarithms."

The third student in the program was introduced not by Corbino or Rasetti but by Segrè himself. His name was Ettore Majorana. A darkly quiet young man, he had a famous uncle, Quirino Majorana, who was a respected experimental physicist at the University of Bologna. Quirino was also a member of the panel of judges for the Rome *concorso* that Fermi won. Ettore was an engineering student and had become friendly with Segrè. Segrè believed that Majorana's mathematical brilliance would be wasted in the engineering field and persuaded him to join the group at Via Panisperna.

Corbino brought Rasetti to Rome earlier in the year as his own special assistant and as the physicist best able to add to the experimental expertise of the Rome group. Rasetti was particularly strong in spectroscopic experimentation, at that point the central concern of experimental physics. Persico, who came in second in the *concorso* for the Rome position, went to Florence. Fermi found himself, to his delight, reunited with Rasetti, his old friend from his days in Pisa. He missed Persico, of course, who was an even older friend, but Persico frequently traveled to Rome to visit the team at Via Panisperna.

The core of the Rome School was now more or less fully formed: Fermi, Rasetti, Segrè, Amaldi, and Majorana. Others joined the group over time, and the group would eventually dissolve as each member went his own separate way, but by late 1927–early 1928 the Rome School was born.

SEGRÈ, MAJORANA, AND AMALDI ALL HAD A PASSION FOR PHYSICS and were each brilliant in their own way. The three students also understood just how special this opportunity was, to work with one of the greatest minds in physics. There was no question as to who led the group. Aside from these elements in common, the three could not have been more different.

Segrè was a profoundly practical man, with a knack for business inherited from his father. He also had a fiery temper, resulting in his nickname "Basilisk," after a mythical creature who could kill with a single glance. He was, in addition, a gifted experimental physicist who in later years shared a Nobel Prize with an American student of Fermi, Owen Chamberlain, for the discovery of the antiproton.

Majorana was by all accounts the brightest of the group, in some ways even brighter than Fermi. His greatest strength was in mathematics; he would often complete a complex calculation quicker than Fermi. Unique among the Rome group, Majorana was a quiet introvert, who spent many solitary hours with his work. The general ambience of the Rome School was highly social. In this he stood apart from the rest.

Amaldi was the youngest of the group, sociable and outgoing, like Segrè an enthusiastic outdoorsman, a strong physicist with a preference for experimental work who readily stepped into the role of Fermi's experimental assistant when the time came. He knew Fermi from childhood and his subsequent 1933 marriage to Ginestra Giovene brought him even closer to the Fermis, because Ginestra was rapidly becoming Laura Fermi's closest friend.

The curriculum these students embarked on was unlike that at any other university and was itself one of the hallmarks of the Rome School. The method of teaching ensured that Fermi made his mark directly on each student. Though there were university lectures on all the standard topics of physics, Fermi's physics students were not required to attend these. Instead, they attended private seminars in Fermi's office. Fermi's formal lectures were reserved for students required to take physics as part of some other degree requirement. The ones who attended the private seminars with Fermi in the afternoon were mainly physics students, although others were occasionally invited. Ginestra Giovene, a general science student at the university, was one such attendee. It was here that she met her future husband, Edoardo. Another was a young man named Gabriello Giannini, who became a successful industrialist and businessman in the United States and later led the Rome group through the complex international patent process for the discovery of induced radiation.

These private seminars were the centerpiece of a Rome physics education. Segrè's description tells us much about the way in which Fermi taught and the secret behind keeping the group as coherent and tightly bound as it was:

> Fermi's seminar was always improvised and informal. In the late afternoon we would meet in his office, and our conversations might give rise to a lecture—for example, if we asked what was known about capillarity, Fermi would improvise a beautiful lecture on its theory. One had the impression that he had been studying capillarity up to that moment and had carefully prepared the lecture. I find in one of my notebooks on the discussions of those years the following topics: blackbody radiation, viscosity of gases, wave mechanics (the establishment of Schrödinger's equation), tensor analysis, optical dispersion theory, gaussian [*sic*] error curve, more quantum mechanics, and Dirac's theory of spin.
>
> In this fashion we reviewed many subjects at a level that corresponded to a beginning graduate course in an American university. Sometimes, however, discussion was on a higher level, and Fermi might explain a paper he had just read. In this way we became conversant with some of the famous papers by Schrödinger and Dirac as they appeared. *We never had a regular course.* [Italics added.] If there was an entire field of which we knew nothing and about which we asked Fermi, he would limit himself to mentioning a good book to read. Thus, when I asked him for some instruction on thermodynamics, he told me to read the book by Planck. But the readings were not always the best, perhaps because he mentioned only the books he himself had studied, which were not necessarily the best pedagogically but simply those he had found in the library at Pisa. After his lecture we would write our notes on it and solve (or try hard to solve) the problems he had given us, or others we thought of.

Ginestra wrote that students learned so much from these sessions because Fermi "brought his problems to the others and worked at them on the blackboard, aloud, with chalk and voice, thus showing how a rational mind reasons, how accidental factors can be discarded and essential ones taken into light; how analogies with known facts help to clear the unknown."

In these seminars Fermi was teaching more than physics. He was teaching his students to think the way he thought, to address problems the way he did, by cutting away all irrelevant factors and focusing on the heart of the matter in order to arrive at the simplest solution. Segrè notes that Fermi would work his way through a problem slowly and steadily, never speeding up when the mathematics was simple, never slowing down when the equations became thornier, much like a steamroller "that moved slowly but knew no obstacles. The final result was always clear, and often one was tempted to ask why it had not been found long ago inasmuch as everything was so simple and natural." This deliberative, almost plodding style of working through a problem almost always resulted in a much deeper understanding of the solution than an approach used by many of Fermi's contemporaries, which glossed over the easier steps and focused all the energy on the more difficult steps.

Fermi, Rasetti, Segrè, and Amaldi also became close socially. Years later Laura Fermi wrote:

> When the four came together, two teachers and two students, they were all young—there were seven years' [*sic*] difference between the oldest and the youngest. They shared a love for physical exercise, a swim in the near-by sea, a climb in the mountains, a long hike, and a game of tennis.

They also shared a "certain playfulness, a naïve love of jokes and silly acting that they brought into their serious work." When Ginestra first encountered the private seminar, Fermi explained that they liked to play a game called "two lire." Anyone in the group could ask anyone else a question. If the person could not answer the question, then that person owed one lire to the one who asked it. If, however, the person who asked the question could not answer the question himself, then he owed two lire to the one whom he had originally challenged. Edoardo Amaldi came up with a question for the bewildered, unnerved Ginestra: if tin has a lower boiling point than olive oil, how can you possibly boil olive oil in a tin-lined skillet? Ginestra was clearly flustered but composed herself sufficiently to think through the solution. She explained that when one cooks with olive oil, the oil is not boiling; rather, it is the water in the food that boils. This was the correct answer and no money changed hands that day.

By this time the *ragazzi di Via Panisperna* (boys of Via Panisperna), as they were known, had each adopted a nickname, giving them a specific relationship to the infallible Fermi, *il Papa* (the Pope). Corbino became *il Padreterno* (Eternal Father) owing to his munificent sponsorship of the group. Rasetti was the *Cardinale Vicario, il Papa*'s right-hand man. Persico was *Prefetto di Propaganda Fede,* Prefect of Propaganda of the Faith, responsible for spreading the gospel of quantum physics wherever he could. *Il Basilisco* (the Basilisk) was the name given to Segrè. Amaldi was *l'Abate* (the Abbot), reflecting his junior position within the team. When young Bruno Pontecorvo came on board in 1933, he too took on a nickname appropriate for his status—*il Cucciolo* (the Puppy).

Eventually, the members of the group (generally excepting Majorana) spent weekends together hiking in the hills around Rome or sunning themselves on the beach at Ostia; they took ski holidays together in the Italian Alps; they dined with each other regularly. Like the "logarithms" before them, they became an intimate circle of friends, Fermi always leading with an easy informality and familiarity that increased group cohesion. They were "family." They became so close, reportedly, that they all began to speak in the same slow, deep, unusually cadenced Italian that Fermi and Rasetti had developed when they were students together. Segrè tells the story of a member of the group who once struck up a conversation with a fellow railway passenger, who immediately asked whether he was a physicist from Rome. The Fermi protégé confirmed that he indeed was and asked how the passenger could tell. The passenger had guessed it from the way he spoke.

In this way, physics in Rome became fun. It is almost as if Fermi, having experienced the dry austerity of physics at Göttingen some five years earlier, had decided that he would build a school along very different principles. It was also a group endeavor, a social endeavor, one in which each person had a role. Even when Fermi was doing theory, he would bring his thoughts to the private seminars and work through them in front of an audience of students and fellow researchers. They might argue over physics by day, but evenings and weekends were spent together, and the group became very close indeed.

In Florence, Garbasso was developing his own group of strong, young physicists. The two groups exchanged visits and met often to discuss their mutual research interests. In addition to Persico, the group included Bruno Rossi, who later became the world's leading expert on cosmic rays; Giuseppe Occhialini, who would go on to major contributions to particle physics, including an important role in the discovery of a particle called the pion; Gilberto Bernardini, a brilliant experimentalist who led the effort to build Italy's first major accelerator lab after the war and later became director of the Scuola Normale; and theorist Giulio Racah, who after the war established modern physics in Israel.

Italian visitors were not the only ones who were at home at Via Panisperna. As word spread that Fermi's group was doing interesting things and having fun doing them, young physicists from around Europe sought post-doc appointments or short visiting stints with Fermi and his colleagues. Sommerfeld's student Hans Bethe stayed for a year and became a good friend of Fermi. Other visitors who ultimately made important contributions included George Placzek, Felix Bloch, Rudolf Peierls, Sam Goudsmit, and Eugene Feenberg. Young Hungarian physicist Edward Teller visited and challenged Fermi to table tennis. Teller was quite good at the game in spite of having a prosthetic foot and always won handily, to Fermi's inevitable frustration.

One problem that Corbino and Fermi faced together early on, as Fermi settled in to his new position in Rome, was how to raise the profile of Italian physics on the international scene. Modern physics always relied on international conferences for the development of new ideas, exchange of information, and thrashing out of major debates and disputes. In the first half of the twentieth century, no series of conferences was as important, or prestigious, as the Solvay conferences. When Fermi arrived back in Rome, no Italian had ever been invited to attend.

The conferences were the idea of Ernest Solvay, a successful Belgian industrial chemist and philanthropist. Beginning in October 1911 and every three years hence, the conferences, under the direction of Dutch physicist Hendrik Lorentz, attracted the greatest names in physics of the day—Planck, Einstein, Sommerfeld, Rutherford, Marie Curie, and Poincaré, among others. Many of the original issues associated with relativity and quantum theory were hashed out at these historic meetings. The fifth Solvay conference, scheduled for late October 1927, would focus on the topic of "Electrons and Photons," involving a variety of complex developments, including Dirac's recent, brilliant work on the quantum theory of the electron. Among the "who's who" of international physicists invited to attend, totaling some thirty-two scientists, not one was Italian.

The absence of any Italian was a source of continuing irritation for the leaders of Italian physics. Not even the famous Guglielmo Marconi received an invitation, probably because he was considered more of an inventor than a scientist. Other distinguished Italians might have merited inclusion, such as Corbino in Rome, Garbasso in Florence, and even Levi-Civita, who, though not a physicist, had developed much of the mathematical framework for Einstein's theory of general relativity. Lorentz ignored them all. Italian physicists were simply not part of the highest-level dialogues in the field.

Feeling shunned, Corbino came up with an idea that he thought would alter the playing field. If Italians could not get invited to Solvay, he would hold a conference of his own. He had two major assets on his side. First, of course, was Fermi, already known to many of those invited to Solvay. The other was the natural beauty of Italy's Lake District. Breathtaking lakes nestled in the valleys of the Italian Alps could, he imagined, provide an attractive place for Italy's own conference on physics. He chose the town of Como for the site of his conference. The hundredth anniversary of the death of famed Italian physicist Alessandro Volta, who invented the electric battery, provided a credible reason to bring the world's great physicists together.

And so it did. Among those who attended the conference in mid-September 1927, just one month before the fifth Solvay conference, were Niels Bohr, Max Born, Walter Bragg, Maurice de Broglie, Werner Heisenberg, Wolfgang Pauli, Max Planck, Ernest Rutherford, Arnold Sommerfeld, Pieter Zeeman, and even Solvay director Hendrik Lorentz himself. The Italians were there in full force, of course, with Corbino and Fermi acting as hosts for the events.

From an historical perspective, the most important presentation was that of Bohr, who took the opportunity to share some initial thoughts on his concept of complementarity, thoughts that he developed in greater detail a month later at the Solvay conference. Fermi, Heisenberg, and Pauli presented no papers of their own, but they did comment on Bohr's paper and on a variety of other issues relating to quantum theory.

On a social level, it was an opportunity for Fermi to meet and greet many of the great names of the field. It was the first time Fermi met Pauli in person. Heisenberg introduced the two of them by saying something to the effect of "May I introduce the applications of the exclusion

FIGURE 7.3. *Left to right:* Fermi, Heisenberg, and Pauli, Lake Como, 1927. *Photograph by Franco Rasetti. Courtesy of AIP Emilio Segrè Visual Archives, Segrè Collection.*

principle to each other." Rasetti memorialized the moment in a photograph he took of the three young quantum theorists sharing a boat ride on Lake Como. In the photo Heisenberg is clearly the happiest, arms folded and beaming. He had just made an historic introduction and was probably quite pleased with himself. Pauli looks happy enough as well. It is Fermi who looks the most awkward; perhaps because Rasetti caught him just at a moment when the wind was whipping his thinning hair up and away from his head. Or perhaps it was the conflict he felt as host, colleague, and competitor with respect to his companions on the boat. Rasetti's photo is one of the iconic images of twentieth-century physics, showing three of its greatest minds at the height of their intellectual powers.

Years later, Rasetti, reflecting on the importance of the Como conference, noted that in general Italian physicists were not aware of Fermi's significance until Sommerfeld underscored the impact of Fermi's work on statistical mechanics at this conference, in particular, pointing to its ability to explain the behavior of electrons in metals. The conference also made clear to the Italian physics community just how far behind it was because many of its senior physicists refused to adopt the new quantum theories of 1925–1926. It was, in Rasetti's view, the moment when Italy began to recognize Fermi's standing.

It was too late to invite Fermi to the fifth Solvay conference just a month later, and it is not clear just how much the Como conference influenced the organizers of Solvay, but they invited Fermi to attend the conferences in 1930, 1933, 1936, and 1939 (although the last one was canceled owing to the outbreak of World War II).

FERMI'S ARRIVAL IN ROME WAS NOT MET WITH UNIVERSAL WELCOME in the corridors of Via Panisperna. In particular, Antonino Lo Surdo, an older experimentalist with offices at Via Panisperna, resented the newcomer. Corbino and Lo Surdo were personal rivals and had been feuding for years before Fermi arrived on the scene. Lo Surdo had made some important discoveries but was overlooked for a Nobel Prize in 1919, perhaps in part because Corbino had worked hard to undermine his scientific reputation in Italy. By the time Fermi arrived in Rome Lo Surdo was a bitter man who resented the appointment of this prodigy to a new professorship at the institute.

Lo Surdo may not have won a Nobel Prize, but he was a member of the distinguished *Accademia dei Lincei* (Academy of Lynxes), the oldest scientific society in the Western world. Founded in 1603, among its earliest members was Galileo. By the early twentieth century, its status as Italy's most prestigious scientific organization was unquestioned. At this point in Fermi's career, it would have been natural for Corbino to nominate his protégé to the *Accademia,* and it would also have been natural for Lo Surdo to block the nomination. But the accepted narrative of these events, as told by Laura Fermi, does not cohere. As she would have us believe, Corbino prepared a nominating letter on behalf of Fermi and, because he would be traveling during the annual election meeting, entrusted this letter to Lo Surdo, who promised to submit it in Corbino's absence. When Corbino returned, however, Fermi was not on the list of those elected. When he confronted Lo Surdo, the latter insisted he had forgotten to submit the nomination letter—an unlikely excuse, to say the least.

It is genuinely difficult to believe that, given their historic enmity, Corbino would have entrusted Fermi's nomination to Lo Surdo. Their feud had lasted over a decade, and Corbino was highly sophisticated and politically adept. Further, the process of nomination was a complex, lengthy one, and many people would have been involved by the time an election was held; members would have been aware that Fermi was up for election well before the election date. The story does not hold up. Corbino probably had a different agenda.

Mussolini had long been planning to establish his own institution to rival the *Accademia dei Lincei* in prestige and importance. The *Reale Accademia d'Italia* (Royal Academy of Italy) was Mussolini's brainchild, and the bill to establish it passed the Senate in 1926. Setting it up was complicated because the fascist dictator wanted to include the greatest cultural and intellectual figures in all of Italy, across all disciplines. He also wanted, somewhat naïvely, to include individuals who were not completely enamored with his fascist regime. Under the circumstances, it took until March 1929 for the government to compile a slate of inaugural candidates for the new academy. Corbino, no fan of the dictator, was nevertheless able to persuade Mussolini to place Fermi on the list of academicians.

Lo Surdo had expected to be named and was deeply disappointed to have been passed over. Corbino had in fact won a double victory. Not only had he influenced the naming of Fermi over Lo Surdo; he had done so at a time when he must have known that, owing to Mussolini's new high-profile academy, the *Accademia dei Lincei* would soon fade in importance. Fermi was delighted. On that date he wrote in large capital letters in his notebook: "INCIPIT VITA NOVA—GAUDEAMUS IGITUR!" (Now begins a new life—let us rejoice!).

Newly appointed members of the *Accademia d'Italia* were celebrated in the press, and ceremonies of great pomp and elegance were held to honor them. Of far greater importance to Fermi, membership included a significant annual stipend of 36,000 lire for life, more than doubling his salary from the university. This enabled him to drop other distracting commitments he had made to supplement his income, most notably an editorship in Treccani's *Enciclopedia*. It entitled him to be known as "His Excellency," a title whose pretentions endlessly amused him. It also included a new uniform designed especially for the honor, with flowing robes and faintly ridiculous headgear. Self-conscious of appearing in public so dressed, Fermi would drive himself in the bébé Peugeot to the meetings of the *Accademia d'Italia* and dress when he got there. On one occasion, he was challenged by a guard who was certain that such a car did not belong at an august meeting of the *Reale Accademia.* Fermi explained—truthfully, he would later recount with a smile—that he was the driver of His Excellency, Enrico Fermi. The guard waved him through.

The public was fascinated. Fermi was by far the youngest academician. Before his induction he was known by physicists throughout Europe for his work in quantum physics. His fame within Italian physics spread as a result of Sommerfeld and others' comments at Como. As a result of his elevation to the *Accademia,* he was now known more broadly. He had become, as he would later say, "a great man."

He was certainly willing to play Mussolini's game, lending his name and his scientific prestige to the new fascist institution. He may well have thought that it was essential for continuing government funding for Italian physics. Corbino, no Fascist himself, probably felt the same way. Although the honors fed Fermi's ego, he considered the various functions and receptions a waste of time. Every hour spent at the *Reale Accademia* was an hour away from his work at Via Panisperna. There was never any doubt about his priorities.

In this way an uneasy symbiosis was established between Fermi and the fascist regime. However readily he accepted the radical ideas of the new physics, he was personally conservative and may at some level have approved of the stability that Mussolini brought to Italy, despite the regime's use of thuggery and violence. Fermi played the game and was trotted out as an example of the brilliant science sponsored by the regime. In return, his work was supported without interference. It was, to be sure, a deal done with the devil, but it served Fermi's purpose. His trips to America, beginning in 1930 and continuing throughout the decade, showed him what opportunities lay abroad under a freer, more prosperous form of government, and he fully appreciated those opportunities. Laura's resistance, however, prevented him from making a move.

In the end, the importance of the Rome period had nothing to do with Fermi's election to the *Accademia d'Italia,* nor to his growing celebrity, nor even to his ability to attract the best young talent to Via Panisperna. The importance of these years can only be measured by the discoveries made by Fermi and his little group perched on a hill overlooking Via Panisperna. The work in Florence had made him famous among physicists. The work in Rome would make him a legend.

CHAPTER EIGHT

BETA RAYS

ENSCONCED AT VIA PANISPERNA, FERMI BEGAN EXTENDING HIS work on statistical mechanics and then turned to quantum electrodynamics, a process that led him to his theory of beta radiation, considered by many today his most important contribution to physics.

Fermi's first important paper of 1928 extended his work on statistical mechanics from a monatomic gas to the much smaller world of electrons around an atom. A monatomic gas is one in which only one type of atom is present, like an idealized helium-filled balloon. Fermi imagined that the cloud of electrons around a nucleus could be viewed as a gas with one type of particle. Fermi could apply his new statistical methods because electrons obey Pauli's exclusion principle—no two can share the same quantum state. Pauli himself had used his exclusion principle to explain the way electrons fill up each orbital and why each orbital can contain only a finite number of electrons. Building on this, Fermi wanted to find an entirely probabilistic description of these orbitals to calculate the probability of finding electrons at any given place within the atom at any given time. Where the probability density was great, one was more likely to find an electron. Where the probability density was low, the probability of finding an electron would be lower. Fermi's approach would also, in principle, give results consistent with Pauli's rules regarding how many electrons could occupy each orbital.

The idea was an example of Fermi's enthusiasm for taking ideas developed in one context and applying them in another. It had some usefulness in analyzing complex atoms and also had applications in calculating how charged particles would behave when moving through matter—the so-called stopping power of matter with respect to these charged particles. Eventually, it gave rise to density functional theory, which has proven sufficiently useful in condensed matter physics and computational chemistry that Walter Kohn and John Pople shared the 1998 Nobel Prize in Chemistry for its development.

There were, however, two problems with Fermi's extension of his statistics to the electrons surrounding an atom. First, the technique as originally presented by Fermi had limited accuracy in a variety of important applications, a problem that was eventually overcome through the development years later of the Kohn-Pople density functional theory. The second problem was that someone else had beaten Fermi to the idea.

Llewellyn Thomas, a young British mathematical physicist who received his degree at Cambridge, read Fermi's 1926 statistics paper and decided to see whether Fermi-Dirac statistics could model the electron's behavior around the nucleus from a statistical perspective. His article, "The Calculation of Atomic Fields," using the assumptions of Fermi-Dirac statistics as a starting point, came out in the *Proceedings of the Cambridge Philosophical Society* in January 1927, some eleven months prior to Fermi's paper published in the journal of the *Accademia dei Lincei*. How Fermi missed Thomas's paper remains a bit of a mystery. Fermi was, at this point, still reading the international journals on a regular basis, and it is difficult to imagine him glossing over Thomas's work. Some have speculated that the physics department did not subscribe to the journal, which is true, but the Istituto Superiore di Sanita, located in the basement of Via Panisperna, certainly did. The journal itself was an important one, publishing the top British physicists and mathematicians of the day, people Fermi followed. It is possible, as some Italian historians point out, that he was indeed aware of the article but considered his own work to be sufficiently different to obviate the need for a direct reference.

In any event, Fermi's attention soon turned to another, far more challenging and far more important issue. In 1927, Dirac continued his exploration into quantum theory and produced something monumental—a quantum theory of the electromagnetic field, known today as "quantum electrodynamics," or QED. Because Dirac's achievement laid the foundation for Fermi's theory of beta radiation, it is essential to understand what Dirac achieved and how Fermi integrated Dirac's accomplishment into his own thinking.

To APPRECIATE THE MAGNITUDE OF DIRAC'S BREAKTHROUGH, IT helps to understand what his predecessors meant by an electromagnetic field. This concept of a field is the great contribution of the experimental physicist Michael Faraday, a high school–educated Englishman who conducted a series of experiments in the 1840s and 1850s to explore the relationship between electricity and magnetism. His concept of a field was quite simple: a region contains a field if objects therein feel a force without a direct physical connection. Observing how magnets and electricity flowing in a current affected each other, he posited a relationship between the electric field and the magnetic field. A Scottish physicist named James Clerk Maxwell subsequently put this relationship into a mathematically rigorous form. The eponymous Maxwell's equations define the characteristics of the electromagnetic field.

Maxwell's equations are classical, with three major characteristics that quantum mechanics would completely undermine. First, if one understands the strength of the field at any point, one can predict with certainty how a charged particle will behave at that point within the field. If one can characterize the field as a whole, one can predict the behavior of a charged particle at any point in the field—if it is moving in one direction at a particular moment, one can predict with absolute certainty where it will be at the next moment, irrespective of how soon that next moment is. Second, in classical electromagnetic theory, a field exerts a force that varies continuously throughout the field, influenced by such factors as the distance between a point and the source of the field, that is, electric currents and charges or magnets. In fact, it is analogous to Newton's laws of gravity, although unlike Newton's gravitational field, Maxwell's equations allow for repulsion as well as attraction between two charged particles or magnets. Finally, in classical electromagnetic theory, the particle itself is not a manifestation of the field. This seems so obvious at first glance that it might hardly be worth commenting on, except that it is not actually true, as Dirac would discover.

The quantum revolution changed forever the way physicists think about the world around us. The world is not perfectly predictable. Though predictions can indeed be made, there is an inherent uncertainty in these predictions that must be taken into account when calculating the behavior of very small objects like atoms or subatomic particles. Furthermore, the world is not continuous. Energy comes in the form of irreducible packets, imparting a granularity to a world previously thought of as smooth. An analogy might be the way we observe a body of water versus a sand dune. Significantly, a particle and its associated field are not independent. A particle is indeed a manifestation of the field with which it is associated. Any conception of an electromagnetic field that accords with the principles of quantum theory would have to incorporate these new insights.

Between 1900 and 1925, the theory of quantum mechanics developed apace, culminating in the 1925 work of Pauli, Heisenberg, and Schrödinger. Dirac followed these developments closely and wrote an astonishingly sophisticated thesis on the subject. Not content to rest there, however, he decided over the next two years to apply his insights into a reformulation of classical electromagnetic theory. In particular, he became interested in how Einstein's theory of the photon—the particle of light that is created or absorbed during the shift in electron energy levels—could be interpreted in a general quantum theory. By 1927, he had done it. In an historic paper published in March 1927, he gave a new account of electrodynamics based on the idea of a quantum field, which treats the particle and the field not as two separate entities but rather as a single system consisting of the energy of the atom, the energy of the radiation field, and a "coupling" factor that connects the two. His mathematics, arcane and difficult even for seasoned physicists to follow, developed a model of the electron's behavior that integrated the matrix mechanics approach of Heisenberg, Born, and Jordan with the wave formulation of Schrödinger. It also incorporated the special theory of relativity developed by Einstein in 1905, necessary because electrons nearest to the atomic nucleus travel at speeds approaching the speed of light. At these speeds, the weird effects of Einstein's theory become relevant.

The Dirac equation described not just the behavior of electrons in a quantum electromagnetic field but also the behavior of all charged fermions, that is, all particles that obey Pauli's exclusion principle. In it, he predicted the existence of antimatter and he explained why particle "spin," as proposed by Goudsmit and Uhlenbeck, was not simply a convenient way of interpreting the behavior of electrons but actually required by natural law. It was a breathtaking tour de force and firmly established Dirac as one of history's greatest physicists. He was a mere twenty-four years old at the time.

Only as theorists explored Dirac's astonishing theory did they come to understand its limitations. The most important one involved the calculation of the "magnetic moment" of an electron, related to the torque an electron "feels" in a magnetic field. The Dirac equation provides a first approximation of the magnetic moment that is reasonably consistent with experimental data, but the calculation requires an iterative process to get greater accuracy. Unfortunately, the mathematics required involves summing a numerical series that does not converge onto a more exact quantity, but, absurdly, approaches infinity. This problem would not be solved either by Dirac or by Fermi or in fact by anyone at all until after World War II, when three brilliant young theorists—Richard Feynman, Julian Schwinger, and Sin-Itiro Tomonaga—cracked the puzzle independently.

Fermi understood the importance of Dirac's work, but it took him until the winter of 1928–1929 before he set about studying it seriously. Much of the mathematics Dirac used was of his own invention, and Fermi, according to Segrè, found it "alien." He began to recast Dirac's theory in his own terms, terms that were mathematically more familiar to him. The private seminars at Via Panisperna became the stage upon which he worked out, aloud in front of an audience consisting of Rasetti, Amaldi, Majorana, Segrè, and Racah, a version of Dirac's work that would be easier to digest. One must also assume, though Segrè and Amaldi are silent on this, that he spent his early morning pre-breakfast hours at home working intensively on the problem.

By April 1929, Fermi had a preliminary version of his presentation ready for lectures he had been asked to give in Paris. He continued working on it over the next year, and by the time of the 1930 summer school in Ann Arbor, he had built up the presentation to an extensive series of lectures, which he presented there and which were subsequently published in 1932. Fermi eased into the subject matter by comparing an atom and the electromagnetic field to a pendulum and a vibrating string connected by a thin elastic thread, which represented the coupling of the two. If the pendulum is at rest and the string starts to vibrate only slightly, the elastic thread will perturb the pendulum only slightly. But when the string vibrates in tune with the amplitude of the pendulum, the elastic thread carries that energy to the pendulum and causes the pendulum to swing in time with the vibrating string, creating a resonance between the pendulum and the string. This description of how an atom and an electromagnetic field interact with each other formed the basis of Fermi's interpretation of Dirac's quantum electrodynamics.

Some seventy years later, Nobel Prize–winning theorist Frank Wilczek would call this description "a masterpiece, instructive and refreshing to read even today. . . . Everything is done from scratch, starting with harmonic oscillators."

Hans Bethe, the young German physicist who spent this period in Rome as a post-doc, recounts the impact the paper had on him:

> Many of you, like myself, have learned their first field theory from Fermi's wonderful article in the *Reviews of Modern Physics* of 1932. It is an example of simplicity in a difficult field which I think is unsurpassed. It came after a number of quite complicated papers and before another set of quite complicated papers on the subject, and without Fermi's enlightening simplicity I think many of us would never have been able to follow into the depths of field theory. I think I'm one of them.

Eugene Wigner, another quantum theory pioneer who was destined to work closely with Fermi during the Manhattan Project, corroborated Bethe's assessment:

> He disliked complicated theories and avoided them as much as possible. Although he was one of the founders of quantum electrodynamics, he resisted using this theory as long as possible. His article on the Quantum Theory of Radiation in the *Reviews of Modern Physics* (1932) is a model of many of his addresses and lectures: nobody not fully familiar with the intricacies of the theory could have written it, nobody could have better avoided those intricacies. However, when he tackled a problem which could not be solved without the explicit use

> of the much disliked concepts of quantum field theories, he accepted this fact and one of his most brilliant papers [on beta decay] is based on quantized fields.

Fermi himself thoroughly understood the Dirac formulation and yet spent two years recasting it in his own terms. As Wigner suggests, he was fully familiar with the mathematical complexity of Dirac's approach, yet he felt uncomfortable with it and sought a simpler way of explaining it. The issue was not mathematical complexity per se. Fermi was a fine mathematician, able to hold his own with some of the greatest mathematicians of his day. Rather, like almost all of Fermi's greatest work, the effort of radical simplification was at least in part pedagogically motivated. He worked it out in front of his colleagues and students, slowly and methodically, making sure that his audience followed at each step of the way. The purpose of this exercise was to make the material accessible to others. It was not simply the physicist's love of simplicity—if he couldn't teach it to someone else, he felt he didn't understand it sufficiently himself. Dirac had no such purpose, writing his papers at his own, highly rarified level. Fermi made Dirac comprehensible to physicists who otherwise might not even have bothered slogging through the eccentric physicist's complex concepts and exotic techniques.

Pauli and Heisenberg both independently wrote papers paralleling Fermi's simplifying approach. Working in the opposite direction, Jordan and Wigner, among others, added an additional layer of complexity to Dirac's initial formulation. Dirac's mathematics explained the creation and annihilation of bosons such as photons in electrodynamic processes, offering a mathematical description of what happens when a photon strikes an electron in the outer shell of an atom (it disappears and the electron "moves" to a higher energy state) and when an electron "moves" to a lower energy state (a photon is created out of "nothing"). Jordan and Wigner, significantly, extended the mathematics describing creation and annihilation of particles from bosons to fermions in what was called a "second quantization." Their contribution was essential to Fermi's formulation of the beta decay paper in late 1933. Segrè suggests that the time lag between the Ann Arbor conference in 1930 and the 1933 beta decay paper was due to Fermi's working through the Jordan second quantization approach so that he felt completely comfortable with it. Once he had thoroughly absorbed the difficult methods pioneered by Jordan and Wigner, however, he determined to use them in ways no one anticipated.

ERNEST RUTHERFORD WAS THE FIRST TO CLASSIFY DIFFERENT TYPES of radioactive emissions by their respective ability to penetrate matter. Emissions that were least able to penetrate he called alpha rays. The most penetrative were called gamma rays. Those with medium penetration ability he called beta rays. Rutherford quickly concluded, through a series of classic experiments, that alpha rays are positively charged, with the charge of two protons and the mass of four protons. Later investigation revealed that the alpha particle consists of two protons and two neutrons, emitted from certain overcrowded nuclei. Similar investigation showed that gamma rays were high-energy photons, of which X-rays were a subset. Beta rays were negatively charged and very light compared with the mass of the alpha particles. They were, in fact, electrons.

For a long time, not much more was known about these forms of radiation, largely because the structure of the atom's nucleus was a mystery. For a time, physicists thought that electrons and protons existed together in the nucleus of the atom, because beta rays seemed to be coming directly from the nucleus of radioactive elements. Many tried to explain the existence of electrons within the nucleus, but those explanations raised more questions than they solved.

A series of very precise experiments in the latter half of the 1920s created a beta ray "crisis," involving the apparent violation of certain laws of conservation that physicists cherish. One of the central conservation laws concerns the conservation of energy. In any physical process, the energy into the process and the energy out of the process must be equal and accounted for. Beta radiation seemed to violate this law. If energy was conserved, the energy of the emitted beta particles should fall within a very narrow range. Every process of beta ray emission was, theoretically, the same as every other one, yet the beta particles emitted fell along a significantly wider spectrum of energy than predicted. That strongly suggested that energy was not being conserved. Imagine baseball batting practice with a machine set to pitch balls at exactly the same speed and in the same direction every time. Yet, suddenly the balls start coming out randomly at different speeds. One would conclude that there is something wrong with the machine.

Some physicists surrendered to the evidence and pronounced that energy was evidently not conserved in this particular interaction. These included Niels Bohr, who ventured this daring, and to physicists distasteful, conclusion in a paper he presented at a major physics conference held by Corbino and Fermi in Rome in 1931. Rutherford, for one, utterly rejected Bohr's conclusion, but it was only the imaginative Pauli who made a suggestion that would explain beta radiation without violating the principle of energy conservation.

In typically audacious terms, Pauli suggested that another particle was being emitted at the same time as the electron, virtually undetectable because it is neutral and of very small (or even no) mass. He first outlined his idea in a letter to physicists attending a conference in Tübingen, Germany, in December 1930. He called this imaginary particle a "neutron" and proposed that it was carrying the balance of the energy not observed in beta decay, thus salvaging the conservation of energy. The particle we today call the neutron had not yet been discovered. James Chadwick's sensational discovery took place some two years later, at which point Fermi and his team proposed the name "neutrino"—little neutral one—for Pauli's hypothetical particle.

Not many people took Pauli's idea seriously. For some, the whole issue of energy conservation in beta radiation was something, in the words of one eminent physicist, "better not to think about . . . at all, like new taxes." Others sided with Bohr and found the idea of a particle that could not be seen and that could penetrate vast distances unimpeded too outlandish to take seriously. Violation of energy conservation was more likely than the existence of such a neutral particle. Further, no one understood the mechanism by which an electron and a neutrino would be emitted. Did they exist in the nucleus all the time, ready to be emitted at the appropriate stimulation? Or were they somehow created within the nucleus and then promptly ejected?

Chadwick's discovery of the neutron in the nucleus of the atom provided a clue. Dirac's quantum electrodynamics, which described the creation and destruction of photons, provided a second clue. Jordan and Wigner's "second quantization" provided the third clue. By mid-1933, Fermi—who took Pauli's proposal quite seriously—used everything he had learned over the previous four years and applied it to the problem of beta radiation. His paper, "A Tentative Theory of Beta Rays," published in late 1933 and early 1934 in scientific journals in Germany and Italy, set out his ideas on the subject. More than eighty years later, it stands as one of the most important achievements of twentieth-century physics.

What Fermi proposed was the existence of a new quantum interaction—an interaction now known as the "weak" interaction because it takes effect only when particles come into extremely close range of each other. This interaction changes neutrons into protons and protons into neutrons. At the very moment that these changes occur, new particles are created and emitted from the nucleus at high energy. When the neutron is changed into a proton, an electron and an antineutrino are emitted. When a proton is changed into a neutron, a positively charged electron (a positron) and a neutrino are emitted. The sum total of the energy of the emitted particles remains constant, but the apportionment of energy among the particles varies according to quantum laws that are a direct consequence of the quantum field theory. The electron and the neutrino (and their antimatter cousins) do not preexist in the nucleus at all. Rather, they are *created* at the moment of emission. The theory allows one to calculate the likelihood of a neutrino or an antineutrino interacting with matter. That likelihood is so low that a neutrino can travel through millions of miles of lead without interacting at all.

The story of the development and publication of the paper is intriguing. Fermi had been working on the paper through the latter part of 1933. Beta radiation was a major preoccupation of the Solvay conference of October 1933, which Fermi attended, and he and Pauli discussed Pauli's neutrino idea at length during the conference. By Christmas of that year, Fermi had progressed sufficiently to feel comfortable presenting the main ideas to his Rome colleagues during a group ski vacation. Segrè reports that soon after this holiday Fermi presented the paper to the British journal *Nature,* because the group at Via Panisperna had decided, following the rise of Hitler in Germany, to boycott German publications even though they remained the most prestigious in the field. According to Segrè, *Nature* rejected the paper, a reviewer claiming that it was too "speculative." In response, Fermi sent the paper to the Italian journal *Nuovo Cimento* and the German *Zeitschrift für Physik,* both of which published it. This story is so central to the legend of the paper that Wikipedia reports *Nature* later publicly regretted having rejected it as one of its most egregious editorial errors.

In fact, no such public statement of regret can be found in any back issues of *Nature*. It is unfortunately impossible to review *Nature*'s archives to find the rejection letter written by the reviewer, because all records were destroyed in a move to new offices several decades ago. Some historians question the entire story. They observe that at that time *Nature* accepted only short notes on these types of subjects and was certainly not a forum for a detailed presentation of new quantum field theories. A more logical British publication would have been the *Proceedings of the Royal Society of London*, which had published all of Dirac's seminal papers on QED and would have been a logical place for Fermi to submit the beta decay paper. These historians suggest that he may have wanted his German counterparts, particularly people like Born, Heisenberg, and of course Pauli himself, to read the paper first. A white lie—that he had tried but failed to get it published by *Nature*—would get him off the hook with his young colleagues who were so opposed to publishing anything in German journals and still achieve his main objective.

Whatever the actual history of the paper, the immediate reaction to it within the physics community was somewhat muted. Pauli and Wigner appreciated the achievement for what it was. Fermi had integrated the entire Dirac quantum field framework into his own thinking and applied it ingeniously to the beta radiation problem. The trouble was that the theory seemed almost impossible to verify experimentally. Neutrinos appeared to be impossible to detect. Fermi himself doubted that they ever would be. Although he could take private satisfaction in having mastered QED and used its mathematics to explain another apparently different phenomenon, only in later decades would the true brilliance of the paper come to be appreciated. It was the first hint of the existence of a new, fourth fundamental force of nature, the weak force, which would ultimately take its place alongside gravity, electromagnetism, and the strong force (the force that holds the nucleus of an atom together). Its exploration has resulted in over a dozen Nobel Prizes and some of physics' biggest surprises.

In the 1970s, Fermi's future University of Chicago student, Chen Ning Yang, who would help to uncover one of the most surprising secrets of the weak force, asked Fermi's colleague and friend Eugene

Wigner what physicists would remember as Fermi's most important contribution to the field. Wigner insisted that the beta radiation paper was Fermi's most important work. Yang disagreed, noting that it was Jordan and Wigner himself who had invented the second quantization with its creation/annihilation operators for fermions. Wigner's reply: "Yes, yes. But we never dreamed that it could be used in real physics." Most people would agree with Wigner's assessment that the beta decay paper remains one of Fermi's true landmark discoveries.

IN THE MEANTIME, SOME NEWS FROM PARIS CAME TO FERMI'S attention. The Joliot-Curies announced that they had induced radiation by bombarding nonradioactive elements with alpha particles. The news gave Fermi yet another idea.

CHAPTER NINE

GOLDFISH

WELL BEFORE HE BEGAN THINKING ABOUT BETA RAYS, FERMI knew that the next big thing would be nuclear physics.

Until about 1930, cutting-edge physics focused at the atomic level. Most of the work at Rome, like that throughout Europe and the rest of the world, tried to understand the structure of atoms and how they behaved. Recognizing the importance of atomic physics, Fermi wrote a comprehensive textbook on the subject in 1928, *Introduzione alla fisica atomica,* for use as a basic introduction for college students throughout Italy. It was yet another way to supplement his university income and provided a much-needed way for Italian physics and engineering students to gain exposure to the new physics.

In late 1929/early 1930, Fermi began to focus his research at the next level down—the nucleus of the atom. At that time, the nucleus was still a bit of a mystery. Physicists knew that it was suspended deep within the inner space of the atom. If a typical carbon atom were magnified to the size of a football field, the nucleus would be a penny in the center of the fifty-yard line and the nearest electrons would be at the goal lines, with empty space between them most of the time. Physicists also knew that the nucleus was positively charged. They knew that it contained the bulk of the mass of an atom, but its inner constituents and structure remained a mystery. One great puzzle was that the mass of the nucleus tended to be about twice as large as it should be, given the charge of the nucleus. Before Fermi's beta decay paper, the emission of beta rays suggested that at least some electrons were also inside the nucleus. No one knew of the existence of the neutron, a neutral particle of almost the same mass as the proton. Furthermore, though many speculated, no one knew how protons could coexist in such close proximity to each other, overcoming the electrostatic force that causes like charges to repel each other.

To Fermi, the nucleus presented an attractive new frontier, so he conspired with Corbino to map out a plan of action. There were a number of major steps in the plan, steps that enabled the Rome School, a few years later, to pounce on a new discovery at exactly the right moment.

First was Corbino's decision to publicly stake out a new direction by making a high-profile speech in September 1929 to a gathering of the Italian Society for the Progression of Science. In this presentation, Corbino baldly stated the new goals of experimental physics: "Italy will regain with honor its lost eminence . . . the only possibility of great discoveries lies in the chance that one might be able to identify the internal nucleus of the atom. This will be the worthy task of physics of the future." With these words he launched a decade-long effort to persuade the fascist state to finance and support Fermi's nuclear work in Rome. With envy, Corbino and Fermi looked across the Atlantic to places like Berkeley, where Ernest Lawrence had built an eleven-inch cyclotron designed to explore the inner workings of the nucleus at high energies. Soon Lawrence would be building even bigger cyclotrons. Equipment like this was expensive, and Corbino concluded that only a national commitment could provide the funding for the expensive equipment required to put Italy at the forefront of this new field.

Unfortunately, Corbino was never able to get the kind of support from the Mussolini regime that he believed the Rome School deserved. The physics program in Rome would be funded for teaching and more modest equipment, but only after World War II would Italy build its own high-energy cyclotron. Mussolini's reluctance to commit the necessary financial resources may well have influenced Fermi's eventual decision to leave Italy for good.

The second step was to get up to speed on the most recent research in the field of nuclear physics. Fermi knew that Rutherford and his team at Cambridge led the field, at least as far as experimental work went, and instructed Amaldi to study the most recent book on radioactivity by the Cambridge physicists and lead a small group, including Fermi, Rasetti, Segrè, and Majorana, through a colloquium on the subject. The massive 575-page book, published in 1930 by Rutherford and his colleagues John Chadwick and Charles Drummond (C. D.) Ellis, summarized all experimental data on the various forms of radiation, with many photos and diagrams. It clearly influenced Fermi's thinking about beta radiation when he turned his mind to the subject in late 1933. The book was essentially an experimenter's treatise and, characteristically for the Cambridge group, it contained little in the way of theoretical speculation, relying heavily on empirical data and experimental technique, a reflection of Rutherford's instinctive distrust of theory. It was exactly the introduction that Fermi wanted.

Third, the team continued to publish and maintain a public profile as it made the transition from atomic to nuclear physics, using spectroscopic analysis to study nuclear spin rather than electron energy shifts. The Rome group may not have had the most up-to-date cyclotrons, but they did have beautiful and precise spectrographs, including one that measured some five feet in length, nicknamed the "crocodile." Rasetti was a master spectroscopic physicist who taught his skills to Amaldi, Segrè, and others who joined the team, most importantly, a young Pisan named Bruno Pontecorvo, who arrived at Via Panisperna in 1933.

FIGURE 9.1. Rasetti's "crocodile" spectrograph. *Photo by Susan Schwartz. Courtesy of the Department of Physics museum, University of Rome, La Sapienza.*

As part of the continuing publication effort, Fermi alone published twenty-six papers between the time of Corbino's speech and the beta radiation paper in 1933, covering subjects as varied as the magnetic moment of the nucleus and the Raman effect, in which the frequency of light changes when bounced off certain molecules. They were all solid, interesting papers, but they were incremental contributions to the field, nothing as significant either as the 1926 paper on statistics or the 1933 beta decay paper.

In another carefully considered step, Fermi sent each member of the team to a different, major foreign lab to learn new experimental techniques and gain insights of researchers who were themselves further along in the transition process. Earlier on, Rasetti went to Caltech to study the Raman effect with the esteemed American physicist Robert Millikan, who won a Nobel Prize in Physics in 1923 for his work measuring the electric charge of a single electron. Segrè visited Pieter Zeeman in Holland to study—not surprisingly—the Zeeman effect. In 1931, Rasetti went to Berlin to study techniques relating to the construction of cloud chambers—the standard particle detector at the time—with experimental physicist Lise Meitner. He also learned how to isolate and prepare radioactive samples for further study. Segrè went to Hamburg, where he studied experimental techniques with Otto Stern, a brilliant experimentalist who would go on to win a Nobel Prize for his measurement of the proton's magnetic moment. Amaldi traveled north to Leipzig, where he spent time with Peter Debye, who won the 1936 Nobel Prize in Chemistry for his work in X-ray diffraction of gases. It is clear that Fermi chose widely differing labs for his team to visit to learn the breadth of skills that he thought would be valuable in future work.

Yet another step, calculated not only to bring the Rome team up to speed on matters nuclear but also to raise Italy's profile in the field, was the convening in October 1931 of an international conference on nuclear physics, sponsored by the *Reale Accademia*. Instead of Como, the site of the 1927 conference, this one was held in Rome, with most of the activities centered on Via Panisperna. Like its 1927 predecessor, it attracted a wide range of impressive scientific names, including Niels Bohr, Marie Curie, Arthur Compton, Hans Geiger, Werner Heisenberg, Lise Meitner, Robert Millikan, Wolfgang Pauli, and Arnold Sommerfeld,

FIGURE 9.2. Group photo, Rome conference, 1931. Marconi stands front, center. To his left on successive steps are Bohr, Corbino, and Fermi, who is enjoying a laugh with his friend Ehrenfest. Persico is standing at the back by the left side of the entrance, under the brass plaque. Arthur Compton can be seen, head down, on the first step directly to Marconi's right. The woman who appears to be dressed in black standing over Marconi's right shoulder is Madame Curie. *From "Convegno di Fisicia Nucleare," Rome: Reale Accademia D'Italia, 1932.*

among others. Representing the Italians, Corbino and Guglielmo Marconi were the copresidents of the conference. In contrast to the Como conference, Fermi now had a formal role as the secretary general of the meeting, responsible for all invitations and organization. It was, in many ways, Fermi's conference. Garbasso from Florence was also there, as was Persico, at this point a professor at the University of Turin. Rasetti attended, as did Tullio Levi-Civita and the young Bruno Rossi.

The papers delivered at the conference covered a wide range of topics in nuclear physics. Ellis from the Cambridge group delivered a paper on beta and gamma rays, summarizing and extending what Fermi and the team learned from the treatise the Cambridge group published in

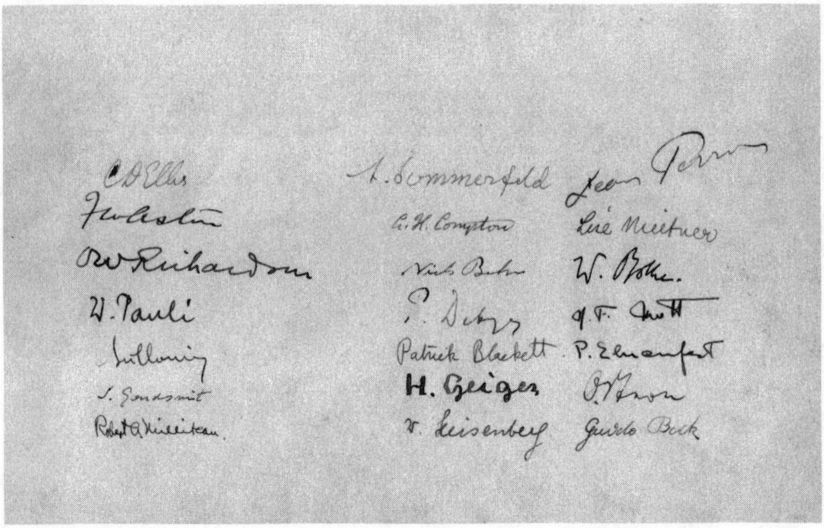

FIGURE 9.3. Signatures procured by Persico of several attendees, including Ellis, Aston,* Richardson,* Pauli,* Brouillin, Goudsmit, Millikan,* Sommerfeld, A. Compton,* Bohr,* Debye,* Blackett,* Geiger, Heisenberg,* Perrin,* Meitner, Bothe,* Mott,* Ehrenfest, ?, Beck.* The asterisks mark those who had won or were to win the Nobel Prize. *Photo by Giovanni Battimelli. Courtesy of the Enrico Persico Archives, Department of Physics, University of Rome, La Sapienza.*

1930. The problems associated with beta decay were on everyone's mind and Pauli spent much of the session chatting with Fermi about them. As we have seen, Bohr's paper ventured the notion that energy was not conserved in beta decay. George Gamow, an ebullient and gregarious Russian theorist who defected to the West two years later at the 1933 Solvay conference, and Cambridge theorist Ralph H. Fowler presented papers proposing theories of nuclear structure. It was a productive meeting, although, as Segrè suggests, it came just a few months too early. In early 1932 American chemist Harold Urey would discover an isotope of hydrogen, deuterium. Even more important, a month later, in February 1932 Rutherford's colleague James Chadwick would announce the discovery of a neutral particle in the nucleus with just a bit more mass than the proton, which he dubbed the neutron. Its existence explained the weird discrepancy between the mass and the charge of the nucleus and also explained the existence of Urey's heavy hydrogen isotope. Later that year, in August, Carl Anderson, an American physicist working with Robert Millikan at Caltech, made a further experimental discovery while studying cosmic rays: the positron, the antimatter counterpart of the electron, predicted by Dirac in 1927.

Two further conferences, one in Paris in 1932 and the Solvay conference of 1933, pushed nuclear physics even further along. Fermi attended both and, immediately after returning from Solvay in 1933, put together his beta decay paper. In February 1934, however, he received the startling news, published in *Nature* and in the French physics journal *Comptes Rendus,* that the French husband and wife team of Irène and Frédéric Joliot-Curie, the former being the daughter of Nobel Prize winner Marie Curie, had made nonradioactive elements like aluminum, boron, and magnesium radioactive by bombarding them with alpha particles from a polonium source. To Fermi, as for the rest of the physics world, this was astonishing news. Scientists had been bombarding elements with alpha particles for some time and had been noting the breakdown into a variety of new isotopes and elements, none of which were radioactive. Though radioactivity was well understood experimentally, the theory behind it was not well developed, and it came as a complete surprise that, with experiments like the ones the Joliot-Curies conducted, radioactivity could be induced in nonradioactive elements. With their experiments, the Joliot-Curies created new radioactive versions of otherwise stable elements.

When Fermi read the Joliot-Curie papers, his critical intuition began to twitch. Because alpha particles are positively charged, he reasoned that they are not particularly efficient as "bullets" for striking the positively charged nuclei of atoms. Positive charges repel each other and, he figured, it would be a lucky alpha particle indeed that would make its way into the nucleus of a target atom. Most would be repelled long before nearing the target. That the Joliot-Curies got any results at all was due to the intensity of the alpha radiation created by the polonium source. Polonium emitted an enormous number of alpha particles per second. Some would be bound to get through. However, if instead of alpha particles, neutrons were aimed at the nucleus of an atom, they would have a much better chance of striking the nucleus directly, causing similar radioactive transmutations, because, being neutrally charged, they were not repelled by the positively charged nucleus. True, the available neutron sources were nowhere nearly as intense as alpha ray sources, but they would not have to be. The better odds any given neutron would have in striking the nucleus would offset the relatively low numbers available.

At this moment, however, the news from the Paris team was public knowledge. Rutherford and his team in Cambridge had been bombarding elements with neutrons for the past year but had not developed sources of sufficient intensity to compete with alpha particles and had so far been unable to do the types of studies that were so successful in Paris. However, they were expert in the experimental techniques being used in Paris to pursue this work. Fermi knew that his real competition would be Rutherford and Chadwick, who would find a way to use neutrons instead of alpha particles to induce radioactivity sooner or later on their own. If Fermi wanted to establish priority in the use of neutrons to bombard nuclei—and the ever-competitive Fermi certainly did—the Rome team would have to work fast.

A new colleague of Fermi's, Gian-Carlo Wick, provided additional stimulus. Wick, a former student of the Ukrainian-Italian physicist Gleb Wataghin in Turin, came to Via Panisperna in 1932 as an assistant to Corbino, when Rasetti, who previously held the post, was promoted to a professorship. Wick was an insightful theorist and observed that the positron emissions seen by the Joliot-Curies were the result of "reverse" beta decay, as Fermi's paper predicted. The idea delighted Fermi.

Fortunately, Fermi knew of a technique to produce high-intensity neutron sources. He and Rasetti had been working on an earlier project that required neutron sources—a spectroscopic study of gamma-ray scattering—and had located a small sample of radium in the bowels of Via Panisperna. The radium belonged to the Institute of Public Health, located in the basement of the building and headed by a prominent public health official named Giulio Cesare Trabacchi, and was being used by the institute for preparations related to cancer treatment. In an act of extraordinary generosity, Trabacchi allowed Rasetti and Fermi to draw off radon gas produced by the radium for the gamma-ray studies. After they pumped the radon gas off the radium and into a glass tube, they dipped the tube in liquid nitrogen, condensing the gas into a liquid and giving them a short time to seal the glass tube before all the radon

FIGURE 9.4. Corbino's boys. *From left to right:* D'Agostino, Segrè, Amaldi, Rasetti, and Fermi. Probably taken in the spring of 1934, during the first neutron bombardment experiments. *Courtesy of the Amaldi Archives, Department of Physics, University of Rome, La Sapienza.*

evaporated. It was a finicky process and often resulted in the liquid nitrogen cracking the glass tube. By November 1933, however, they had more or less perfected the technique. In the wake of the news from Paris, they decided that a mixture of radon gas and beryllium would provide exactly the intense neutron source they required to see whether neutrons could induce radioactivity in otherwise stable elements.

The team included Fermi, Rasetti, Segrè, and Amaldi and a new member, a radiochemist named Oscar D'Agostino, who had been working with Trabacchi in the basement of the building and who was at that moment studying radiochemistry separation at the Joliot-Curies' lab in Paris. Fermi arranged for a division of labor. He and Rasetti would prepare the neutron source. Fermi and Amaldi would expose the target elements to the neutron source and would measure the resulting radiation with Geiger counters they built by hand. Segrè would help out as needed in either of these processes and would also use his considerable entrepreneurial skills to scour Rome and procure target elements to expose. D'Agostino would analyze the by-products of the bombardment using newly developing techniques of radiochemistry. Trabacchi was also considered an honorary member of the group, having loaned the team the radium from which they obtained the radon gas.

A seventh would be added to the team during the year. Bruno Pontecorvo, from a wealthy Jewish family long associated with the textile trade in Pisa, arrived at Via Panisperna in 1933 and participated in Fermi's gamma-ray studies that year. He was strikingly handsome and a fine athlete. He was also decidedly left-wing, verging on communist. At this point, though, his family's social standing and his own involvement with Fermi's team inoculated him against attack by the aggressively anticommunist fascist regime. He was, it turned out, a gifted researcher, destined to make a singular contribution to the story of neutron bombardment.

So much has been written about the period from March to October 1934 at Via Panisperna and so much of what has been written comes from the memories of participants well after the fact that historians must be cautious in accepting any particular participant's narrative at face value. One example is the story, told by Laura Fermi, of how Rasetti was away in Morocco on an extended vacation when the work on neutron bombardment started and that Fermi sent a cable asking him to return so he could participate in the experiments. In fact, Rasetti was in Rome giving lectures on spectroscopy during the period of initial work, although he did not participate in these initial experiments. On March 20, 1934, when Fermi first induced radioactivity through neutron bombardment, Rasetti was delivering the final lecture of the course. He left for a conference in Morocco, not an extended vacation. Another story, told by Segrè and Rasetti, is that the project began by exposing elements to neutron bombardment, going systematically through the periodic table of elements starting from the beginning of the table. The lab notebook for this initial period, lost for decades and found in

2006 by professors Francesco Guerra and Nadia Robotti in the estate of D'Agostino, suggests that Fermi started with the element fluorine, number nine in the periodic table.

Fermi's experimental design was complicated by several constraints. First, the measurement of radioactivity by Geiger counters had to take place in an area that was not affected by the intense radioactivity of the radon gas itself, so they placed the Geiger counters in a room at the farthest end of the lab's corridor. Since the half-life of some irradiated targets was very short, measured in mere minutes, getting the target to the counters involved running up and down the corridor at high speed. For the next few years, while work on neutron bombardment continued, distinguished visitors arriving at Via Panisperna were astonished to find Fermi, Amaldi, and others in lab coats running back and forth along the second floor corridor of the institute, with Fermi, as was his nature, always in the lead, carrying irradiated samples.*

Second, the geometry of the experiment needed careful consideration. Glass tubes of radon-beryllium radiated neutrons in all directions with equal intensity. Understanding this, Fermi decided to form the target materials into cylinders and drill holes vertically through them, into which glass tubes containing the neutron source were inserted. When the exposure was complete, the tube was removed and the sample quickly transported to the room with the Geiger counters.

On March 25, 1934, in the first of a series of ten short reports he sent to the journal *Ricerca Scientifica,* the journal of the Italian National Research Council, Fermi reported on the induced radiation in fluorine and in aluminum. For each of these reports, he sent preprints to physics labs throughout Europe and the United States to establish priority over the discovery of induced radiation through neutron bombardment and to provide data for others to replicate should they wish to do so. This time, in contrast to the muted reaction to the beta decay paper, the physics

---

* Those who speculate that Fermi's later terminal stomach cancer might have been caused by radiation exposure point to these experiments as possible evidence. The entire team was involved, and only Fermi developed stomach cancer. Given Fermi's hands-on experimental style, it is likely that he did most of the running up and down the corridors himself, holding the irradiated targets close to his chest. Throughout his life, whether running on the beach or in a lab corridor, Enrico Fermi was always out in front.

community paid immediate attention. In receipt of the first reports at Cambridge, Rutherford drafted a generous reply to Fermi:

> Your results are of great interest, and no doubt later we shall be able to obtain more information as to the actual mechanism of such transformations. It is by no means clear that in all cases the process is as simple as it appears to be the case in the observations of the Joliots.
>
> I congratulate you on your escape from theoretical physics! You seem to have struck a good line to start with. You may be interested to hear that Professor Dirac also is doing some experiments. This seems to be a good augury for the future of theoretical physics!

During the summer, Amaldi and Segrè traveled to England to deliver to Rutherford a paper on the work for publication in the *Proceedings of the Royal Society*. When they met in Cambridge with the legendary experimentalist, so much of whose work laid the foundation for the Rome team's efforts, Segrè asked whether it would be possible to arrange for speedy publication. Rutherford replied, with characteristic wit, "What did you think I was president of the Royal Society for?"

Leading up to the summer academic holidays, Fermi and the team continued through the periodic table, finally coming to the heaviest of elements, thorium and uranium. With these heavy elements, the expectation, shared widely throughout the physics community, was that neutron bombardment would result in the creation of even heavier elements, so-called transuranic elements. Nevertheless, Fermi insisted that the team be thorough, and D'Agostino went through the process of trying to identify any lighter by-products, moving down the periodic table until he came to lead, assuming there would be nothing lighter. Finding none, he gave up. The tentative conclusion the team reached—somewhat reluctantly, since D'Agostino was unable to get a clean separation of by-products—was that new heavier elements might actually have been created. Only a German physicist, Ida Noddack, suggested that the transuranic hypothesis was wrong, that Fermi had actually caused the uranium nucleus to split into two much smaller pieces, lower than lead in the periodic table. Her suggestion was ignored, largely because neither she nor anyone else could come up with a possible mechanism for explaining such an event.

The tentative conclusion became a definitive one, however, at the beginning of the summer break when Corbino, in a premature but enthusiastic speech before the *Accademia dei Lincei,* publicly announced the discovery of transuranic elements by Fermi and the rest of the team. The speech made headlines in Italy and around the world. Corbino had not consulted with anyone beforehand, and Fermi was devastated that an uncertain conclusion had been presented as final. Meticulous in his conservative approach of announcing results only when he was absolutely sure of them, he worried that his reputation might be ruined, particularly if it was proven not to be true. He spent a sleepless night wondering what he should do. Corbino was his mentor and key supporter in Italy, and there were limits to how openly Fermi could differ with him. The next morning Fermi approached Corbino directly with his concerns. Corbino understood his mistake immediately and tried to downplay the story, but the damage had been done. The story was just too exciting to go away of its own accord. Perhaps because he wanted to believe it or perhaps because he did not want to embarrass Corbino, Fermi himself never absolutely repudiated it. Five years later, the Nobel committee awarded Fermi the physics prize for this work, citing slow neutrons and the discovery of transuranic elements. At that same moment, the work of the brilliant team of Lise Meitner, Otto Hahn, and Fritz Strassmann, laboring away in Berlin, uncovered the truth of what Fermi and his team had actually done. They hadn't discovered transuranic elements at all. They had split the uranium atom.

Work on neutron bombardment paused during the summer, while team members went their separate ways. Fermi spent much of that summer lecturing in South America and came back through London, where he attended a conference and delivered a full report on the neutron work. In the fall, work continued under his direction. At this point, he invited Pontecorvo, who had been at the institute for a year, to join the team.

One of the problems Fermi was trying to solve was the difficulty of getting reproducible results from specific irradiations. The level and type of radioactivity induced seemed to differ from experiment to experiment. The best the team could do was to categorize levels of induced radioactivity as strong, medium, or weak. Fermi wanted to see whether a more quantitative standard could be developed, and he asked Amaldi and Pontecorvo to give it a try.

Very occasionally, Mother Nature decides to give us a peek behind the curtain at what is really happening. On October 18, 1934, Amaldi and Pontecorvo were given just such a peek.

They began by irradiating silver, with a known half-life of 2.3 minutes. They wanted to establish this as a standard against which other quantitative measurements would proceed. The problem they encountered, however, was that the effect of the neutron source on the silver target depended not only on the distance from the source to the target but also on the table on which the source and the target were placed. When they placed the silver on a marble table, the level of radioactivity was markedly lower than when placed on a wooden table. This was perplexing, to say the least. Why should the level of induced radioactivity change depending on the table on which the source and target were placed? Amaldi and Pontecorvo continued measurements throughout the next day. By Saturday, October 20, 1934, this strange phenomenon refused to go away, and they approached Fermi with the puzzle.

Fermi had been preparing a wedge made of lead to place between the neutron source and the silver target. What he did now is best described in his own words, related to his good friend Subrahmanyan Chandrasekhar after World War II:

> I will tell you how I came to make the discovery which I suppose is the most important one I have made. We were working very hard in the neutron-induced radioactivity and the results we were obtaining made no sense. One day, as I came to the laboratory, it occurred to me that I should examine the effect of placing a piece of lead before the incident neutrons. Instead of my usual custom, I took great pains to have the piece of lead precisely machined. I was clearly dissatisfied with something: I tried every excuse to postpone putting the piece of lead in its place. When finally, with some reluctance, I was going to put it in its place, I said to myself: "No, I do not want this piece of lead here; what I want is a piece of paraffin." It was just like that with no advance warning, no conscious prior reasoning. I immediately took some odd piece of paraffin and placed it where the lead was to have been.

Fermi's memory may not be reliable. His notebooks suggest that the lead wedge was actually used before he decided to use paraffin. In any case, Fermi conducted the paraffin experiment with Amaldi and Persico in the morning, while Segrè and Rasetti were engaged in supervising exams in another part of the building. At about noon, he repeated the experiment in front of the whole team, along with Persico and Bruno Rossi, both of whom were visiting. The results were astonishing. The level of induced radioactivity in the silver was much higher than it had been without the paraffin—indeed, much higher than any levels the team had yet measured.

Having established the effect of the paraffin, Fermi decided—for the first, but not the last time, in the midst of a crucial experiment—that the team should break for lunch. He was, as always, a man of habit, but this break also gave him time to mull over the extraordinary effect they had witnessed. By three o'clock, when the team returned to Via Panisperna refreshed and ready to pick up where they had left off, Fermi understood the phenomenon and was ready to share his insights.

The first observation Fermi made was that paraffin has a high proportion of hydrocarbons in its makeup, which means that much of paraffin consists of hydrogen. The second observation was that a hydrogen nucleus has very nearly the same mass as a neutron, in contrast to heavier nuclei, where the mass is two, three, fifty, or even a hundred times the mass of a neutron. The effect of a neutron bouncing around against hydrogen atoms was to slow down the neutron considerably. As an analogy, it is helpful to think of balls on a billiard table. When a cue ball hits another ball, the kinetic energy is redistributed between the two balls and the cue ball bounces away from the target ball at a slower speed because much of the cue ball's kinetic energy is transferred to the target ball, which also travels considerably as a result of the impact. Now imagine that instead of a cue ball, a ping-pong ball is driven into a bowling ball on that same billiard table. The bowling ball will hardly move; little of the kinetic energy of the ping-pong ball is transferred because the ping-pong ball is so much lighter than the bowling ball. The ping-pong ball will hardly slow down upon impact but instead will carom around the billiard table at about the same speed as it had before hitting the bowling ball.

In this way, hydrogenous substances could slow down neutrons in a way substances rich in heavier elements could not. The next question was: Why would slowing down neutrons boost the radioactivity of the target elements? This was the final part of the puzzle that Fermi figured out during the lunch break. At high speeds the neutron is likely to spend less time inside a nucleus it enters. A slower neutron has a higher probability of entering the nucleus, bouncing around inside, and coming to rest there, thereby causing the instability that gives rise to radioactivity. It was, in fact, just the opposite of the conventional wisdom, which suggested that higher energy neutrons would induce greater radioactivity in the target.

Fermi's ideas explained the results of the paraffin experiment. A wooden table has more hydrogen atoms than does a marble table, leading to the anomalies seen by Amaldi and Pontecorvo. When a hydrogenous filter like paraffin was placed between the source and the target, far more of the neutrons were slowed down.

What was needed to verify the observation was an experiment with a substance that has an even higher concentration of hydrogen at room temperature. Fortunately, there was water on the premises: a goldfish pond in the rear garden of Via Panisperna. In what may well be an apocryphal tale, the group supposedly traipsed outside and watched as Fermi repeated the experiment in the pond, using water as the medium to slow down the neutrons. The effect of the water was even stronger than that of the paraffin. History does not record how the goldfish were affected.*

That evening the team repaired to Amaldi's apartment. Amaldi's wife, Ginestra, had a job at the National Research Council and could bring a report of the discovery to work with her on Monday morning and hand it to the editors at *Ricerca Scientifica*. That evening her job was to type up the report as it was dictated by Fermi, with vigorous interruptions and arguments from the rest of the team. At times the group was so noisy that the Amaldis' maid later asked Ginestra whether the group

* There seems to be some question as to whether the goldfish pond incident really took place. Laura Fermi described it as above, but it is mentioned neither by Segrè nor by Amaldi. Ugo Amaldi reports that his father did not recall the incident and doubted that it occurred. Ugo Amaldi, interview with author, June 8, 2016.

had had too much to drink. The report is dated October 22, 1934, the date on which Ginestra submitted it for the team.

Fermi's explanation of his decision that Saturday to use paraffin illuminates as much about him and his thinking as it does about the problem itself. The actual mechanism by which neutrons induced radioactivity was still being understood and Fermi was himself seeking a better grasp. He had been thinking about nuclear reactions for several years, the first result of which was his theory of beta radiation. When he considered a problem, he thought about it continually, using the early morning hours of quiet to set out his thoughts and continuing in spare moments at the office and in his private seminars. He had thought about the problem in such depth that it had burrowed into his subconscious. In this way, he was perhaps the best prepared person in the world to react when confronted with the anomalies that Amaldi and Pontecorvo had discovered accidentally on October 18 and 19, 1934. The team's inability to standardize the impact of neutron bombardment had baffled him, but now everything fit into place. Somehow the accumulated data knocked something loose in his subconscious and he knew instinctively to reach for the block of paraffin. When the results were so dramatic, he was in a position to withdraw for a few hours and put together a definitive explanation of the phenomenon. Perhaps others might have been able to do this and we certainly owe much to Amaldi and Pontecorvo, who took note of the anomalies and brought them to Fermi's attention. Yet, in the end, it was Fermi who created the definitive experiment and explained its results to the satisfaction of his team and to the physics community at large.

Corbino became aware of the discovery that Monday and immediately understood that it was important, although he, like the rest of them, could not anticipate just why it would become an historic milestone. He was thinking mainly that a method of enhancing induced radiation would have commercial value in the production of radioactive substances for use in cancer treatment and other medical applications. He insisted that the method of slow neutron–induced radiation be patented, and Fermi immediately began work filing a patent for the process.

The Italian patent, which was granted a year or so later, names Fermi, Amaldi, Pontecorvo, Rasetti, and Segrè as the inventors of the slow-neutron process. An agreement was reached whereby D'Agostino and Trabacchi would share equally in any commercial benefit. Fermi's former student Gabriello Giannini had moved to the United States and now offered to shepherd the idea through the US patent office and manage the process internationally. The issue of the patents for the slow-neutron process would arise again after the war, with results that would be disappointing to the team. For now, though, Fermi and the others members felt that they had a lock on the new concept and had certainly established priority over competing teams in Cambridge, Paris, and Berlin.

Immediately upon the discovery of the slow-neutron effect, the group commenced redoing all the work that had been done since March 1934, seeing how exposure to slow neutrons irradiated each element in different ways. By the end of 1934, Fermi was confident that he understood the effects. In February 1935, he submitted a rather lengthy paper to the Royal Society summarizing the work in slow neutrons. He also began to analyze neutron-nuclei collisions using a primitive, paper-based form of simulation he would later pursue at Los Alamos with the first generation of computers, in which the course of a neutron's travel through a particular target would be simulated according to the probabilities of different outcomes at each stage of the journey. Using pencil and paper, he could repeat the simulations over and over again to analyze the distribution of outcomes given the underlying probabilities. This method was later christened the "Monte Carlo" method, reflecting the role of chance in the range of possible outcomes, much as it would be in a casino. It was one of Fermi's lasting and most broadly useful analytical contributions. Oddly, Fermi did not seem to think that this new analytical technique was a significant development, perhaps because he did not anticipate the subsequent advent of electronic computers that made Monte Carlo simulations far easier to do. He only told Segrè about his attempts with this kind of calculation years later, during the war.

THE YEAR 1934 IS REGARDED AS A HIGH WATER MARK FOR THE ROME School. In a period of little more than seven months, Fermi and his team explored radioactivity and the atomic nucleus with poor financial backing and primitive equipment, certainly nothing of the sort available to Lawrence's team at Berkeley. Yet they made the astonishing discovery that slowing down neutrons enhanced the radioactivity induced by neutron bombardment. They developed a detailed body of data that could be used by other teams throughout the physics world to replicate and study. By pushing hard on the experimental program, Fermi achieved results of lasting significance ahead of all his competitors, including the formidable teams in Berlin, Paris, Berkeley, and Cambridge. In the process he set the stage, quite unwittingly, for an historic drama some five years later, a drama in which he would find himself a major player. Fortunately for the world at large, neither Fermi, nor the team around him, realized at the time what they had actually accomplished.

CHAPTER TEN

PHYSICS AS SOMA

THE PERIOD BETWEEN EARLY 1935, WHEN FERMI SUBMITTED HIS long report on slow neutrons to the Royal Society in London, and November 1938, when he left Italy for good, was marked more by a consolidation of knowledge rather than by any new, revolutionary discoveries. In part this reflects a natural cycle in scientific inquiry, particularly for people like Fermi who, having made a significant discovery, tried to understand everything about that discovery before moving on to other research interests. In part, though, it reflects the increasingly political environment in Italy under fascism, culminating in the Italian aggression against Ethiopia and the gradual slide from domination over Nazi Germany to a humiliating subservience to the regime in Berlin. It also reflects developments in the Roman academic world that complicated Fermi's situation and worked to restrain his own research agenda.

THE WORK ON SLOW NEUTRONS CONTINUED UNABATED. SEVERAL members of the team moved on, a natural result of professional pressures and opportunities. Segrè was offered a position at the University of Palermo in Sicily, where Corbino had done some of his early research in physics, and from the summer of 1935 onward was no longer a fixture at Via Panisperna. Next to leave was Bruno Pontecorvo, who moved to Paris in 1936 to work with the Joliot-Curies. D'Agostino returned to his job with Trabacchi at about this time, as well. Fermi continued to work with Rasetti and Amaldi, and together the three of them did exhaustive research on slow-neutron bombardment. Of the twenty-one papers Fermi published between the submission to the Royal Society in March 1935 and his departure from Italy in December 1938, only three focused on matters unrelated to slow neutrons: eulogies for Corbino, Rutherford, and Marconi, in that order. Of particular scientific importance, at least in retrospect, was a series of papers on the diffusion of neutrons in matter. These papers became a foundation for work done in the United States in 1939 and 1940, when Fermi began to explore the possibility of controlled fission reactions. Nothing the team discovered as they repeated experiment after experiment was as significant or as exciting as the original discovery of slow neutrons, but that did not matter to Fermi. His instinct for thoroughness drove him to complete these measurements in spite of—or perhaps because of—the deteriorating political situation. Burying himself in work was natural for him, but it also provided an escape from a thoroughly depressing political landscape.

Elsewhere in the world of physics, the main development in this period came not from Europe but, interestingly, from Japan. Until 1935, there were no satisfactory theories as to how a nucleus with many protons held together, despite the electrostatic repulsion of positively charged particles with each other. Something was holding protons and neutrons together in a tightly bonded mass at the center of the atom. The nature of the force was still a mystery.

At the moment Fermi was preparing the 1935 report on slow neutrons for the Royal Society, a Japanese physicist named Hideki Yukawa proposed a quantum field theory to explain how the nucleus is held together against electromagnetic repulsion. As part of the theory he proposed the existence of a massive particle, which he called a "mesotron," which would act as a sort of glue to bind the protons and neutrons within the nucleus. In the history of scientific nomenclature, the mesotron's story is particularly confusing. Its name was soon shortened to meson. Then it was called a pi-meson, and Fermi, in a typical attempt to find a catchy name for a new particle, shortened the name to "pion," a label that stuck. Most of the work on Yukawa's theory would have to wait until after World War II, when particle accelerators became powerful enough to produce pions as a product of bombarding nuclei with high-energy protons. For his theory of nuclear forces, Yukawa won the 1949 Nobel Prize.

Fermi played a major role in these pion studies after the war, but if he was aware of the theory as early as 1935, he certainly understood that the equipment at his disposal was completely inadequate to meaningfully explore Yukawa's ideas about the atomic nucleus. He was, under the circumstances, compelled to continue his neutron studies, hoping for the day when he would have access to a powerful cyclotron particle accelerator.

THE INTERNATIONAL SITUATION AND THE GROWING RELATIONSHIP between Mussolini and Hitler certainly had an impact on events at Via Panisperna.

In December 1934, skirmishes broke out between Italian and Ethiopian military forces along the border between Italy's colonial protectorate in Somaliland and Haile Selassie's ancient independent kingdom. The skirmishes gave Mussolini the excuse he was looking for to launch a dramatic military intervention that would be the fascist regime's crowning military glory. Threatening to invade Ethiopia, Mussolini whipped his population into frenzied patriotic fervor, at the height of which he called upon all married women to trade in their gold wedding rings for rings of steel. The government's fiscal state was precarious at the best of times and unable to finance a major military campaign without additional financial reserves. Mussolini understood the power of public ritual and made sure that women throughout the country offered up their nuptial gold in public ceremonies that had the dual purposes of raising the country's gold reserves and pushing war patriotism to a fevered pitch. Even Laura Fermi was impressed with the outpouring of devotion to the cause and offered up the gold ring Enrico had bought for her in 1928.

Mussolini's increasingly aggressive stance lost the support of Britain and France at the League of Nations and thrust him into the arms of the one major power that would support him—Nazi Germany.

During the run-up to hostilities, the atmosphere at Via Panisperna was deeply affected. Segrè noted an inexplicable anxiety among his colleagues during the spring and mentioned it to Fermi. Fermi told him to go to the institute library and consult a large book in the middle of a table in the library. It was a world atlas and it opened up without effort to the well-thumbed pages on which were displayed the countries of the Horn of Africa. Segrè may have been oblivious, but clearly Fermi was not. He certainly discussed international events with Corbino, who as a member of the Senate was thoroughly informed of the government's thinking on the subject. In the summer of 1935, Fermi went to the United States to attend the summer session at Ann Arbor, leaving Laura behind, and again was struck by how attractive the United States was as a place to work. At this point, the main thing keeping him from accepting a job in the United States was Laura's refusal to leave Italy.

He spent the next two summers in the United States, as well. In 1936, George Pegram, head of the Columbia physics department, invited him to give a series of summer lectures on thermodynamics. The results, compiled later that year into a short book, are a model of clarity and simplicity and still worth reading for anyone interested in a brief but pithy introduction by one of the true masters of the field. While there, Fermi received an offer from Pegram to join the faculty on a permanent basis, but knowing of Laura's resistance to leaving her beloved Rome, he politely declined. In 1937, he spent the summer in California, lecturing at Berkeley, where he spent time with the inventor of the cyclotron, Ernest Lawrence, as well as with a German-trained American theorist named J. Robert Oppenheimer. He also lectured at Stanford University, where he renewed his friendship with a young Swiss experimentalist named Felix Bloch. Bloch had completed a post-doc in Rome in 1932 and soon after, with Hitler's rise to power, left Europe for Stanford.

Bloch and Fermi drove back across the United States together from California to New York, during which time they found the ubiquitous Burma Shave billboards greatly amusing: "At crossing roads don't trust to luck / The other car may be a truck! Burma Shave." Bloch accompanied Fermi back to Rome and joined Laura and Enrico on a trip to the country home in Tuscany in a new, more elegant car than the bébé Peugeot that served as the Fermis' first car. The highway was now lined with billboards displaying menacing yet faintly ridiculous fascist slogans authored by Mussolini himself: "Mussolini is always right!" "To win is necessary. But to fight is more necessary." Fermi and Bloch passed the time by converting these to Burma Shave slogans: "Mussolini is always right. Burma Shave!" The amusement they found in this subversive humor was, of course, short lived. Bloch returned to California and Fermi stayed behind, trying to stay out of trouble under an increasingly delusional fascist regime.

The Tuscan villa Laura's aunt and uncle owned was the Fermi family's summer refuge, and Enrico would join Laura and the children whenever he returned from his summer stints in the States. There they would spend the last few weeks of the summer enjoying the relative tranquility with Nella and Giulio, still young children. In the Tuscan countryside they could escape the constant barrage of fascist propaganda that Mussolini pumped out in an effort to keep the nation together under his increasingly erratic and paranoid rule. Whether they argued over the possibility of a move to the United States remains unrecorded, although it would be odd if the subject did not come up at least occasionally. Still, Laura refused to consider leaving Rome and Tuscany behind.

THE ITALY FERMI RETURNED TO IN THE FALL OF 1935 WAS ON WAR footing, and in October Mussolini decided to act, launching a full-scale invasion of Ethiopia. The Italian victory, inevitable though it was, took some seven months to secure and involved Italian use of chemical weapons against a poorly equipped Ethiopian military. Mussolini had his military victory, but little did he appreciate that its price would be ever-increasing reliance on economic, military, and diplomatic support from the ever more powerful Nazi regime in Berlin. Once the dominant member of the fascist club, Mussolini now found himself playing number two to an increasingly powerful and bellicose Hitler.

To avoid the stress of the daily headlines and the constant war propaganda, Fermi and Amaldi drove themselves at an even more aggressive pace. Weekend jaunts to the beach or the mountains became less frequent, and the easy camaraderie, the occasional pranks and jokes that defined the early years of the Rome School became difficult to maintain. "Physics as soma," Amaldi later called it, using the term from Aldous Huxley's 1932 *Brave New World*, which described a drug that induced a zombie-like state. It was their way of avoiding having to think about the increasingly troubling direction of the Italian regime.

Physics in Rome was no longer pure fun. Indeed, for Fermi, who had found unalloyed enjoyment in physics since his early youth, physics would not delight him again for at least a decade. Even then, in the aftermath of World War II and his own role in ending it, physics was never again for him the entirely carefree delight that characterized his early career.

When Fermi returned to Rome in 1936, the fascist regime had built him a brand new institute.

In the early 1930s, Mussolini decided to bring all the disparate departments of the University of Rome together to one location, a relatively open area south of the old medical center, northeast of Rome's Termini railroad station.

The building that would house the physics department, designed by one of Mussolini's favorite architects, Giuseppe Pagano, was a functional, horizontal, five-story building, faced with glazed mustard-colored brick, far better suited to its purpose than the old rabbit warren on Via Panisperna. Never particularly sentimental, Fermi did not miss the old lab and appreciated the convenience of the new campus, a short bicycle ride from home and an easy lunchtime commute.

The team continued to disperse. The brilliant Ettore Majorana, who speculated presciently on some of the fundamental characteristics of neutrinos, received a position at the University of Naples in 1937, notably without having to enter a competition. He then disappeared without a trace the following year, on a boat ride from Palermo to Naples. The mystery of what happened to him remains unsolved but has been the subject of continuing conjecture to this day. Although new Italian students were admitted to the physics program at the University of Rome, including Oreste Piccioni and Eugene Fubini, the flow of post-docs and more senior visitors from the United States and Europe slowed, in large part because of political tensions and the flight of Jewish scientists from Germany and Austria to safe havens in the United States, the United Kingdom, and Canada. They were going to places like New York and Berkeley and London, not to fascist Rome, irrespective of the special treatment accorded Fermi and his team. In 1927, Italy could attract great European and American scientists to an international conference in Como. By 1931, however, some invitations to the Rome conference were discreetly declined owing to reservations about the character of the fascist regime. By 1937, doubts about the regime had vanished and, even if he had tried, Fermi would have found it difficult to attract the best international minds.

The biggest blow to Fermi and the dwindling Rome team was Corbino's relatively sudden death from pneumonia in January 1937. Of all the people Fermi had encountered in his professional career in Italy, Corbino was clearly the most important and certainly one of the most extraordinary. In the early 1930s the regime required almost every academic in Italy to take an oath of allegiance to the Fascist Party. Corbino notably did not comply and, unlike Fermi, he never joined the Fascist Party. Corbino's judgment, administrative genius, and seniority protected him. He identified Fermi as his star understudy early on, believing that the twenty-three-year-old would be able to push Italy into the forefront of international physics. Fermi never disappointed him, and the only serious recorded conflict between the two—when Corbino made his premature speech in 1934 regarding the discovery of transuranic elements—was quickly smoothed over. Corbino was more than Fermi's boss. He was the younger man's mentor, protector, and father figure, even standing in for Fermi's late father at Fermi's 1928 wedding. The sudden and unexpected loss of Corbino dealt a serious blow not only to Fermi but to the group as a whole.

Fermi wrote a eulogy for Corbino that not only focused on his scientific achievements but also touched on his important character traits, reflecting the debt Fermi felt for Corbino's help and support. Fermi acknowledged Corbino's influence on his career with his method of radically simplifying the study of apparently complex phenomena. The two physicists shared this love of simplicity, reflecting a conviction that the beauty of the discipline lay in its ability to describe complexity through the elucidation of simple underlying relationships. In their long and fruitful relationship, surely Corbino learned this from Fermi as much as Fermi did from Corbino.

The loss of his mentor and friend was compounded by the elevation of Lo Surdo to director of the institute. Fermi would not have sought this post, an administrative one with little advantage in terms of pursuing research, but it must have rankled Fermi that Corbino's political opponent in the world of Rome physics would now run the institute that Fermi called home. To make matters worse, Marconi—head of the CNR and a strong supporter of Fermi and his team—died several months later. The early part of 1937 was not a happy time for Fermi.

MEANWHILE, ITALY AND GERMANY GREW EVER CLOSER, WITH Germany increasingly the dominant partner. During a May 1938 state visit to Rome, Hitler received a hero's welcome. A new railway station, Ostiense, was built for the arrival of the German dictator. Street lamps were torn down and replaced with lamps topped with Nazi eagle iconography. Mussolini prepared an elaborate welcome, and Hitler seemed pleased with the reception. On a grand tour of the city, the two leaders passed through the beautiful central park of Rome, the Villa Borghese. The Fermis' nanny was walking with young Giulio at the time, and the event seemed to register in the mind of the three-year-old. A few years later he would, rather innocently, step up to the defense of the two dictators while at elementary school in Chicago, causing not inconsiderable consternation for his parents, who were keenly sensitive about their immigrant status and their privileged position in the war effort.

The Villa Borghese was a short walk from the new Fermi home. In early 1938, the Fermis made a decision to upgrade their living accommodations. The old apartment on Via Belluno was not small by Roman standards, but the family could afford a nicer one, with Enrico's double salaries from the university and the *Reale Accademia,* not to mention royalties from Enrico's textbook and Laura's independent income from her family. Perhaps at the urging of Laura's close friend Ginestra Amaldi, the Fermis found an apartment on an elegant side street called Via Lorenzo Magalotti, less than a block from the Amaldis' home and a short walk from the Villa Borghese, and moved there in May 1938. The upper-middle-class neighborhood was leafy and quiet, with expensive apartment buildings and small villas lining the Via dei Parioli, where the Amaldis had moved not long before. On Via Magalotti, the apartment was fitted out with a green marble-lined bathroom. Laura loved it. "It satisfied my ambitions of grandeur," she later recalled, "which had been rising as Enrico's position had steadily grown better. . . . I felt rich, well established, and firmly rooted in Rome."

The timing speaks to Fermi's thoughts regarding potential emigration. In spite of everything, in early 1938 the family had no immediate plans to leave Rome. Events over the next two months would, however, radically change the family's calculations.

HITLER HAD BEEN PRESSING MUSSOLINI FOR SOME TIME TO CRACK down on the ancient and assimilated community of Italian Jews. Under the Nazi regime, Hitler had been able to carry out an increasingly aggressive anti-Semitic social policy in Germany, first excluding Jews from professional positions at universities and hospitals and banks and law firms and then broadening the restrictions, limiting the ownership of property and greatly reducing their rights as citizens of the Third Reich. Worse was soon to come.

To Hitler's increasing annoyance, Mussolini had done nothing of the sort. Though he remained a virulent and violent Italian nationalist, his views on the Jewish population in Italy were complex and not entirely negative. The beautiful Margherita Sarfatti, one of the ideological drivers of Italian fascism, was Jewish. She served as Mussolini's confidante, biographer, muse, and mistress for well over a decade. Jewish families were considered in every sense Italian, as well they might have been, given that their presence in Rome predated the unification of the country by some two thousand years. Ironically, many Jews wholeheartedly supported the fascist regime. Many, like Laura Fermi's family, hardly viewed themselves as Jewish in any but a remote ethnic sense of the term. They were Italians first and foremost. Laura had readily acceded to Enrico's request to raise the children as Catholics.

While he was the senior partner in the relationship, Mussolini was easily able to fend off Hitler's demands for him to crack down on Italy's Jews, but by 1938 *il Duce* realized this was no longer possible, or at least no longer worth the trouble. In July, the regime issued new racial policies, followed in September by promulgation of laws that began to restrict Jewish access to jobs and positions, to places at universities, and to the financial and real estate markets. The news came as a shock to the Fermi family, who followed these events by radio in the quiet of the Tuscan countryside. It was during discussions over the balance of this anxious summer that Laura finally gave way to Enrico's persistent requests and agreed to move to the United States.

They assumed, correctly as it turned out, that they were under surveillance by the fascist secret police. Fermi was one of the very few figures of international prominence in fascist Italy, and Mussolini would, Fermi knew, take extraordinary measures to keep him in Rome. Indeed, every time Fermi went abroad he had to apply for permission to do so, under the watchful eye of Mussolini's domestic and foreign spies, and clearly his wife's unwillingness to travel with him to the United States ensured his ultimate return to Rome. Fermi's file with the fascist political police, though not very large, indicates that he had been under surveillance for some time, at least since 1932. Now Laura and the children were going to come with him. He would need to be discreet.

Fermi drafted letters to several American universities, explaining that the circumstances that had held him in Italy no longer applied and inquiring whether a position might be available. Together, Laura and Enrico drove north from Tuscany into the Alps and dropped each letter in a different rural post box. They hoped that they would evade detection by the regime, and they were right. We do not know the names of all the universities queried, but we do know that George Pegram at Columbia, who had offered him a faculty position two years before, replied quickly and unreservedly. Given Fermi's prior travels in the States, we might surmise that the University of Michigan, Stanford, and Berkeley were among the others. By this time, his good friend Hans Bethe had been at Cornell for three years, and it is certainly possible he also sent a letter there. In any event, Fermi decided to accept the offer from Columbia and began to plan for departure in the first quarter of 1939. This gave Fermi some time to quietly sort out his affairs at home and leave his research program in safe hands. The cover story would be that he was taking a six-month sabbatical. He hoped that the Italian authorities would believe him.

It remains an open question as to whether the racial laws promulgated beginning in September 1938 would have affected Laura. In deference perhaps to his own ambivalence, Mussolini incorporated an opt-out provision in the racial laws. One could apply to the regime for exemption from the laws, a provision designed for Jewish families who had proven useful to the regime and upon whom the regime relied for prestige and legitimacy. Laura's father, Admiral Capon, applied on behalf of his whole family and his application was accepted in early 1939, meaning that Laura could have returned to Italy at any time during the Mussolini years without endangering herself or her family. Things changed, however, when Mussolini fell in 1943 and the Germans effectively took direct control over much of Italy. Jews, including Laura's father, were rounded up systematically and sent off to concentration camps, where many, including Admiral Capon, perished.

Certainly, Italy's increasing isolation and closer alignment with Germany troubled Fermi and constrained his work. Fermi was a vital part of an international community, but that community was vital for him as well, and he knew that isolation from this community would inevitably damage his ability to stay at the cutting edge of the field. He also was never able to obtain the funding he felt he needed and deserved for the experimental research program he envisioned in Rome. After Corbino's sudden death, that funding became ever more problematic. These factors, as much as the racial laws, would have weighed on him. In short, there were, by mid-1938, many reasons for Fermi to leave Italy.

In mid-October 1938, however, a conversation in Copenhagen changed his carefully laid plans.

CHAPTER ELEVEN

THE NOBEL PRIZE

In October 1938, Fermi was in Copenhagen for a conference hosted by Niels Bohr. At one point during the conference, Bohr asked Fermi whether they might have a private word. Pulling him aside, Bohr asked Fermi an astonishing question: If Fermi were to be awarded a Nobel Prize, would he be in a position to accept?

The story of Fermi's Nobel Prize is one of the more fascinating tales of intrigue surrounding the world's most prestigious scientific award. It begins, however, not in October 1938 but in 1936 and involves not the physics prize but the more controversial peace prize.

In 1936, the peace prize was awarded retroactively to a German peace activist named Carl von Ossietzky, who in 1935 revealed the secret of German rearmament to a stunned world. For this act the Nazi government charged him with high treason and sent him to a concentration camp, where he ultimately died. In the meantime, to garner support for his release, the Nobel Peace Prize committee awarded him the 1936 prize. Predictably, Hitler flew into a rage and decreed that no German national would ever again be allowed to receive the Nobel Prize in any field. The whole episode was a public relations catastrophe, one the Nobel Foundation determined not to repeat.

The Nobel Foundation delegates the task of naming the winners in physics and chemistry to the Royal Swedish Academy of Sciences, a small committee that decides in strict secrecy on the basis of recommendations from a select group who are canvassed confidentially for their views. Only recently have files been opened to scholars interested in the nomination process and even then only in instances when the winners have passed away. Anyone can submit a Nobel Prize nomination—nominators need not have won to be invited to submit nominations. For example, Fermi nominated Heisenberg several times before Heisenberg won the prize in 1932, prior to Fermi's award. Dirac and Schrödinger shared the prize the following year. Records show the surprising fact that Dirac had received only three nominations prior to the award, whereas Schrödinger had received forty-one, including two from Einstein and three from Bohr. The award is not, and never has been, simply a popularity contest. The great German theoretician Arnold Sommerfeld was nominated over eighty times but never won.

None of this, of course, would have been known to Fermi at the time. He is the first and, as far as records go, only person to have been approached beforehand to see whether he would be able to accept the award.

The concern of the Swedish Academy was that, with Italy growing ever closer to Germany, Mussolini might feel compelled to follow Hitler's lead and prevent Fermi from attending the ceremony and accepting the prize. However, they need not have worried. Fermi may not have known Mussolini well at all—he met him once and, like many Italians, he came away more impressed than he had anticipated—but he knew enough about the fascist regime's need for international acceptance that he could say with confidence that it would not be a problem. Bohr dutifully reported back to the academy.

Fermi received thirty-six nominations over the course of his career, of which three were in chemistry and the rest in physics. Heisenberg nominated Fermi in 1936 and 1937. The de Broglie brothers, Louis and Maurice, nominated Fermi repeatedly. His old friend Persico, though never a laureate himself, was actually the first to nominate him. James Franck from Germany and Arthur Compton from the University of Chicago also nominated him. Max Planck was another supporter. Absent from the list, ironically, given the sequence of events, was Bohr himself. In 1938, Bohr nominated no one. In previous years, he nominated Schrödinger and Heisenberg as well as Franck and Otto Stern. Bohr had enormous respect for Fermi, but perhaps he limited his nominations to those with whom he had worked directly. For all their interactions at conferences and meetings and the visits Fermi made to Copenhagen, the two never actually worked together. In addition, they had very different approaches to physics: Fermi utterly disinterested in anything other than physics itself; Bohr almost mystically fascinated by what physics revealed about the nature of underlying reality.

In any case, Bohr's question posed a dilemma for Fermi. The prize came with a twenty-three-karat gold medal weighing a bit more than six ounces and a significant cash award, some 155,000 Swedish kroner in 1938, worth more than $500,000 in today's money. Italy had imposed currency restrictions that would prevent Fermi from taking the money with him if he brought it back to Rome and then left when he had originally planned, in the first quarter of 1939. Upon his return to Rome from Copenhagen, he discussed this with Laura and they quickly changed their plans. They would take with them to Stockholm as many of their belongings as they could discreetly manage, receive the prize, and go directly to their new home in New York, bringing the medal and prize money with them.

So, when the call came from Sweden to the Fermi household, on November 10, 1938, it was not entirely a surprise. As Laura describes the events of the day, an initial call came through from an international operator early that morning before Enrico left for work informing them that they would be receiving a call from Stockholm at six o'clock that evening. Laura took the call, woke Enrico, and they realized immediately what the afternoon call would be about. When Laura proposed that Enrico should skip work and spend the day with her to celebrate, he demurred. He understood that it was quite likely that the call would deliver the news he had anticipated after his conversation with Bohr in Copenhagen, but as someone who always calculated the odds, he also understood that he might be sharing the prize with another scientist. Dirac shared it with Schrödinger in 1933 and the prize was split again in 1936 and 1937. Fermi was not willing to go on a spending spree to celebrate without further information, and he also worried that buying too many transportable objects of value might give away their plans to leave permanently. In the end, Laura persuaded him to join her on a shopping trip, but he limited himself to purchasing a watch—"inconspicuous and useful" as he described it.

Laura spent the day with her husband, absorbing the sights and sounds of the Rome she loved, the Rome that had been her home since birth. As she later described:

> I was determined to be of good cheer and to chase away the nostalgia that came over me at the familiar sight of the Roman streets; of the old, faded buildings that had preserved their full charm; of the clumps of ancients trees that everywhere disrupted the monotony of the streets, rising above a discolored wall or behind an iron fence, silent and monumental witnesses to human restlessness; of the numberless fountains of Rome, which indulged the opulence of water, shot it toward the sky, and let it come back in cascades of diamond-like droplets, in rainbow patterns. I was going to enjoy these sights and give thanks to God for thirty years of life in Rome.

It would be hard to find a more wistful, elegiac tribute to Rome, what it meant to her, what she would lose by leaving it behind.

Later that afternoon, they sat by the phone anxiously awaiting the call. Several times an impatient Ginestra Amaldi phoned to see whether they had heard from Stockholm yet, but six o'clock came and went without the call they were expecting. Enrico turned on the radio to provide a distraction, only to find the airwaves filled with news of yet another wave of anti-Semitic laws promulgated by the increasingly desperate regime, laws with real bite: prohibiting Jews from attending public schools, firing Jewish teachers, restricting the practices of Jewish professionals to Jewish clients only, withdrawing passports, and even dissolving firms and commercial enterprises owned by Jews. By the time the call came, the Fermis were in need of some good news. They got it.

Fermi had been awarded the Nobel Prize on his own. He would share the prize money with no one. The award would cite Fermi "for his demonstrations of the existence of new radioactive elements produced by neutron irradiation, and for his related discovery of nuclear reactions brought about by slow neutrons." Within the Fermi circle of friends, the good news traveled fast, and soon a crowd descended on the apartment at Via Magalotti, Ginestra leading the way, organizing an impromptu celebratory dinner. For the time being at least, Fermi set aside his reservations about the discovery of transuranic elements and joined in the festivities. Even little Giulio and Nella participated, two-year-old Giulio climbing impishly on the legs of the men who towered over him and seven-year-old Nella trying with little success to keep him in line.

THE ORIGINAL PLAN DID NOT INCLUDE A STOPOVER IN STOCKHOLM. The new plan did, and unfortunately that meant traveling by train through Germany. Although there was no other sensible way to get to Stockholm, a trip through Germany posed a problem. Enrico and the children would carry passports that indicated they were Roman Catholic, but Laura's would identify her as a Jew—that is, if Italian authorities decided not to revoke her passport as part of the new racial laws. A passport with a stamp indicating Laura was a Jew could cause serious problems at the German border. As a practical step, the solution was clear. Laura had to convert. In the modest San Roberto Bellarmino church on Via dei Parioli, the Monsignor Ernesto Ruffini, who was later elevated to cardinal and who became an ardent opponent of church reform in the early 1960s, baptized Laura into the Catholic faith and conducted a quiet religious marriage between Laura and Enrico to seal the deal. Only the Amaldis were in attendance. Laura never mentioned her baptism or second marriage ceremony in her memoir, suggesting a certain discomfort with this aspect of their departure. Yet one can understand her actions, even though, like her husband, religion played no part in her life. This act of apparent hypocrisy enabled her passport to read the same as her husband's and her children's. There would, presumably, be no problems at the border with Germany.

The final preparations were fraught. How much to bring with them was an issue—they told Italian officials that they were taking a six-month sabbatical, so bringing the entire contents of the apartment at Via Magalotti was out of the question. In fact, bringing any furnishings at all would have tipped off the government to their real plans. In the end they decided to bring a minimum of clothing and only one of their household help. Laura would shoulder most of the burden of child care herself, at least until they had settled sufficiently and could make new arrangements in New York. The situation was especially tense for the small group of intimates who knew the full plan; for obvious reasons they could say nothing to anyone else. Of all their close friends, it was Ginestra who was most critical of their decision to leave. She believed Fermi was abandoning students and colleagues who had no choice but to stay and cope with increasingly difficult circumstances. On the platform at Rome's Termini station, where they gathered to say goodbye on December 6, 1938, Ginestra gave voice to bitterness and resentment. Laura recalled the moment vividly some fifteen years later. "Enrico's departure," Ginestra said, "is a betrayal of the young people who have come to study with him and who have trusted him for guidance and help." Ginestra's husband was quick to defend his mentor and friend: "Fascism is to be blamed, not Fermi." Laura, for her part, was stung by Ginestra's words, even as she tried to find some enthusiasm for the onward journey. She felt a host of moral contradictions, reflecting her closest friend's rebuke. Her particular concern was what she perceived to be her responsibility as a daughter: "And, perhaps the most perturbing of all moral contradictions, should a woman forget her responsibilities as a daughter to follow the call of those as wife and mother? . . . questions that have not found an answer over the centuries cannot be settled in the few minutes before a train departs." She was thinking of her sisters and most especially her father, who remained convinced that the regime would not target him. She was not so sure, perhaps because the option to opt out of the racial laws had not yet been formulated. In the end, however, there was no question what Laura would do.

Even the normally talkative Rasetti seemed downcast. "I hope I'll see you soon" was all he could muster.

The train departed. The greatest Italian scientist since Galileo was on his way into self-imposed exile. He would return only twice after the war, never for more than a few weeks.

THE TRAIN RIDE WAS STRESSFUL, PARTICULARLY WHEN A GERMAN border official had trouble finding the transit visas that Enrico had obtained for the whole family, but the confusion was settled quickly and the family passed through Germany without any further problems. In

Stockholm they were put up at the Grand Hotel, which richly merits its name, and were received by the Italian ambassador, a sophisticated man from an aristocratic family. Laura liked him.

The ceremony, on Saturday, December 10, 1938, followed roughly the same format as all Nobel ceremonies before or since. It started at the Royal Opera House, where Enrico Fermi and Pearl S. Buck, the American author and only other recipient that year, sat on stage waiting to be introduced to the king. We don't know what Buck was thinking. Perhaps, as Laura Fermi suggests, she was still wondering why she had been selected for this honor. We know from Laura that Enrico was obsessing about his evening wear. He may have been familiar with the elaborately ridiculous uniform of the *Reale Accademia,* but he was not used to wearing white tie and tails. The vest of the outfit was, as is traditional, highly starched and prone to popping up into the face of the person wearing it. He was also a bit worried about standing up to receive the award itself. As newsreels show, recipients in those days were required to approach

FIGURE 11.1. A quiet moment in Stockholm, December 1938, during the Nobel Prize ceremonies. Pearl Buck is on the left, and Fermi is smiling at a young girl, identity unknown. *Courtesy of the Pearl Buck Estate.*

the king and return to their seat facing the king at all times. Turning one's back on the Swedish sovereign was simply not done. In any case, in later years Fermi would proudly boast about how he counted out the steps carefully as he approached the king, who shook hands with him and handed him the medal and the diploma that accompanies it, following which Fermi executed a backward return to his seat without a hitch.

A huge banquet in the town hall followed the ceremony, with the guests of honor and their spouses seated with the king at the center of the long main table. A formal ball finished the night off, and Laura was thrilled to dance with King Gustav V, although she noted later that this was the second European monarch with whom she had danced. Accompanying Enrico to the 1933 Solvay conference, she had danced with the king of Belgium. All in all, it was a heady few days, capped with a private dinner the next evening at the palace for recipients and their spouses.

In what remains a tradition to this day, the recipients each made a speech reflecting on their work. Fermi chose, understandably, to focus on the efforts for which the Nobel committee had recognized him, the development of the slow-neutron method of bombardment and the results of that endeavor. It was a relatively short speech, not very technical, but it did specifically mention the discovery of two transuranic by-products: element 93, "ausonium," and element 94, "hesperium." He also noted that the German team of Otto Hahn and Lise Meitner had subsequently discovered elements through number 96. If he harbored doubts about the reality of these transuranic discoveries, he kept them to himself. This would come back to haunt him a month later.

The Italian press was furious with Fermi's decision to shake hands with the king rather than give the fascist salute. Fascist newspapers, increasingly aware of the growing relationship between Mussolini and Hitler, had downplayed the announcement of the award in November, keen to have an Italian honored by the Nobel Foundation but wary of aggravating Hitler. Now they pounced on Fermi, charging that his failure to deliver the fascist salute reflected subversive tendencies on the part of this vaunted member of the *Accademia d'Italia*. That he shook hands with the king only compounded the sin. Mussolini and his regime disparaged the shaking of hands as feminine. The public furor in Rome eventually died down, but the watchful eyes of the fascist regime were more alert than ever.

The celebrations and parties behind them, the family proceeded south and west, first to Copenhagen to pay respects to Bohr, then across the English Channel and onto a ship at Southampton destined for New York Harbor. There they set sail for the New World on December 24, 1938. Across the Atlantic, a new life beckoned.

PART THREE

THE MANHATTAN PROJECT

CHAPTER TWELVE

THE NEW WORLD

THE TRANSATLANTIC JOURNEY ON THE CUNARD LINER *FRANCONIA* was uneventful, and as the ship pulled into New York harbor on January 2, 1939, Fermi grandly announced that they had now established the American branch of the Fermi family.

Neither he nor Laura had much to do with their fellow passengers, among whom were the great French composer Nadia Boulanger, the German stage director Erwin Piscator, and some seventy members of the famed D'Oyly Carte Opera Company of London, traveling to New York for a six-week run of Gilbert and Sullivan performances. Fermi was not particularly interested in music unless it formed the backdrop to a party and dancing.

Perhaps of greater importance, it was not the happiest moment for the Fermi family. Fermi had uprooted himself from colleagues he had worked with for over a decade, men and women who were close friends as well as professional associates. Laura was even less happy, having left behind the city she called home all her life for a big and sometimes brutal city that Enrico had decided would be their new home. Passing through New York briefly on the way to the 1930 Ann Arbor summer school, she was impressed with the vitality and size of the city but also found it strange, crude, and unattractive. Now she also had two small children in tow and was under no illusion that her husband would make time to share the burdens of parenthood in a new and unfamiliar city.

She would have to find Nella a school—she ended up at Horace Mann, a progressive, highly prestigious private school in the Riverdale section of the Bronx—and teach the younger Giulio English herself. She did, however, travel with a maid who could help her with the transition.

Pegram met them on shore, along with Fermi's former-student-turned-American-businessman Gabriello Giannini, and the two of them escorted the Fermis uptown to a small hotel on the Columbia campus, the King's Crown on West 116th Street, where they were installed until they could find long-term lodgings. Its cramped and dark corridors gave the family an added impetus to find more permanent living quarters.

A month had elapsed from the time they left Rome until they arrived in New York City. During this time, he was uncharacteristically out of touch with the vibrant, almost hyper-communicative scientific community in Europe and was unable to correspond with the many scientists on whom he relied for information about cutting-edge developments. This was, in retrospect, fortunate. During the month of December, dramatic events unfolded in Germany, proving that Fermi had colossally misunderstood the results of his neutron experiments, the very experiments for which he had just received the Nobel Prize. He would never quite get over the embarrassment of having been so wrong. When the news of those experiments eventually crossed the Atlantic, Fermi and others found themselves on a runaway train careening down the tracks in a direction no one could foresee.

While Fermi was en route to the States, scientists in Germany, conducting experiments almost identical to those conducted by Fermi and his Panisperna boys in 1934, reported that they had split the nucleus of the uranium atom.

Otto Hahn was a cautious and careful man who, over a decades-long career, had built a reputation as one of the finest chemists in the world. Born in 1879, he had been engaged in intensive study of the chemistry of radioactive substances from the early 1900s, and as such is considered the father of radiochemistry, the chemical study of radioactive materials. He was particularly expert in chemical separation of radioactive elements from one another.

In 1907 he met Lise Meitner, a young woman physicist who had been working with Max Planck. Born in Vienna in 1878 and raised as a Protestant, but with a Jewish grandfather, Meitner was the second woman to gain a doctorate in physics at Berlin, and Hahn was immediately drawn to her intellect and drive. With her physicist's understanding of the radioactive processes inside the nucleus, Meitner collaborated with Hahn in the chemistry labs of the Kaiser Wilhelm Institute in Dahlem, Berlin, for the next thirty years.

Hahn the chemist and Meitner the physicist were, in many ways, the ideal collaborators, able to complement each other's skills as the team sought to understand radioactivity. It was, however, primarily a chemistry project, focusing almost entirely on the chemical separation and identification of the by-products of radioactive decay. Because of their work at the heavy end of the periodic table, among the class of elements called "actinides," they were immediately taken with Fermi's announcement in 1934 of the results of the initial neutron bombardment experiments on uranium, an actinide itself, in particular Fermi's observation that one of the four by-products of uranium bombardment seemed to be a transuranic element, element number 93.

The two colleagues began intensive work on neutron bombardment of uranium and over the next four years focused almost exclusively on identifying one after another of what they thought to be transuranic elements and their isotopes. By 1938, they believed they had identified elements 93 through 98 in the periodic table, but definitive confirmation continued to elude them.

For these experiments, they brought on board a third chemist, Friedrich "Fritz" Strassmann, some twenty years younger than Hahn and Meitner, who had a reputation as a strong analytical chemist. An active anti-Nazi, in 1933 he resigned from the Society of German Chemists over its ejection of Jewish members, although he was what the Nazis would have considered a "pure" German. As Hahn and Meitner's assistant, he was the crucial third member of the team.

The German team had competition. Rutherford, an enthusiastic fan of Fermi's experiments in 1934, was pursuing similar experiments at his Cavendish Laboratories at Cambridge University, working with colleagues like James Chadwick, who discovered the neutron in 1932. Irene Joliot-Curie, Marie Curie's daughter, whose work creating artificial radioactivity by bombarding nuclei with alpha particles in 1933 inspired Fermi in his neutron bombardment experiments, was working with her colleague Pavel Savitch to identify by-products of uranium bombardment. A bright graduate student of Ernest O. Lawrence at Berkeley, Philip Abelson, was a third competitor. In the end, though, it was Hahn and Strassmann, with significant help from Meitner and her young nephew, Otto Frisch, who made the critical discovery.

Hitler's annexation of Austria in March 1938 had a direct impact on the Berlin team. Meitner, one-quarter Jewish, was a citizen of Austria and thus untouchable by the Germans as Nazi anti-Semitic laws hit hard within the German academic community. Suddenly, with the stroke of a pen, she was a German citizen, without the protection of a sovereign foreign government. In July she escaped to Holland with the help of two Dutch physicists. Hahn gave her his mother's wedding ring with which to bribe border guards if necessary. A promised position at the University of Groningen fell through, and she wound up in Stockholm, working with Manne Siegbahn, the head of the physics department at the Royal Swedish Academy of Sciences, whose members were about to elect Fermi as the next recipient of the Nobel Prize.

Research back in Dahlem continued without Meitner, although Hahn corresponded frequently with his former partner. Hahn and Strassmann continued to analyze by-products of neutron bombardment of pure uranium and were struck by results produced in similar experiments done by the Joliot-Curies in Paris—results that neither the French nor the Berlin team could understand. They replicated these experiments themselves and eventually, having eliminated every conceivable alternative explanation, Hahn and Strassmann came to a shocking conclusion: there were no transuranic elements at all in the by-products of neutron bombardment of uranium. The by-products were, instead, much lighter elements, barium and krypton. *The only way such by-products could have been produced was in the splitting of the uranium nucleus into two much smaller pieces.*

One of the main reasons Hahn had such trouble understanding what he had done was the lack of a theoretical framework to explain fission. It might be easy for us to imagine how fission occurs in a general way, but for a physicist or a chemist, the details are important, and Hahn's knowledge of the nucleus and how it behaves was insufficient to provide a sound explanation. What seems obvious in retrospect was hardly obvious at the time. Before concluding that the by-products were barium and krypton, in early November Hahn traveled to Copenhagen to consult with Bohr, whom he believed would certainly have an answer to the conundrum of the perplexing results. Meitner and her nephew, a bright young physicist named Otto Frisch, were also there and the four of them discussed the puzzle at length, but came to no conclusions. Now, desperate for some sort of justification for his conclusion that the uranium nucleus had been split, he sent a private letter to Meitner, who was back in Sweden. Frisch happened to be visiting his aunt when the letter arrived.

Together they spent an afternoon discussing Hahn's astonishing news. Meitner's insight—Frisch always credited his aunt—was that the nucleus of the uranium atom was held together in a manner analogous to the way surface tension holds together a drop of water. Under the right circumstances, forces can overcome the surface tension of a water drop (which is an electrostatic phenomenon) and cause the drop to push apart, or to "burst," as Meitner put it, under electrostatic repulsion. The impact of the neutron on the uranium nucleus produced a similar effect, overcoming the force holding the nucleus together and pushing the nucleus apart into two distinct pieces, in the process releasing enormous energy. Meitner calculated that the energy released each time a uranium nucleus split would be on the order of several hundred million electron volts (MeV), tiny in absolute terms, but almost incomprehensibly large relative to the energy released in chemical reactions.

She communicated her theory back to her former colleagues in Dahlem, who published their results on January 6, 1939. Frisch conducted a confirming experiment on January 13, 1939, detected the ionization pulse Meitner had predicted, and published his results four days later. Meitner and Frisch followed up a month after with an article explaining the theory behind fission, but by then everyone in the physics world knew of their work. For the second time in less than five years, a small scientific team, closeted away in a quiet corner of a highly oppressive totalitarian society, came up with pathbreaking, historic research.

Frisch informed Bohr of the breakthrough in early January 1939. Bohr's reaction: "Oh what idiots we have all been! Oh but this is wonderful!" Frisch extracted a promise from Bohr, who was about ready to sail to America to visit Einstein at Princeton and attend a physics conference in Washington, to postpone announcement of the news until

Frisch could do his experiment. So when Bohr arrived in New York harbor aboard the steamship *Drottningholm,* word of uranium fission had not yet reached American shores, even though Frisch had completed his experiment. If Fermi had been in Rome at the time, Hahn would almost certainly have informed Fermi of his conclusion out of professional courtesy, knowing that it completely repudiated the transuranic hypothesis. As it was, he only became certain after Fermi was well on his way to the United States.

LAURA SPENT TIME LOOKING FOR APARTMENTS IN THE COLUMBIA neighborhood, eventually identifying one on Riverside Drive just around the corner from the university campus. Enrico settled into new offices in Pupin Hall, the home of the Columbia physics department. Pegram, who had been instrumental in bringing Fermi to Columbia, found him space on the seventh floor but left it at that. Fermi was reduced to scrounging around for laboratory equipment.

Back at the King's Crown, Fermi ran into another of the hotel's residents, a brilliant Hungarian scientist named Leo Szilard. Although he did not have a formal faculty position at Columbia, Szilard hung around the physics department, developing relationships with some of the young stars that Pegram had drawn to the department, men like I. I. Rabi and Willis Lamb. Szilard was a true eccentric, a brilliant polymath who flitted from one idea to another with the rapidity of a brightly colored butterfly in a bed of flowers. He had no visible means of support and an extravagant lifestyle, leading those who watched the fascinating, entertaining, and sometimes irritating Hungarian to conclude that he was either independently wealthy or that he had the support of wealthy friends. The latter explanation was no doubt closer to the truth. Szilard understood chemistry and physics and later in his career contributed to biology, as well. He was also an inveterate inventor. He worked with Einstein to develop and patent an ingenious refrigerator. Like many of his inventions, it amounted to nothing.

One invention that was to become useful, however, was what Szilard called a "chain reaction." In 1933, reportedly while standing at a central London street corner waiting for the light to change, he thought of a physical process in which radioactive emission of neutrons might strike the nuclei of neighboring atoms, causing them to emit more neutrons, which would strike other nuclei and so on and so forth in a geometrically increasing cascade, releasing enormous energy in the process. He did not yet know how to create such a chain reaction, but he patented the idea in England, where he was living at the time. Concerned even at this early date with secrecy and the risk of Nazi Germany gaining access to such dangerous technology, he assigned the patent rights over to the British Admiralty, who, not knowing what to do with it, promptly shelved and forgot it.

Szilard was one of those unusual Hungarian refugees who arrived in the United States prior to the war and whom Szilard himself had christened "the Martians," because they seemingly came from another planet, with an intelligence that surpassed most Earthlings. They had all known each other as youths in Budapest: Szilard, Edward Teller, John von Neumann, Eugene Wigner, Theodore von Karman, Paul Halmos, George Polya, and Paul Erdos. Fermi loved Szilard's label. He once quipped: "Of course, they [extraterrestrials] are already here among us: they just call themselves Hungarians."

Fermi and Szilard were polar opposites—the former, thorough, calm, methodical, frugal in his personal habits, and focused entirely on physics with few interests outside the field; the latter, highly excitable, easily distracted, jumping from inspiration to inspiration, crossing disciplinary boundaries with ease, a product of high European culture, an epicure if ever there was one—and the two geniuses would never have had much to do with each other had fission never been discovered, but news of Hahn's discovery was about to reach the American shores, creating an historic and quite extraordinary, if momentary, case of strange bedfellows.

CHAPTER THIRTEEN

SPLITTING THE ATOM

Exactly two weeks after his arrival in New York, on January 16, 1939, Fermi and his wife traveled back to the Manhattan piers, this time to meet Niels Bohr, who was arriving on the Swedish-American Line's SS *Drottningholm*. Also there to meet the great man was John Wheeler, a twenty-seven-year-old Princeton physicist whose job it was to take Bohr and his son Erik and fellow passenger, Belgian physicist Leon Rosenfeld, to the Princeton campus.

Bohr was in town for two reasons. He wanted to visit Einstein at Princeton to discuss physics and the increasingly dire political situation in Europe. He also wanted to attend the fifth annual Washington Conference on Theoretical Physics, cosponsored by the Carnegie Institution, represented by Merle Tuve, and George Washington University, represented by Fermi's old friend Edward Teller and Russian émigré George Gamow. The conference was scheduled for January 26, 1939.

The Fermis persuaded Bohr and his son to spend the day with them before heading to Princeton. Bohr, keeping his promise to Frisch, did not mention the Hahn-Strassmann-Meitner results to Fermi. He may also have wanted to spare Fermi the inevitable embarrassment the news would cause. Fermi would find out soon enough. Why should Bohr be the one to tell him? As it was, Fermi and Bohr spoke of other things and the Fermis were glad to see their Danish friend, a friendly and familiar face from across the Atlantic.

In the meantime, Wheeler whisked Rosenfeld off to Princeton. On the train ride, probably unaware of Bohr's promise to Frisch, Rosenfeld broke the news to an astonished Wheeler. The next day, on January 17, 1939, Rosenfeld gave an impromptu talk on fission to the weekly physics department meeting, at Wheeler's urging. I. I. Rabi, visiting from Columbia, was in attendance but, for reasons that are not entirely clear, did not bring the news of fission back to Columbia. On Friday, another Columbia physicist, Willis Lamb, visited Princeton and brought the news to Fermi the next morning.

There is no direct record of how Fermi reacted to the news. Lamb reported with presumed understatement that Fermi received the information with great interest. Fermi must have been mortified. His embarrassment that he might have been the one to discover fission some five years earlier was compounded by the fact that the Nobel Prize citation mentioned the now discredited discovery of transuranic elements. Additionally, Fermi must have felt that, given the depth of his understanding of nuclear physics, he should have been the one to have had the Meitner insight. It would have been natural for the man to kick himself. As the years passed, he developed a sense of humor about his mistake. After the war, when reviewing the construction plans for a new nuclear science institute at the University of Chicago, he and his colleagues were speculating on the identity of a vaguely defined human figure shown in bas-relief above the entrance. Fermi quipped he was "probably a scientist not discovering fission."

On the other hand, none of the great scientists who followed his lead in March 1934, when he began the neutron experiments, came up with fission as a possibility, and these included truly brilliant experimentalists like Rutherford and Joliot-Curie. Fermi later told his wife that the Via Panisperna team was not good enough in chemistry, and he may have privately blamed D'Agostino, the team's chemist. Such blame would be unfair, considering that the Dahlem team was arguably the best in the world in radiochemistry and that it took them four years of intensive work to unlock the secret. Confronted with the puzzling results of the Dahlem team as early as November 1938, even Bohr was unable to process it correctly. No one had seen the solution until it was too obvious to ignore. Segrè lays part of the blame for the Rome team's failure on their experimental design. Shielding the uranium the way they did prevented them from seeing the ionizing pulse of energy, which would have naturally caused them to look for explanations leading to the discovery of fission.

In retrospect the true culprit was what psychologists call "cognitive dissonance." First identified by academic psychologist Leon Festinger, cognitive dissonance occurs when we are confronted with empirical data at odds with the way we "know" the world to work. To resolve this discrepancy, we choose to ignore data or try to fit the data into our preconceived belief structure. Sometimes, there is a crisis and the belief structure eventually crumbles. Hahn, Rutherford, Joliot-Curie, Bohr, Fermi, and other major scientists involved in the analysis of uranium bombardment believed that particles in the nucleus were like bricks and mortar, rigid and incapable of deforming into the shape required for fission to occur. It was only when confronted repeatedly with evidence to the contrary that they realized they had all been wrong. The discovery of fission is a classic case of cognitive dissonance in action.

Soon after his visit to Princeton, Bohr was back in New York. Wandering around Pupin Hall, he ran into a young Columbia physics graduate student by the name of Herbert Anderson. Squirreled away in his lab, Anderson had missed the commotion over fission, and Bohr eagerly filled him in. Anderson, who had been looking for a way to introduce himself to Fermi, now saw his chance. When Bohr left, Anderson immediately went to Fermi's seventh-floor office and introduced himself, mentioning the conversation with Bohr about fission. Characteristically, Fermi jumped at the opportunity to teach this young stranger something: "Let me tell you about fission!" He outlined the theory for Anderson, walked him through some ideas for experiments, and by the end of the conversation Anderson was offering Fermi the modest equipment he had in his lab. Anderson had figured out a way to connect an ionization chamber, capable of detecting an ionizing pulse, to an oscilloscope, invented in 1932 in Britain to display electrical signals on a small cathode-ray screen, similar to a television. Because uranium fission was accompanied by an ionizing pulse of some two hundred million electron volts, Anderson's equipment seemed perfect for the job.

Fermi had found an eager collaborator and, even more important, someone he could teach. Anderson had found a mentor, and would play an important role in the Americanization of Fermi. The partnership would last for the next fifteen years.

ANDERSON WAS A SLIM, ATHLETIC, ATTRACTIVE TWENTY-FIVE-YEAR-old graduate student who had been helping John Dunning, a physics professor and head of the team, build Columbia's first cyclotron, working out some of the kinks in the project. Anderson's previous degree in electrical engineering enabled him to solve some thorny engineering and design problems that arose as they built the new machine. He was also ambitious and understood that befriending the legendary physicist could prove beneficial to his own career. He was right.

Fermi was happy to throw himself back into work after the revelation of the Hahn-Strassmann-Meitner breakthrough, eager to repeat the neutron bombardment experiment with proper instrumentation to see the ionizing pulse himself. With Anderson, Dunning, and several other Columbia physicists, Fermi set up the experiment, using two sources for neutrons, the Columbia cyclotron and a Rome-type radon-beryllium glass bulb, and Anderson's ionizing chamber/oscilloscope for detection. Fermi had to leave for Washington before the results of the experiment came in, but Anderson assured him they would relay the results to him at the conference.

The Washington Conference was already quite important. At the previous one in 1938, Hans Bethe was so stimulated by the discussions during the conference that he used his train ride back from Washington to Cornell to work out the fusion cycle of the sun, effectively demonstrating why and how stars shine and create new elements in the process, work that won him a Nobel Prize some thirty years later. So it is not surprising that the attendees of the January 26, 1939, Washington Conference were a who's who of contemporary theoretical physics in the United States and abroad: aside from Bohr and Fermi, the fifty-one participants included current and future luminaries such as Bethe, Gregory Breit, George Gamow, Maria Goeppert, I. I. Rabi, Edward Teller, Merle Tuve, George Uhlenbeck, and nuclear chemist Harold Urey.

The conference was supposed to discuss low-temperature physics. From the beginning, however, Bohr and Fermi stole the show. Bohr took the podium to announce that the uranium atom had been split. He then turned it over to Fermi, who discussed the theory behind the

FIGURE 13.1. The Fifth Washington Conference on Theoretical Physics, January 1939. Fermi, who has just explained fission to the group, is smiling in the front row, second from the left. Bohr is fourth from the left. Hans Bethe is directly above Bohr. Maria Mayer and Edward Teller are sitting together, two seats to Bethe's right. *Courtesy of the Department of Physics, George Washington University.*

phenomenon, using Meitner's work and the liquid drop analogy. This left the conference in an uproar. The Columbia experimental results came through during the conference and were duly reported by Fermi. The team in New York measured the ionization at 90 MeV, smaller than theoretically expected but within the error range of the equipment they were using. They had seen fission. Shortly after the conference, teams at Copenhagen, Johns Hopkins, the Carnegie Institution, Berkeley, and elsewhere climbed on to the experimental bandwagon, each one repeating the experiment and detecting the strong ionization pulse.

If his colleagues had lost any respect at all for Fermi due to his 1934 oversight—and they had little reason to do so, considering that all of them were guilty of the same error—they quickly regained it through his frankness about it and his enthusiastic acceptance of the Hahn-Strassmann-Meitner results. At the Washington Conference he took center stage, a place where he naturally belonged. His colleagues agreed.

In the group photo at the end of the conference, Fermi sat in the front row, beaming.

While Fermi was in Washington, Leo Szilard was battling a cold.

On January 16, 1939, one week before the Washington Conference, on the same day that Bohr arrived in New York, Szilard took the train from New York to Princeton to visit his old friend Eugene Wigner, who was in the hospital suffering from a serious case of jaundice. The courtly and quiet Wigner, whom Szilard had known since their childhood in Budapest, was already one of the most respected quantum theorists in the world. After looking in on his friend, Szilard settled himself in Wigner's apartment.

Wigner learned of Rosenfeld's presentation at Princeton while still in the hospital and mentioned it to Szilard, who immediately grasped its significance. If, during the fission of uranium, neutrons were emitted, uranium could be the basis of the chain reaction that Szilard first envisioned in 1933.

While in Princeton, Szilard left Wigner's apartment poorly protected during a heavy rainstorm and came down with a nasty cold. It is unclear whether Szilard had been invited to the Washington Conference—it was supposedly on low-temperature physics, something he knew a bit about. Szilard remained cooped up in Wigner's apartment in Princeton for about ten days owing to his cold and spent the time alone obsessing on the chain reaction idea while the Washington Conference was underway without him.

By the time Szilard managed to get himself back to the King's Crown hotel in Manhattan, the only thing he wanted to do was find Fermi, who would understand how fission and the chain reaction were related and who would appreciate the importance of the moment. Distressed by the news that Bohr and Fermi had already briefed some fifty physicists in Washington about fission, he also wanted to plead the case for secrecy over all future work on uranium. "Since it was a private meeting," Szilard later recalled, "the cat was not entirely out of the bag, but its tail was sticking out." Hahn had done his work in Berlin, and so Hitler now had access to the most dangerous information on the planet. No sane person would willingly make it easier for the Nazis to create a fission bomb. If

Hitler succeeded in making a fission weapon, the war in Europe, now seemingly unavoidable, would be over before it even began.

Szilard visited Pupin Hall looking for Fermi but could not find him. He ran into Rabi and asked him to pass along his plea for secrecy to Fermi. Rabi agreed. The next day, still on the hunt for Fermi, Szilard looked in on Rabi and asked if he had spoken to Fermi. Rabi told him he had. Szilard asked Rabi for Fermi's reaction, and Rabi reported that Fermi, proudly showing off his newly acquired American slang, had said "Nuts!" Szilard asked Rabi to elaborate, but Rabi declined, suggesting instead that they should walk over to Fermi's office and hear it from the man himself.

When the two of them confronted Fermi, he seemed unsympathetic to Szilard's insistence on secrecy. He explained that he understood the concept of the chain reaction quite well, but in his view secrecy was not necessary because the probability of a successful chain reaction occurring in uranium was too remote to be of any practical concern.

Rabi pressed him to define "remote possibility" to which Fermi, consistent with his view of the world, replied, "Well, ten percent."

Rabi and Szilard were astonished. Rabi shot back, "Ten percent is not a remote possibility if it means we may die of it. If I have pneumonia and a doctor tells me that there is a remote possibility that I might die of it, and it's ten percent, I get excited about it."

When Fermi said he thought 10 percent was remote, he meant it. Some years later he was in conversation with a number of physicists when the subject of faster-than-light travel came up. Someone asked what the odds were that physicists would discover that the speed of light could be exceeded. Teller estimated one in ten million. Fermi estimated it at 10 percent. In his experience, 10 percent probabilities never happened, reflecting a deep-seated and somewhat unique view of how the world worked and what constituted a rare event.

Perhaps Fermi also dismissed the possibility of a chain reaction because he understood what it might lead to and was reluctant to start on the road toward a nuclear weapon. He may also have judged the project too difficult for Germany to pursue successfully. In any case, Fermi's position was totally at odds with the usually sober Rabi, not to mention the usually volatile Szilard. Szilard later recalled this moment:

> From the very beginning the line was drawn; the difference between Fermi's position throughout this [period] and mine was marked on the first day we talked about it. We both wanted to be conservative but Fermi thought that the conservative thing was to play down the possibility that this may happen and I thought the conservative thing was to assume that it would happen and take all the necessary precautions. . . . I personally felt that these things should be discussed privately among the physicists of England, France, and America, and that there should be no publication on this topic if it should turn out that neutrons are, in fact, emitted and that a chain reaction might be possible.

While Fermi considered Szilard's plea, the Hungarian also wrote to Joliot-Curie in Paris with the same request, but the French physicist was less accommodating than Fermi and flatly refused to be bound by secrecy.

FERMI AND SZILARD NOW SET OUT INDEPENDENTLY TO CONDUCT experiments to determine whether neutrons were emitted during uranium fission, and if so, how many. The results of these experiments would determine whether a chain reaction was possible. If on average more than one neutron was emitted each time a uranium nucleus split, then the type of cascade that Szilard had dreamed of for five years would be possible. If the average was either one neutron or even less, then the cascade would not occur and an explosion would not result.

Fermi brought Anderson under his wing. They placed a neutron source in a glass bulb and immersed it in a cylindrical tank of water, three feet high and three feet in diameter, measuring the induced radioactivity in a rhodium foil placed in the tank at various distances from the bulb. Then they put uranium oxide in the bulb along with the neutron source and compared the results with those of the bulb without the uranium oxide. They found a 6 percent increase in the radioactivity of the rhodium when the uranium was present—suggesting that the fission of uranium had resulted in neutron emission. By his calculations, Fermi believed that if the neutron emissions were from fission alone, this would suggest that two neutrons were produced for each fission reaction. A chain reaction was thus theoretically possible.

In parallel to Fermi, Szilard decided to do a similar experiment with a slightly different neutron source, producing only slow neutrons, thus eliminating the possibility that fast neutrons had merely knocked neutrons off the uranium nuclei without splitting them. He negotiated to "rent" a gram of radium from the Radium Chemical Company with a loan of $2,000 from a businessman friend and a letter of reference from Wigner himself, who assured the company that the experiments would be conducted under the auspices of a university. Not deterred by his lack of a faculty position at Columbia—or anywhere else, for that matter—the irrepressible Szilard then persuaded Pegram to authorize him to conduct experiments over a three-month period and recruited an eager young Columbia physics faculty member, Walter Zinn, to work with him. Zinn, a thirty-three-year-old Canadian who joined the Columbia faculty as an instructor in the early 1930s was happy to help, although this was the last time he was to work directly with Szilard, probably because of a difference in their experimental styles. He would soon become one of the core members of Fermi's first nuclear reactor team.

Szilard's experiment, a bit more complex than Fermi's, used a target with a series of carefully nested boxes containing uranium and various moderating media such as paraffin and demonstrated neutron emission from fission. With this result in hand, Szilard offered up his source to Fermi and Anderson. When they substituted it into their experiment, there was a 30 percent increase in the detected radioactivity. In the paper he published with Anderson, Fermi was careful to note that direct comparisons could not be made between the two sources because the geometric configurations were so different, but once again he concluded it was probable that at least two neutrons were emitted for each fission reaction.

The results of these two parallel, relatively primitive experiments, completed in early March 1939, were not enough to determine the practicality of a chain reaction, but they were sufficient to persuade Szilard that the US government needed to be alerted. A recent Nobel laureate, an exceptionally clear lecturer, and the acknowledged expert on all things nuclear, Fermi was the natural person to present the discovery to the government, but sensing Fermi's reluctance to enter the breach, Szilard arranged a meeting in Pegram's office of Pegram, Fermi, and himself. He asked Wigner from Princeton to join them as well.

The discussion began with the enormous implications of the experimental results. Fermi was probably still of the view that an actual weapon would be impractical—no calculations, for example, had been done on how much uranium would be needed for a weapon—but up against a three-on-one assault, Fermi could not resist. Neither Wigner nor Pegram were excitable men, so their palpable alarm about the possibility of a German nuclear breakthrough must have impressed Fermi. As a foreign national, Fermi was naturally anxious about briefing the US military on such a sensitive subject only three months after arriving in New York. Szilard could be ignored, but Pegram and Wigner simply could not.

The conversation ended with a discussion of secrecy. Wigner joined Szilard in pressing for secrecy about work on the chain reaction. Fermi respected Wigner as much as he respected anyone, but he remained opposed to any restrictions on publication, deferring to Pegram to make the final decision. Pegram opted for sending papers on the two experiments to the *Physical Review*, but with the extraordinary request that their publication be delayed while the community considered the implications of the secrecy debate. In particular, Joliot-Curie in Paris was resisting any attempts to keep chain reaction research under wraps.

The next day Enrico Fermi met the US Navy.

CHAPTER FOURTEEN

FERMI MEETS THE NAVY

YEARS LATER, IN WHAT SHE DESCRIBED AS A "FIT OF HOUSE-cleaning enthusiasm," combing through a filing cabinet that stored family papers, Laura Fermi found a copy of the letter Pegram sent to Admiral Stanford Hooper at the Office of Chief of Naval Operations. Dated March 16, 1939, it begins:

> Dear Sir:
>
> This morning I had a telephone conversation with Mr. Compton in the office of the Assistant Secretary of the Navy [Charles Edison, Pegram's friend and son of the inventor], who has doubtless reported the conversation to you. It had to do with the possibility that experiments in the physics laboratories of Columbia University reveal that conditions may be found under which the element uranium may be able to liberate its large excess of atomic energy and that this might mean the possibility that uranium might be used as an explosive that would liberate a million times as much energy per pound as any known explosive. My own feeling is that the probabilities are against this, but my colleagues and I think that the bare possibility should not be disregarded and I therefore telephoned to Mr. Edison's office this morning chiefly to arrange a channel through which the results of our experiments

might, if the occasion should arise, be transmitted to the proper authorities in the United States Navy.

Professor Enrico Fermi who, together with Szilard, Dr. Zinn, Mr. Anderson and others, has been working on this problem in our laboratories, went to Washington this afternoon to lecture before the Philosophical Society in Washington this evening and will be in Washington tomorrow. He will telephone your office and if you wish to see him will be glad to tell you more definitely what the state of knowledge on this subject is at present.

Professor Fermi, formerly of Rome, is Professor of Physics at Columbia University. In December last he was awarded the Nobel Prize in Physics in 1938 for the work he did on the artificial creation of radioactive elements by means of neutrons. There is no man more competent in this field of nuclear physics than Professor Fermi.

Professor Fermi has recently arrived to stay permanently in this country and will become an American citizen in due course. He is very much at home in this country, having visited here often to lecture at the University of Michigan, Stanford University and at Columbia.

Professor Fermi will be staying tomorrow with Professor Edward Teller of George Washington University.

Sincerely yours,
George E. Pegram
Professor of Physics

Across Fermi's copy of it, Pegram scrawled a note. "Dear Fermi—This may prepare the way for you a little better than Mr. Compton's explanation to Adm. Hooper."

Laura was mystified by the letter, having never seen it before and blissfully unaware that her husband met with the Navy on the subject of a potential atomic bomb in March 1939. In what was to become a habit, he had not spoken to her of his initial research relating to the potential for a nuclear explosive. When she confronted him about it, he explained that he had kept it as a kind of insurance policy. In December 1941, when the Axis declared war on the United States, he felt he might need some proof of loyalty to his new country and set it carefully aside in a manila folder. If any authorities challenged him, he would pull it out.

The interchange between Laura and her husband reveals much. Even as early as March 1939, Fermi was maintaining a certain veil of secrecy around his uranium work—this in spite of his disagreements with Szilard over the whole issue of secrecy. The secrecy issue was undecided, but with characteristic caution he decided to say nothing to Laura. The other point, perhaps more telling, is that at the moment the United States entered the war, the Fermis became, albeit for a short time, enemy aliens. Fermi himself was not disposed to take his importance in the war effort for granted and felt the need for documentary proof of his loyalty to the United States. He needn't have worried. By the time the United States entered the war, he was one of the key players in the effort to build a nuclear weapon and few, if any, doubted his loyalty. At the time, however, he was keenly aware of his status as a foreign national of a potential enemy power and felt the need to have an insurance policy tucked away in his filing cabinet.

Reading the letter from Pegram, Admiral Hooper may have wondered what all the fuss was about. Pegram had to tread carefully. The idea of creating an explosive out of uranium would have struck Navy technical staff at first blush as preposterous, and Pegram believed that overselling the concept would have resulted in doors being shut in Fermi's face. The letter was designed in part to give the admiral, who may never have heard of Fermi, a sense of the man's stature within the scientific community. Pegram also went out of his way to assure the admiral that he and his Navy colleagues would be able to understand this foreign national, that Fermi was "very much at home" in America and would make a good presentation to the presumably unworldly Navy staff. In retrospect, tentative though it was, Pegram drafted the perfect letter to initiate contact between the scientific community and the US government on the potential for an atomic bomb.

March 17, 1939, was a cool day in Washington, and the high of forty-nine degrees suggests that the cherry trees around the tidal basin, a 1912 gift from the Japanese people, had not yet blossomed. The basin would have been visible from Navy headquarters, housed in an enormous and exceedingly ugly edifice on what is now beautiful parkland just north of the long, narrow reflecting pool on the west side of Constitution Mall.

Fermi arrived to a tepid welcome in the nondescript board room and overheard an unenthusiastic staff member announce to the assembled group, "There's a wop outside."

In the conference room, Hooper had assembled a range of technical experts from various offices in the Navy, officers responsible for ordnance, engineering, construction, and repair, as well as a team from the Navy Research Labs headed by Dr. Ross Gunn. Although it was a Navy meeting, a technical team from the Army had been included as a courtesy. Fermi spoke for about an hour providing an overview of the physics of nuclear fission, the potential for the development of nuclear energy, and the prospect of a weapon based on the fission of uranium. According to notes taken by Captain Garrett L. Schuyler, later chief of the Research Division of Ordnance, Fermi described the principles of slow-neutron fission and the importance of neutron emission from fission to create a chain reaction. In summarizing the experiments he and Szilard had just completed, Fermi concluded that "the excess in the number of released neutrons is not very great and has not yet been demonstrated *absolutely* beyond the possible limits of experimental error." He added that new experiments were planned in the next months to make more definitive measurements and if "these experiments show more neutrons are released from the uranium atoms than are necessary to split them up, continuous release of energy in a mass of uranium is theoretically possible." He gave a clear description of critical mass: "In the small samples used so far . . . the released neutrons are possibly not all effective because some will too rapidly escape; but in a sufficiently large mass of uranium, they necessarily will be all trapped and available in time."

At this point Captain Schuyler piped up with a question: "What might be the size of this critical mass?" Fermi smiled and gave an answer consistent with his strategy of downplaying the possibility of a nuclear weapon: "Well," he replied, "it just might turn out to be the size of a small star." Although Fermi might have been deliberately modulating the practicality of a fission bomb, it is also true that no one—neither Fermi the pessimist nor Szilard the optimist—actually knew what the critical mass might be. Not nearly enough was known to even begin such a calculation.

Fermi turned to the issue of uranium isotopes. He explained to the assembled military scientists that natural uranium consisted of a mixture of the two isotopes, about 99.3 percent of which was uranium 238 (U-238) and 0.7 percent, uranium 235 (U-235). On the basis of theoretical work done the previous month at Princeton by Bohr, Wheeler, and Czech émigré George Placzek, scientists now believed that the far rarer U-235 isotope was responsible for the fission reaction. The only clean way to build a fission weapon would be to separate the two isotopes, but at this point no one knew how to do this.

In summary, Fermi made it clear that it *may* be possible to unlock atomic energy through fission and that this possibility, with all its attendant uncertainties, should be brought to the attention of the military.

Had he witnessed his colleague's performance, Szilard would have been disappointed. Fermi was disinclined to stir the Navy into a frenzy of action, and he didn't. A natural reticence to make bold statements about science before he had clarified the facts for himself, combined with a feeling that the less said about the terrible possibilities the better, drove him to impart enough information for the military to decide on next steps without recommending any specific course of action. Years later, Szilard dismissively suggested that nothing came of the meeting.

This was not entirely true. Fermi's lecture fired up the Navy's head of research Ross Gunn, who immediately saw that uranium could provide a source of energy for submarines. He launched a long, frustrating, but ultimately successful effort to develop nuclear-powered naval vessels. Before acting, however, he had to find out more about this odd, understated little man with the strong Italian accent. He called Merle Tuve, the Carnegie Institution physicist who was one of the cosponsors of the January conference at which Bohr and Fermi had presented. "Who is this man Fermi? What kind of man is he? Is he a Fascist or what? What is he?" Tuve assured Gunn of Fermi's impeccable credentials. That was enough for Gunn. Unfortunately for the Navy scientist, he could not get his own project off the ground until after the war, when national priorities had shifted away from the Manhattan Project.

Though the briefing did not exactly spur the Navy into action, it did result in a check to Columbia for $1,500 for continued research into fission. Who knew? Might there be something to this bizarre, science fiction idea of a new explosive based on nuclear fission? It seemed worthwhile, for a small expenditure, to keep tabs on the research being done at Columbia.

Fermi was sailing in uncharted waters when he arrived on Constitution Mall in March 1939. He came bearing outlandish ideas, ideas that an institutionally conservative military was unprepared to accept. The military was not accustomed to funding private scientific research. In retrospect, it is impressive that he received even a small grant for his experiments.

BACK IN NEW YORK, FERMI BEGAN TO SETTLE INTO LIFE IN HIS NEW country.

The family home at 450 Riverside Drive was one of a row of apartment buildings overlooking the Hudson River, built in the early 1900s to house Columbia faculty. It was reasonably comfortable, but in the winter the wind howled up from the river and up the hill on West 116th Street. Walking the children up the hill or back down against the wind was not one of Laura's favorite activities. In spite of this, she gradually adjusted to life in a new city, making the best of her situation.

Fermi plunged into classroom work. He taught three courses that spring term, including a course on geophysics, one of his favorite subjects, with a group of rather fortunate undergrads, as well as higher level courses on quantum mechanics and applied quantum mechanics.

He was also getting to know other members of the Columbia faculty, people who would become close colleagues and friends over the next decade. Rabi was one of them. An irascible, punchy personality with a wicked sense of humor, Rabi had met Fermi during his years as a postdoc in Europe. In the early 1930s, Rabi began experimental work that eventually resulted in his discovery of the nuclear magnetic resonance effect that is the basis of today's MRI scanners. Fermi and Rabi hit it off right away. Years later, Rabi would tell his biographer that aside from Einstein, he considered Fermi the greatest physicist he had ever known.

Another fast friend was Harold Urey. Slightly older than Fermi, Urey won the Nobel Prize in Chemistry in 1934 for his isolation of deuterium, the isotope of hydrogen with a proton and a neutron in its nucleus. Urey befriended Enrico and Laura and spent a good deal of time selling them on the joys of living the American dream in the suburban town of Leonia, New Jersey, where Urey himself lived. Within a year he succeeded, and the Fermis became suburban Americans with front and back lawns and a makeshift workshop in the basement. Fermi never quite got the hang of suburbia—he and Laura were city folk at heart—and their front lawn was often the least well manicured on the block. It was, however, the beginning of a lifelong friendship with the Ureys.

As the family adjusted to life in the States, making new friends and settling into what they hoped would be a quiet domestic life, nuclear fission and Szilard's obsession with chain reactions continued to preoccupy Fermi.

FERMI HAD TOLD HIS NAVY AUDIENCE THAT HE WAS PLANNING another experiment to clarify some of the uncertainties surrounding fission and the possibility of a chain reaction. For this experiment, Fermi and Szilard collaborated as principal partners. Given their radically different work styles, it is not surprising that this was also their last direct collaboration.

Creating a fission chain reaction using natural uranium taken from the ground poses some real problems. Natural uranium is composed of two isotopes. U-238, which is extremely difficult to split, accounts for 99.3 percent of natural uranium. U-235, which readily splits and is ideal for creating a chain reaction, accounts for only 0.7 percent of natural uranium. Thus, a chain reaction, if feasible, would require many tons of natural uranium, if not separated, to have enough of the fissionable isotope U-235 needed for the reaction. Although no one yet knew how to separate the two isotopes (U-235 and U-238) from each other, Fermi's Columbia colleague John Dunning doubted the feasibility of a chain reaction based on natural uranium and so pressed for first solving the separation problem, purifying natural uranium so that the relatively small amount of U-235 needed for a chain reaction could be isolated. Doubting that the techniques for isotope separation could be developed quickly, Fermi and Szilard preferred using natural uranium, instinctively sensing that a chain reaction could be developed with a large enough supply of it. After much debate, Fermi's proposal won out. Fermi recognized that isotope separation would eventually need to be solved but believed that an initial demonstration using natural uranium would provide proof of the chain reaction concept.

Beginning in April 1939, Fermi and Szilard, along with Anderson, modified the water tank experiment, filling the tank with a 10 percent solution of manganese sulfate, a substance that becomes radioactive in proportion to the neutrons that hit it. Into this solution they placed a matrix of fifty-two tin cans, two inches in diameter and two feet high. In the middle of the tank they placed Szilard's neutron source.

Using four hundred pounds of uranium oxide the ever-resourceful Szilard had "borrowed" in one of his famous sleights of hand from the El Dorado Radium Corporation, which owned significant deposits of uranium ore in Canada, they measured a 10 percent increase in the radioactivity of the manganese with the uranium oxide present inside the cans, confirming the results of the two previous experiments.

Then they set about trying to calculate the ratio of neutrons emitted per neutron absorbed during fission, what they called the "reproduction factor." A cascade of fission reactions, as Szilard originally envisioned, required an average reproduction rate greater than one—even the slightest amount greater than one would eventually work. The experiment resulted finally in a measurement of about 1.5 fast neutrons emitted for every neutron absorbed during fission. They reported that "a nuclear chain reaction could be maintained in a system in which neutrons are slowed down without much absorption until they reach thermal energies and are then mostly absorbed by uranium rather than by another element." They also suspected that the water they used in the experiment was absorbing too many neutrons: "It remains an open question, however, whether this holds for a system in which hydrogen is used for slowing down the neutrons." This was the first conceptual outline of what was to become the "pile," the world's first nuclear reactor in 1942.

Anderson later noted several important points about the experiment.

First, the measurements they took indicated that plain water would probably not be a good moderator for a fully functioning reactor, because hydrogen atoms in the water had a tendency to absorb slow neutrons into their nuclei, taking them out of the chain reaction.

Second, the team became aware of the importance of a phenomenon known as "resonance absorption." U-238, which makes up the vast majority of natural uranium, tends to absorb slow-ish neutrons without undergoing fission, thus taking them out of the chain reaction. Fermi estimated that this phenomenon was responsible for a 20 percent decrease in the average number of emitted neutrons, an estimate based as much on Fermi's intuition as on the data. To solve this problem, Fermi decided that lumping uranium into smaller chunks would reduce the tendency of the fast neutrons emitted in fission to slow down and become absorbed by the U-238.

Anderson went out of his way to explain why this was the last time that Fermi and Szilard collaborated directly on an experiment:

> This was the first, and also the last, experiment in which Szilard and Fermi collaborated together. Szilard's way of working on an experiment did not appeal to Fermi. Szilard was not willing to do his share of the experimental work, neither in the preparation nor in the conduct of the measurements. He hired an assistant to do what we would have required of him. The assistant, S. E. Krewer, was quite competent, so we could not complain on this score, but the scheme did not conform with Fermi's idea of how a joint experiment should be carried out, with all the work distributed more or less equally and each willing and able to do whatever fell to his lot. Fermi's vigor and energy made it possible for him to contribute somewhat more than his share, so that any dragging of feet on the part of the others stood out more sharply in contrast.

Fermi was never so explicit. During his January 1954 lecture to the American Physical Society about this period at Columbia, all he would say about his brilliant but frustrating collaborator was that he was "a very peculiar man, extremely intelligent," a description that brought hearty laughter from the audience. Whatever reservations Fermi had about Szilard's willingness to get his hands dirty in the lab—and uranium oxide is very dirty—he retained enormous respect for Szilard's ability to think creatively about difficult scientific problems.

Fermi, Szilard, and Anderson submitted their experimental results for publication in *Physical Review* in early July 1939. Soon Fermi and the family were off to Ann Arbor, Michigan, for another summer school session, where he lectured on the absorption of cosmic rays in the atmosphere and in solids. He also met up with an old acquaintance from his days in Germany, and they had a conversation that lingered in Fermi's mind for the next six years.

IN LATE JULY 1939, WERNER HEISENBERG ARRIVED IN ANN ARBOR to spend a week participating in the Goudsmit-Uhlenbeck summer session, seeing old friends and discussing the state of the world.

The state of the world was grim. To anyone reading the daily reports coming out of Berlin, Moscow, Paris, and London, it was clear that European powers were preparing for an outbreak of hostilities in the very near future. The betting was that Germany would invade its ideological enemy, the Soviet Union, within the month.

At that moment, Heisenberg arrived on the scene in Ann Arbor.

Max Dresden, a young student tending bar at a party hosted by refugee physicist Otto Laporte, witnessed an encounter between Fermi and Heisenberg. Dresden, who would go on to a distinguished career at Stanford, describes the evening:

> There was actually not much to do, so we could pay close attention to the conversations. There was really only one central topic. Fermi had just left fascist Italy to come to the US; Heisenberg had decided to return to Nazi Germany. The crucial part of their argument was whether a decent, honest scientist could function and maintain his scientific integrity and personal self respect in a country where all standards of decency and humanity had been suspended. Heisenberg believed that with his prestige, reputation and known loyalty to Germany, he could influence and perhaps even guide the government in more rational channels. Fermi believed no such thing. He kept on saying: "These people [the Fascists] have no principles; they will kill anybody who might be a threat—and they won't think twice about it. You have only the influence they grant you." Heisenberg didn't believe the situation was that bad. I believe it was Laporte who asked what Heisenberg would do in case of a Nazi-Soviet pact. Heisenberg was totally unwilling to entertain that possibility: "No patriotic German would ever consider that option." The discussion continued for a long time without resolution. Heisenberg felt Germany needed him, that it was his obligation to go back. Fermi did not think there was anything anyone could do in Italy (or Europe); he was afraid for the life of his wife (her father was later killed); and so he felt it was better to make a fresh start in the US. But none of the decisions had come easy. The role of physics

and physicists was mentioned off and on. After the party was over everybody left in a state of apprehension and depression.

Some three decades later Heisenberg recalled another one-on-one conversation with Fermi, at Fermi's Ann Arbor apartment. Fermi started out on a positive note, describing how his move to the United States was liberating, how the United States had been home to European refugees for generations, and how stimulating it was to start all over again in his new homeland. "Here, in a larger and freer country, [Europeans] could live without being weighed down by the heavy ballast of their historical past. In Italy I was a great man; here I am once again a young physicist, and that is incomparably more exciting. Why don't you cast off all that ballast, too, and start anew?" Heisenberg replied that he understood the attraction, but that abandoning Germany now, he would feel himself a traitor, particularly to younger physicists who did not have the ability to emigrate and to find work wherever they wanted. For Fermi, however, any responsibility he felt for the students he left behind was outweighed by the many compelling reasons he had for leaving Italy. In later years, Fermi's notably generous treatment of his American students may have been an effort to compensate for lingering feelings of guilt he had over abandoning his Italian students.

Fermi pressed on, explicitly referring to the possibility of using the discovery of atomic fission to create a bomb. He warned that Heisenberg would be expected to work on such a project. Heisenberg expressed doubt that such a weapon could be built, at least not quickly. Fermi then asked, "Don't you think it possible that Hitler may win the war?" Heisenberg expressed doubt, given the balance of technological resources available on each side. Fermi was incredulous that under the circumstances Heisenberg still wanted to return to Germany. Heisenberg explained that patriotism was a stronger factor for him than doubt about the war's outcome.

Fermi ended the conversation, noting, "That's a great pity. Let's just hope we will meet again after the war."

Fermi may have had reservations about the possibility and wisdom of pursuing an atomic weapon. He had soft-pedaled the idea in his meeting with the Navy in March 1939 and continued to have doubts about the technological feasibility of the weapon. His encounter with Heisenberg dramatically altered his perspective. He had known Heisenberg since 1922 and, though Fermi might not have particularly liked the man, he had followed his career and his contributions with great interest. He had even nominated Heisenberg for a Nobel Prize. Heisenberg would now return to his home country and when the inevitable war broke out would be tasked by Hitler with developing a weapon based on nuclear fission. Fermi had enough respect for Heisenberg to know the serious threat he posed. Whatever reservations Fermi had about the project, he would now have to pursue it with vigor. He had no real choice in the matter.

THE RESULTS OF THE SPRING EXPERIMENT WITH SZILARD AND Anderson were never far from Fermi's mind. Over the summer he corresponded with Szilard on a central problem: if water was not a suitable moderator for the chain reaction, was there another substance that would be suitable?

Their thoughts turned to a form of carbon called graphite.

CHAPTER FIFTEEN

PILES OF GRAPHITE

IN VIA PANISPERNA, CORBINO'S BOYS DISCOVERED THAT LIGHT nuclei could slow down neutrons. The paraffin block experiment was a perfect demonstration of this phenomenon. At the time the Rome team did not know they were splitting atoms and thus had no interest in the neutrons that might be emitted from fission reactions. Now, as he worked to create a controlled chain reaction, the behavior of neutrons themselves mattered greatly to Fermi.

If hydrogen was not suitable for sustaining the chain reaction, what light nucleus would work? The periodic table of elements is organized from lightest to heaviest—hydrogen, with an atomic number of 1, is first and, in 1939, uranium, the heaviest naturally occurring element, ended the table with an atomic number of 92. It was quite natural to examine elements sequentially along the periodic table from hydrogen onward for the next best neutron moderator, especially for someone as methodical as Enrico Fermi.

Hydrogen has two heavy isotopes, deuterium and tritium, both of which would be less likely to capture neutrons, but these isotopes are rare in nature and difficult to manufacture. Moving up the periodic table, helium is naturally found as a gas, and liquid helium is so cold that it needs special handling. Lithium, beryllium, and boron are the next in line, but the first two are relatively dangerous to work with and boron was not readily available—the major source of all boron is Turkey. Fermi later discovered that boron absorbs neutrons and would not make a suitable moderator. Next after boron is carbon.

Carbon is safe, plentiful, and comes in a variety of solid forms. Coal, of course, is one such source, but it is too soft to be machined with precision. Diamond is another form, but it is far rarer, and as the hardest substance on the planet it is virtually impossible to machine. Graphite, a crystalline form of carbon, is not quite as plentiful as coal but still easily obtained in nature in large quantities and quite easy to machine with precision. Every common pencil contains a piece of graphite that has been machined down to a thin rod. Making graphite bricks is relatively easy.

Fermi and Szilard came to the idea of substituting graphite for water and spent much of the summer corresponding about the possibility of using graphite as a moderator. Carbon atoms are about twelve times heavier than hydrogen atoms, but both scientists believed that carbon might absorb a sufficient amount of a neutron's kinetic energy to slow it down for the purpose of a fission chain reaction. Perhaps it would not grab neutrons out of the chain reaction, the way hydrogen did. The scheme that Fermi had in mind would require a lot of graphite, and Szilard was a man who knew how to get it.

While he was corresponding with Fermi, Szilard was also scheming with his old friends Edward Teller and Eugene Wigner to kick-start the US government's interest in uranium chain reaction research. The story of how Szilard and his fellow Hungarians persuaded the most famous scientist in the world, Albert Einstein, to sign a letter to President Roosevelt on August 2, 1939, urging the president to initiate large-scale research into the possibility of a nuclear weapon is well known. The image of a carload of Hungarian geniuses, chauffeured by a New York investment banker friend of Szilard, hunting for Einstein's house in the wilds of Long Island's North Fork is one of the more vivid of this entire period. The letter galvanized American research in fission weapons. Fermi was neither directly involved in writing the letter nor in getting it signed by the great man, but he was mentioned in the famous first paragraph, drafted by Szilard himself:

> Some recent work by E. Fermi and L. Szilard, which has been communicated to me in manuscript, leads me to expect that the element uranium may be turned into a new and important source of energy in the immediate future. Certain aspects of the situation which has arisen seem to call for watchfulness and if necessary, quick action on the part of the Administration. I believe therefore that it is my duty to bring to your attention the following facts and recommendations.

Events that summer in Europe only served to heighten the sense of urgency. In diplomatic, military, and intelligence circles, rumors swirled that Germany was poised to invade Poland in a lightning attack. The negotiations between Germany and the Soviet Union that led to the Molotov-Ribbentrop Pact remained secret for much of August—they were underway when Heisenberg dismissed out of hand Laporte's speculative question at the Ann Arbor party about a Nazi-Soviet alliance—but the two governments sprang their surprise alliance on an incredulous world on August 23, 1939. Hostilities began a week later. The United States was not yet involved, and many influential politicians and public figures remained opposed to US involvement, but Roosevelt was already engaged in quiet efforts to bring American public opinion and industrial might in line behind eventual engagement in a European war on the side of the Allies.

Roosevelt finally received the letter in October 1939 and authorized work on fission as an immediate priority. The national effort launched by Roosevelt eventually evolved into the largest, most complex military-scientific program ever conceived. At that moment, however, the Manhattan Project was limited to Fermi's work at Columbia's Pupin Labs.

Beginning in the fall of 1939, Fermi wrote forty-seven papers describing the experimental work leading to the creation of the world's first controlled self-sustaining chain reaction on December 2, 1942. The work was methodical, demanding, and sometimes dangerous and involved a growing group of physicists. Fermi continued relying on Szilard's skill in obtaining increasingly pure batches of graphite. Szilard also served as a sounding board and a sometimes irritating cheerleader for the project. Pegram, as head of the physics department at Columbia and a dean of the college, threw his considerable weight and sound judgment behind the project. Anderson and Zinn lent their experimental knowhow and attention to detail, and Fermi added a bright young Columbia graduate student, George Weil, to his team. Other students were to follow, including Albert Wattenberg and Bernard Feld.

Given the importance of the chain reaction experiments, one might assume that they completely preoccupied Fermi, but he continued to pursue other research interests along the way. For example, he lectured on the geophysics of iron in the core of the earth at the 1940 Washington Conference, the one subsequent to the conference at which Bohr and Fermi sprang fission on an unsuspecting world. In the spring of 1940, Fermi went to Berkeley to give the annual, highly prestigious Hitchcock Lecture on "High Energies and Small Distances in Modern Physics." His archival notebook demonstrates the effort he made to prepare for these lectures. His Ann Arbor work on cosmic rays passing through gases and solids continued into the fall at Columbia, resulting in laborious and frustrating calculations on the relationship between the density of a medium and the speed with which an ionized particle slows down. Apropos of this, Fermi quipped to Anderson that "he could calculate almost anything to an accuracy of ten percent in less than a day, but to improve the accuracy by a factor of three might take him six months." Colleagues sometimes noted his tendency to get frustrated if he could not immediately solve a problem. This was a good example.

He also continued with a full teaching load, including courses on geophysics and quantum theory.

Like all Americans, he followed the war in Europe. On the first of every month, he and a group of faculty colleagues—a "Society of Prophets" Laura called them—would meet at the faculty club to predict developments over the coming month, writing down answers to ten yes-or-no questions. Laura reports that by the time the society dissolved, when the Fermis departed for Chicago in the summer of 1942, Enrico had established himself as *the* Prophet, having predicted successfully 97 percent of the time. He did this, she writes, using the most conservative algorithm imaginable: the next month would look almost exactly like the previous month. He did, however, miss one prediction—the surprise German invasion of the Soviet Union. The game was ideal for someone of Fermi's temperament, invariably conservative and skeptical of any predictions of quick or revolutionary change.

In the lab, the bulk of his efforts focused on fission and chain reactions and, within a few months of returning from Ann Arbor, he and his team were making progress.

THE CONCEPT OF A PILE DIFFERS LITTLE CONCEPTUALLY FROM THE configuration of the water tank experiments that Fermi, Szilard, and Anderson carried out during the winter and spring of 1939. Effectively, Fermi decided to substitute graphite bricks for the water and to build up, as well as out.

The water tank experiments were not designed to analyze the way neutrons diffused within the moderating medium. Realizing how crucial it would be to understand this diffusion process with graphite, Fermi devised a series of experiments to do just that. In rooms at Pupin and later in the basement of nearby Schermerhorn Hall, he and his colleagues—with the occasional help of burly members of the Columbia football team who were press-ganged by Pegram—stacked graphite bricks into square columns several feet thick, placed rhodium foil at key locations throughout the stacks, set a neutron source at the top of the stacks, and studied how neutrons made their way through the pile. Once the foils were exposed, they were quickly extricated from the stacks and run down a corridor so the radioactivity could be measured by Geiger counters. Fermi and Anderson raced down the corridors to take advantage of rhodium's short, forty-four-second half-life, re-creating scenes from Via Panisperna in 1934, most likely with Fermi in the lead. They built one stack after another, getting covered in fine black graphite dust that made them look more like coal miners than experimental physicists. As the stacks grew in height, to over ten feet, the physicists required ladders to get to the top of the stacks and place the neutron source. Many years later, when Fermi recalled these experiments at a public lecture, he drew a laugh from the audience as he described it as "the first time when I started climbing on top of my equipment because it was just too tall—I'm not a tall man." These diffusion experiments were critical in establishing how neutrons slowed down during their voyage through the graphite and gave some sense of how often neutrons might be absorbed by the graphite.

These experiments began in the spring of 1940 and continued throughout much of the rest of the year. Anderson, chasing after Fermi running down the corridor with rhodium foil, played Amaldi's role from six years earlier at Via Panisperna. Szilard kept up with the increasing demand for ever larger quantities of graphite, playing the same procurement role as Segrè did in Rome. Szilard had help from the "Committee on Uranium," a group of senior scientists and military officers established by Roosevelt to provide guidance and coordination for the new effort authorized by the president. It was an "all-American" group. As foreign nationals, Fermi, Szilard, Teller, and Wigner were formally excluded but met frequently with the committee to provide input into their decisions. One of their first decisions was to allocate $6,000 to buy what Fermi described as "a huge amount" of graphite.

As the diffusion experiments continued through 1940, Fermi and Szilard began to suspect that impurities in the graphite were absorbing neutrons at a rate that would reduce the probability of a successful chain reaction. A visit to one of the manufacturers, National Carbon in Cleveland, resulted in the identification of trace quantities of boron in the graphite as the culprit. In 1941, Szilard worked closely with the engineers at National Carbon to develop a processed form of graphite with fewer impurities.

A second experimental project took place alongside the diffusion studies, the purpose of which was to test out Fermi's idea of placing the uranium in lumps throughout the graphite pile, to reduce the likelihood that neutrons would be absorbed by U-238.

These experiments pushed materials science to new frontiers. Industrial graphite had too many impurities to make it useful as a moderator for the chain reaction. Szilard and the engineers at National Carbon worked hard to develop methods to remove boron and produce a suitable graphite moderator. In doing so, they created the world's first "nuclear graphite," a form of pure graphite now used throughout the world in graphite-moderated nuclear reactors.

Another aspect of materials science that got a boost through these experiments involved the production of uranium. Before the war, uranium was valuable only insofar as it was used in scientific experiments, and the most readily available form of the element was uranium oxide, not ideal for the purposes of the "exponential" pile.* There being little economic need for uranium metal, production was still relatively primitive. The best that industry could do was to produce the metal in a powder form, which had a tendency to spontaneously combust when exposed to air. Monitoring these experiments with great interest, the Committee on Uranium worked with a variety of manufacturers to improve uranium metal production techniques.

During this period, the team continued to grow, with Wigner, Wheeler, and others from Princeton joining Fermi's team.

BY SEPTEMBER 1941, FERMI BELIEVED THAT THE TEAM HAD MADE sufficient progress to build a true working exponential pile. The Committee on Uranium allocated the considerable sum of $40,000 for the purchase of massive quantities of uranium and graphite, and Szilard negotiated the purchases with industrial vendors. The Columbia football players were called back into service, and the pile began taking shape in the basement of Schermerhorn Hall. Fermi later spoke with awed amusement at the ease with which the Columbia athletes packed heavy cans of uranium oxide—uranium was and remains one of the heaviest elements found in nature—and hoisted them into a lattice whose structure Fermi determined as he balanced theoretical calculations and educated guesswork with the practicalities of working with the materials at hand. Large-scale electronic computers not having been invented, it was impossible to do a full-blown calculation as to what size the lattice would have to be to produce a self-sustaining chain reaction. What *was* possible to measure was the performance of a pile of a specific size and specification and extrapolate whether such a structure, if extended infinitely, would produce such a chain reaction.

The actual pile they built grew to a stack of graphite bricks eight feet on each side and eleven feet high. Within the stack of graphite, square tin cans of uranium oxide, eight inches on each side, were distributed in a three-dimensional matrix of some 288 cans. Slits were placed strategically to allow for insertion and removal of iridium foils to measure

* In a "critical" pile the reactions are self-sustaining; in an "exponential" pile neutron production grows geometrically.

radioactivity. The neutron source would be placed in a bed of paraffin at the bottom of the pile.

This first pile perfectly reflected Fermi's experimental style. Its design was partly a product of sophisticated theoretical considerations, in particular the lumping of uranium throughout the graphite in a "lattice" framework. Yet the design also reflected basic practicalities, such as the dimensions of the graphite bricks themselves. Fermi's design proved easy to build and lent itself to systematic measurement and evaluation through a series of carefully controlled experiments. He played an active part in its construction, piling graphite bricks and cans of uranium oxide alongside the rest of the team. Unfortunately, however, the results of the experiment were disappointing. By Fermi's calculation, even extending the structure they built into infinity, the performance would be 13 percent below what would be necessary for a self-sustaining chain reaction.

Undaunted, Fermi and his colleagues were convinced that adjustments in the structure of the lattice and improvements in the purity of the materials could squeeze more excess neutrons out of the process and deliver the desired result. The Committee on Uranium seemed to agree, as did the various bodies that were now coordinating and directing all national work on fission. Central to this effort was Vannevar Bush, an MIT-trained engineer with a skeptical, Yankee demeanor. Bush, an extraordinarily energetic administrator, reported directly to the president.

As the project grew in organizational complexity, research into fission became highly secretive. Secrecy in fission research had begun as an informal agreement among physicists working in the United States and Britain to avoid publishing experimental results that might help German scientists, but had morphed into a formal statutory edifice of security classification. Many of the scientists most deeply involved with this sensitive work—men like Fermi, Szilard, and Wigner—were foreign nationals and found themselves excluded from deliberations within the committees organized to guide and develop fission research. Political leaders all solicited their views, but the decisions took place behind closed doors without the foreign-born scientists.

Others may have treated the need for secrecy differently, but Fermi took it quite seriously and never discussed his work with his wife. Throughout the war, from New York to Chicago to Los Alamos, Fermi was silent about his activities and his role. Laura Fermi only learned anything substantive about her husband's role in the Manhattan Project in August 1945 after the bomb was dropped, when Fermi handed her a copy of an unclassified US government report on the project.

The work at Columbia on the pile, as well as important technical progress elsewhere, gave Bush a sense of optimism that a fission bomb was indeed possible. On December 6, 1941, he announced an "all-out" effort to pursue a fission weapon. The next day Japan attacked Pearl Harbor and within a few days the United States was at war against Japan, Germany, and Fermi's home country of Italy. Bush reorganized the project's leadership once again, creating a new independent organization, called S-1, to replace the Committee on Uranium. Others on the executive committee included Harold Urey, Ernest Lawrence, and Arthur Compton. Lawrence, the inventor of the cyclotron and the leader of the experimental physics group at Berkeley, would be responsible for directing research on plutonium and for developing methods of isotope separation based on his cyclotron experience. Urey, the brilliant Columbia physical chemist who befriended Fermi in early 1939, would be responsible for chemical separation issues associated with plutonium production and would also pursue promising lines of work on isotope separation. Compton, who sat on various oversight committees during this period, would direct further research into the properties of uranium and plutonium under the aegis of the newly created Metallurgical Laboratory at the University of Chicago.

Eager to prove his loyalty to his new country, Fermi was in an awkward position. He was an enemy alien working at the heart of the US government's most sensitive and secret military project. Change was afoot.

CHAPTER SIXTEEN

THE MOVE TO CHICAGO

THAT CHANGE WOULD INVOLVE A MOVE TO CHICAGO, TO WORK under the supervision of Arthur Compton, a physicist he respected but hardly knew.

In December 1941, Vannevar Bush tasked Compton with an enormous challenge: the coordination and management of more than a dozen uranium research teams across the country. The fear of German progress on a fission weapon exerted enormous pressure. Now that the United States was formally in a war against the Axis powers, time was at a premium.

Compton was a fine choice for the job. He had been involved in science policy for several years prior to the US entry in the war and knew the senior Manhattan Project leadership well. They, in turn, respected his scientific abilities and his sound judgment. Bush hoped that a disparate team of physicists, many of whom had never before worked in large teams under a single leader, would follow his lead. Compton came to prominence in the early 1920s with a series of X-ray scattering experiments that supported Einstein's hypothesis that light consisted of particles, called photons. The importance of the work was immediately recognized, and he was awarded the 1927 Nobel Prize. He was one of the very few Americans invited to the Solvay conferences before the war. Before the 1927 Solvay conference, he also attended Corbino's 1927 conference at Lake Como, where he first encountered the young Fermi.

Tall, athletic, handsome, with a thick head of dark hair slicked back fashionably, Compton looked like a leader. Born into an Ohio family of academic high achievers, Arthur shared and cultivated his family's strong Christian faith, eventually serving as a deacon in a local Baptist church. He was one of only a handful of scientists in the Manhattan Project who spoke openly about his religious beliefs, which hardly endeared him to the largely irreligious group he was leading. On one occasion while he was managing the Met Lab, he brought a Bible to a fractious meeting and tried to establish his authority by quoting from it. This was not necessary and did not work. His authority did not come from his adherence to biblical principles but rather from his undoubted scientific achievements, his sense of judgment and fair play, and a direct line to the country's wartime political leadership.

In January 1942, Compton brought the various teams together for a series of meetings in Chicago and New York to thrash out a strategy for centralizing the research effort. Two key decisions came out of these discussions. One was a timetable for the development of the bomb: determination of the feasibility of a chain reaction no later than July 1, 1942; achievement of a controlled chain reaction by January 1943; production of plutonium for the bomb by January 1944; and a working bomb by January 1945. The second was a decision on how to centralize the work. After considerable debate, he made a Solomon-like decision. Lawrence and his team at UC Berkeley would remain in California, but all other teams, including Fermi's, would come to Chicago. Once the decision was made, Compton immediately informed Fermi by phone, because Fermi was unable to attend the final Chicago meeting owing to a bad cold he had caught not long after the first Chicago meeting. (He sent Szilard in his place.) Compton reports that Fermi immediately agreed, no doubt unwilling to object, given his enemy alien status. Fermi was not, however, a happy man. His team at Columbia was working well together and he had complete control over the project. The move would be inconvenient personally and professionally. He would be working with some of the same people—Anderson was sent to Chicago immediately to coordinate the project, while Zinn stayed behind with Fermi to begin organizing the move—but he would soon be working with many new people, untested and unknown, under the leadership of Arthur Compton.

His status as an enemy alien is surely one of the most bizarre aspects of the entire Manhattan Project story. Here was an Italian national and a member of the Fascist Party at the very center of one of the most secret projects of the US war effort. Travel restrictions were only part of the story. Those who knew him, who knew how lucky the United States was to have him on the side of the Allies, had no doubts about his loyalty, but many of those involved—particularly military officers who were increasingly important in the organization of the project—did not know him at all. The FBI was suspicious of Fermi, but it was more concerned about Szilard. He traveled widely, had no visible means of support, had a lavish lifestyle that often caused him financial problems, and was a high-profile eccentric—exactly the type to attract the FBI's attention.

The initial FBI report filed on Fermi, dated August 13, 1940, stated:

> He is supposed to have left Italy because of the fact that his wife is Jewish. He has been a Nobel Prize winner. His associates like him personally and greatly admire his intellect. He is undoubtedly a Fascist. It is suggested that, before employing him on matters of a secret nature, a much more careful investigation be made. Employment of this person on secret work is not recommended.

Further FBI investigation at Columbia, dated October 22, 1940, confirmed the view of Fermi as an outstanding scientist and loyal "as long as Fascist Party retains control in Italy," although Professor LaMer from the Columbia chemistry department felt it best not to grant clearances to any foreign national, irrespective of whether a specific individual might be trustworthy. Fortunately, the government cleared him for secret work. After all, much of what was secret in the Manhattan Project originated in Fermi's brain.

During this period, Enrico and Laura began to feel sufficiently vulnerable that they dug a hole in the floor of their basement in Leonia and buried a tin can of cash for use in case of an emergency. They never had need for the can, of course, and Fermi's loyalty was never seriously in question. Nevertheless, his enemy alien status had a powerful psychological impact on him. Combined with his natural inclination to defer to authority figures, this probably explains Fermi's reluctance to make a fuss when Compton asked him to move to Chicago. It also may explain

Laura's apparent willingness to make the move without complaint, because she believed that the move would be temporary. Later on in Chicago, however, when Fermi discovered that his mail—along with that of everyone else on the project—was being read by US Army censors, he complained vociferously, but with little effect.

In the first years of the war, when the German onslaught made it seem possible that they might actually win the war, the Fermis and their Columbia colleagues Joseph and Maria Mayer talked about what they would do if America became a fascist state. The two couples had met at the Ann Arbor summer session in 1930 and had become fast friends. When the Fermis arrived in New York in 1939, they were delighted to find that the Mayers had also arrived at Columbia. If the Nazis conquered the United States, the four of them agreed they would settle on a desert island in the South Pacific, far from political danger. Their division of labor would reflect their relative strengths: at sea, Joe would captain and Enrico would navigate, and on the island Enrico would farm, Maria would curate a small but essential collection of books, and Laura would keep everyone clothed. A doctor and a few others would join them to fill out the little community. It was a ridiculous fantasy that they may not have taken seriously and yet it can be seen as another example, shared by both refugee couples, of underlying uncertainty and insecurity regarding their new homeland.

THERE WAS JUST ENOUGH TIME BETWEEN THE CALL FROM COMPTON and the scheduled move to Chicago to build one more pile at Columbia. This one was more successful.

Fermi redesigned the structure, using cylindrical blocks of compressed uranium powder provided by a new, young member of the team, John Marshall, who was given the task of "sintering" the uranium powder that was provided for the project. Sintering is the process of compressing powder to such an extent that it becomes a solid, similar to the compression of charcoal briquettes. Marshall's sintering press produced solid slugs of uranium oxide powder about three inches high and three inches in diameter, weighing about four pounds apiece—still not as dense as pure uranium metal, but denser than previous preparations. All told, some 2,160 slugs of uranium, weighing a total of more than four tons, were embedded in a pile of graphite eight feet on each side and some eleven feet, four inches high. It completely filled the space in the basement of Schermerhorn.

The second modification involved trying to extract as much air out of the pile as possible. Fermi and the team worried that nitrogen in the air between the bricks might be reducing the reproduction factor. Any improvement in that factor, however small, would be important. Fermi used the analogy of canned food—then quite popular because of rationing for the war effort—and searched for someone to build a "can" around the entire pile. He found a workman employed by Columbia who hardly spoke English but whose soldering technique was outstanding. He built a tin can around the pile, with a valve set into one side attached to a vacuum pump. The can held its integrity and the air was vacuumed almost completely out of the pile, replaced by carbon-dioxide.

The results were about 4 percent better than the previous pile, but still well below the absolute level required for a sustained reaction. Fermi was still optimistic, believing that further improvements in size, geometry, and purification would bring about the desired result. He was also able to run a series of tests to determine how impurities in either the uranium or the graphite might affect the reproduction factor. For example, he determined that cadmium was an extremely effective absorber of neutrons. This knowledge came in handy over the next few months. He did not, apparently, test for xenon. This also became relevant later in the project.

During these experiments in early 1942, two accidents occurred, underscoring the hazards of working with the unstable and dangerous materials required to prepare the pile. In one incident, Zinn was working with powdered thorium to test its ability to absorb neutrons and, though he had taken the precaution of wearing goggles and gloves, the powder exploded in his face when he opened its airtight container. He suffered severe burns on his hands and face, but the goggles saved his eyesight.

Another accident involved Pegram, Fermi, and Anderson, but it was Anderson who would suffer the consequences. Like thorium, powdered beryllium has a tendency to catch fire if not handled carefully. The three of them received a shipment of powdered radium and beryllium for use in preparing a neutron source. They found the preparation to be slightly damp, placed it on a hot plate to dry it out quickly, and left the lab room. When they returned, the powder was on fire. Anderson rushed in to put out the blaze and no one seemed to be the worse for the incident. Years later, however, Anderson began to have breathing problems, which were traced back to inhalation of the beryllium powder during this incident. The illness, called berylliosis, eventually killed him at the age of seventy-four.

In April 1942, the pile was dismantled, removed from the Schermerhorn basement, and shipped to Chicago.

At just about the time Compton was making the decision to bring Fermi and his team to Chicago, Fermi was having a conversation with his old friend and colleague Edward Teller.

Teller immigrated to the United States from Europe in 1935, when he was offered a job at George Washington University in Washington, DC. He was in attendance when Fermi and Bohr explained uranium fission to the electrified audience at the March 1939 conference in Washington and then moved to Columbia in the fall of 1941 to help Fermi and the team with their graphite pile experiments. The two dined regularly at the Columbia faculty club. "Walking back to the laboratory after lunch one day," Teller relates, "Fermi posed the question: 'Now that we have a good prospect of developing an atomic bomb, couldn't such an explosion be used to start something similar to the reactions in the sun?'" In what must count as one of the greatest understatements in the history of science, Teller continues, "The problem interested me."

The idea that the sun is powered by fusion reactions was proposed in the late 1920s by the British physicist Robert Atkinson and a German named Fritz Houtermans, later elaborated upon by the eccentric, fun-loving Russian George Gamow, and finally worked out in detail by Hans Bethe in 1938. By January 1942, physicists had a complete understanding of the basic processes underlying the way the sun and other stars work. In the hot, dense core of a star, protons are moving so fast that they break through the "Coulomb barrier"—the electromagnetic repulsion keeping two protons apart. The protons fuse, creating helium nuclei and eventually a range of other heavier nuclei. Each fusion results in radiant energy, in the form of photons and neutrinos. The physics of the process is quite complex, and Fermi's beta decay theory helped Bethe work it out in detail. Per unit of mass, the energy released in fusion, particularly of hydrogen, is far higher than that released in uranium fission.

Fermi's calculations suggested that the temperatures achievable in fission weapons might well be sufficient to set off a fusion reaction in hydrogen. In sharing this idea with Teller, he unwittingly changed the course of Teller's life. Teller became a man possessed. He realized that because hydrogen was so plentiful and so stable, a fusion weapon could have virtually unlimited destructive power, the only limit being how much hydrogen the bomb might contain. That summer Teller arrived at Berkeley to join J. Robert Oppenheimer's team exploring the feasibility of a fast-neutron chain reaction for the bomb—the slow-neutron reactions that Fermi was studying in his pile experiments were not appropriate for the bomb itself—but Teller could think of nothing other than a fusion device. When Teller arrived in Los Alamos the following year, his obsessive work continued.

In that fateful walk across the Columbia campus, Fermi revealed something else of great importance, although Teller does not explicitly comment upon it in his memoirs. By early 1942, even though the pile experiments had been less successful than he had hoped, Fermi had come to the conclusion that fission weapons would work. He may from time to time have expressed doubts, but from that point onward his presumption was almost certainly that fission weapons were feasible.

THE PROCESS OF MOVING THE COLUMBIA PILE PROJECT TO CHICAGO took five months, during which time the graphite bricks and uranium slugs were sent by special shipment across country, Zinn in New York, Anderson in Chicago, and Fermi shuttling between the two. Fermi initially took up residence in Chicago at International House, an independently run Gothic-style dormitory a short walk from the center of campus. He brought two young Columbia graduate students with him, Albert Wattenberg and Bernard Feld. Wattenberg recalls playing chess frequently with Fermi during this period. Fermi could beat him in chess, but always lost in tennis. Laura stayed behind with the two children, now eleven and six, to join her husband when school let out in June.

The University of Chicago was an extraordinary place, a center for scholarship and learning that rivaled schools far older and more prestigious. By the time Fermi arrived in mid-1942, Robert Maynard Hutchins, the *enfant terrible* who became president in 1929, had built a true academic powerhouse in the Hyde Park neighborhood on the south side of Chicago. The physics department was located in two adjacent buildings, Ryerson and Ekhart, almost squarely in the middle of campus. Compton gave Fermi an office in Ekhart, from which the new arrival directed activities.

As soon as the material arrived, work began on new pile experiments. Joining Fermi's team were several key individuals who became close colleagues for the rest of his life. One was an accomplished experimental physicist named Samuel Allison. A little older than Fermi, Allison spent much of the 1930s involved in X-ray scattering studies with Compton. The two wrote the standard textbook on the subject and over the years at Chicago Allison became a trusted associate of Compton. Prior to Fermi's arrival at Chicago, Compton asked Allison to develop a reactor pile working with beryllium as a neutron moderator, and by mid-1942 he had actually achieved better results than Fermi, even though beryllium was a more dangerous substance with which to work. Allison and Fermi became close colleagues during the war and remained so afterward.

Another colleague was a young PhD student working on a thesis under Chicago physicist Robert Mulliken. Her name was Leona Libby.* Tall, athletic, and attractive, she was the only female member of Fermi's Chicago team. She lived with her sister near the university and early on became friendly with the Fermis. Laura would often cook meals for Leona and Herb Anderson, who took a room on the third floor of Arthur Compton's spacious home nearby. Anderson and Libby soon discovered a mutual love of swimming and would take time off every afternoon for a dip in the freezing fresh water of Lake Michigan. Fermi, who was a passionate and exceptionally powerful swimmer fond of a peculiar dog-paddle stroke, often joined them. Harold Agnew, then a student working with Fermi at the Met Lab, recounts an outing in which Fermi challenged a group of younger colleagues to a swim from 55th Street

* When she first met Fermi, Leona was unmarried and her maiden name was Woods. She would marry twice during her lifetime—first to physicist John Marshall, and then to chemist Willard Libby. Because her memoir was written during her marriage to Libby, she is referred to hereafter as Leona Libby, to avoid any confusion.

north for about a mile. The group set off and Fermi quickly took the lead. Over his shoulder, he could see the members of the group lagging and turned back to give them encouragement, swimming in circles around them, egging them on. They finally reached their destination at 47th Street, and the group climbed ashore exhausted. Fermi gleefully announced he would swim back. His wet and weary colleagues decided to return by foot.

Others from the Columbia team came and settled in Chicago. Szilard spent much time there and with Zinn was instrumental in pressing manufacturers to find new ways of purifying graphite and uranium. John Marshall, who joined Fermi's team in mid-1941 and worked out the process of sintering uranium powder, also came with the Columbia team, sharing the third floor of Compton's home with Anderson. Marshall met Leona Libby soon after arriving in Chicago and within a year they were married.

Leona Libby was obviously charmed by Fermi. In later years she would write of this time:

> Fermi would like to show superendurance, to swim farther, to walk farther, to climb farther with less fatigue, and he usually could. In the same way he liked to win at throwing the jackknife, pitching pennies, or playing tennis, and he usually did. These qualities of gaiety and informality of his character made it easy for the young members of the laboratory to become acquainted with him. He was an amazingly comfortable companion, rarely impatient, usually calm and mildly amused.

It was a heady experience for her, for Marshall, and for the other young members of the Met Lab. Already a legend among physicists for his work in Europe, already the subject of growing mythology, here was Enrico Fermi in the flesh, and they found him to be unassuming, approachable, informal, and fun. He had a healthy respect for his own abilities, but that was based on an empirical fact—he was just that much better a physicist than anyone else. Fermi had an enormous personal impact on his colleagues, with whom he collaborated and also went swimming and hiking. They would remember these days for the rest of their lives.

For those first few months in Chicago, Fermi enjoyed working with his new young colleagues and was reminded of the early Rome years. He was no longer working under the cloud of fascism, and his eager, positive new colleagues enjoyed his infectious sense of fun, a sense that had been entirely absent in the final years in Rome. He would stay close to these colleagues for the rest of his life.

Over the course of 1942, however, many others joined the Met Lab, working under Compton's watchful, forceful leadership. Perhaps most important, the Princeton team—Wigner and Wheeler, in particular—arrived in Chicago soon after Fermi. They delved deep into the theory of the pile and were instrumental in the further development of the pile concept at later stages of the Manhattan Project. In the end, some forty physicists were in Chicago working in secret to create the world's first controlled uranium chain reaction. As Compton grew increasingly confident in Fermi's abilities, he put Fermi in charge of ever more elements of the project, until Fermi complained privately to Segrè that he felt like he was doing physics "by phone," perhaps referring to the dramatic increase in his administrative responsibilities, which took him away from the lab and pure physics. Fermi disagreed with some of the directives the team received from project leaders in Washington. Szilard recalls that Fermi once complained, in frustration, "If we brought the bomb to them ready made on a silver platter, there would still be a fifty-fifty chance that they would mess it up." He sometimes felt that he was a cog—an important cog, but a cog nevertheless—in an increasingly large and unwieldy machine.

Nuclear reactors today are major engineering projects involving careful planning and reams of design drawings, all carefully vetted and reviewed at every step. In contrast, the first operational nuclear reactor was planned in Fermi's head, based not on extensive engineering drawings but on his sense of how the neutron flow would develop within the heart of the pile and make its way from one uranium slug to the next. He gave general instructions to machinists and his fellow physicists and let them do the rest. He did not have access to computers to calculate what the geometry of the lattice should be, how big the pile would have to be before it went critical, or how hot it might get as it ran. All these calculations were done either in his head or on his ever-ready slide rule, with the more junior physicists at his side providing back up. By the time work actually began on the final Chicago pile in November 1942, Fermi and his team had built twenty-nine experimental piles testing various aspects of material and configuration. These experiments gave Fermi an intuitive sense of how the pile should be constructed and led to the apparatus that took shape in the squash court under the stands at the abandoned Chicago football stadium.

In later years, Fermi told his wife that the overall structure of the pile came to him in May 1942 while walking with colleagues along the Indiana dunes on Lake Michigan's southern shore. Like the previous piles, it would be modular, constructed of bricks of graphite embedded with uranium and interspersed with bricks of pure graphite. It was basically quite simple. Modern reactors, built with cooling mechanisms, multiple redundant safety mechanisms, elaborate diagnostics and designed to produce electricity, are highly complex. Fermi's piles were brutally simple, with only two objectives in mind. One was the proof of concept—a controlled, self-sustaining nuclear fission chain reaction. The other was to serve as a machine to produce plutonium.

If natural uranium is exposed to neutron bombardment, the U-238 in the uranium sometimes undergoes a series of transformations through beta decay into a new element, plutonium 239 (Pu-239). Studying Pu-239, physicists concluded that it might also be used as a material for weapons.

In 1940, a UC Berkeley team led by chemist Glenn Seaborg produced a small amount of this new element by bombarding natural uranium in the Berkeley cyclotron. This process by its very nature could produce only minute quantities of the new element, nowhere near enough for a weapon, but enough to study its properties. Experiments at Berkeley subsequently demonstrated that, as theory predicted, it would be a suitable alternative to U-235 for a fission weapon. It might also prove easier to produce in substantial quantities than U-235. If Fermi's exponential pile could be made to work, perhaps it could be scaled up to become a plutonium factory. Traditional chemistry could be used to separate Pu-239 from reactor by-products. Hence the priority placed on plutonium research.

So the project Fermi undertook to complete when he arrived in Chicago had two distinct purposes. The creation of a self-sustaining fission chain reaction was clearly important to demonstrate the chain reaction concept. If one could be created, then in principle uranium fission weapons could be built, although enormous challenges would remain. It would also create the possibility of plutonium production, providing a second possible route to a fission weapon. At this particular juncture the success of the Manhattan Project depended almost entirely on Fermi's ability to achieve a self-sustaining reaction. If he felt any pressure at all, he did not show it, perhaps because by this time he felt certain he could make it work.

LAURA ARRIVED WITH THE CHILDREN IN SEPTEMBER 1942 AND they moved into a grand, old, three-story house at 5537 South Woodlawn that had been vacated by its owner, investment manager Sydney Stein Jr., who moved to Washington to help in the war effort at the Bureau of the Budget. As enemy aliens, the Fermis were not allowed to keep the large floor-standing radio that came with the living room furnishings. After consulting with the FBI, the landlord removed it. On the third floor lived two Japanese exchange students, stranded in Chicago when the war broke out. With an Italian family occupying the rest of the house, the landlord decided—presumably also in consultation with the FBI—to evict the students.

The Fermis soon began to entertain at the house on a regular basis. With new physicists arriving in Chicago almost daily, Laura believed that she could help out, in spite of all the secrecy surrounding her husband's work, by providing an active social life for the newcomers. Libby recalls attending parties with some of the most distinguished scientists of the day, watching as Enrico led the group in some of his favorite parlor games from his Rome days—games he was always determined to win. These frequent parties forged social bonds within the team and gave wives who were not privy to their husbands' actual work a sense that they were doing something useful for the war effort. The bonds forged in Chicago and cemented during the later period at Los Alamos were to last for decades.

THE IDEA THAT BEGAN TO FORM ON THE INDIANA DUNES WAS different from the idea that drove the geometry of previous piles. The Columbia piles were squared-off towers of graphite and uranium, rising as high as the Schermerhorn ceiling permitted. Now he began to play with another shape: a flattened, roughly spherical, ellipsoid shape. It was clear to Fermi that such a shape allowed for neutron diffusion that would be more optimal for the reproduction factor. Surface area was Fermi's enemy, because as neutrons escaped from the pile through the surface contact with the air they were lost from future fissions. A cube of a given volume has a greater surface area than a sphere of the same volume. The smaller the surface area for a given volume, the more likely it would be for neutrons to stay inside the pile. Thus, a spherical shape was better than a squared-off shape. Before he could begin to build anything, however, the Chicago team would have to solve two major problems. One of them involved the purity of materials for the new pile. The other was to figure out how to make a spherical shape sit stably on the floor.

The more he thought about it and the more he discussed the matter with Szilard, Wigner, Allison, Wheeler, and others, it was clear that purity of the materials used in constructing the pile would be a central issue. Impurities could absorb neutrons in unpredictable ways, slowing the process and constraining the reproduction factor. Even if they did not absorb neutrons themselves, under neutron bombardment they might transform into new isotopes that would be neutron absorbers. Szilard and Zinn continued working with producers of graphite and uranium to get materials of sufficient purity to increase the reproduction rate to an acceptable level. Throughout the summer and fall, Fermi and the team built one small experimental pile after another, testing newly arrived materials: new graphite, better quality uranium oxide powder, and uranium metal, cast into egg-shaped lumps by a team at the University of Iowa. Fermi began to sense how the different materials reacted, how different lattice structures produced different intensities of neutrons, how the reproduction factor varied with material and configuration. These experiments continued during the late summer and early fall of 1942.

The ellipsoid shape created a problem of its own: how to build it so that it would be absolutely stable on the floor of the lab. The pile would grow layer by layer from the ground up, with the bottom layer laid out in a rough circle. It would gradually grow wider until it reached a certain height above the floor and then begin to reverse its growth symmetrically. From the side it would look like a flattened sphere. How would all these bricks be held in place? The solution was to create a wooden frame that would, they hoped, hold the layers stable as the pile rose from the lab floor. It required strength and stability, because the pile would be quite heavy and the team could ill afford an accident involving the pile sliding into a messy mountain of uranium and graphite on the lab floor. Fermi found a master carpenter employed by the university who was up to the task.

Instrumentation also needed to be considered. To monitor the reactions, Fermi's favorite iridium foils could be placed deep inside the pile, removed, and tested for radioactivity. The problem with this process was that it was cumbersome and unsuitable for monitoring the reactions while they were taking place. Volney Wilson, a long-time Compton collaborator, would be in charge of the team responsible for developing this new instrumentation. Working with Wilson were Herb Anderson and Leona Libby. The new instruments would click loudly when either a neutron or a gamma ray was detected and would drive an electronic pen to graph the level of neutron activity on paper tacked to a circular drum, much the way an earthquake detector traces seismic motion.

Another matter deeply concerned Fermi: safety. His studies in New York told him that cadmium was a highly efficient neutron absorber. To control the fission reactions and make sure the pile did not run out of control, perhaps leading to an explosion (the term *meltdown* had not yet been invented), he decided to insert cadmium-covered wooden strips at strategic points throughout the pile. With all the cadmium strips in place, the pile could not go "critical," their term for a sustained chain reaction. There would also be a fail-safe mechanism: if during the course of operation the pile became too reactive and at risk of blowing up, a rope could be cut that would allow all the cadmium strips to drop back into the pile simultaneously, bringing the reaction to an abrupt halt. This mechanism would come to be called SCRAM. Though debate continues as to what the acronym came from, its meaning is painfully obvious.

Once a critical mass of material had been assembled—once the reactor had a reproduction rate that grew exponentially—the way to turn the reactor on would be to remove all the cadmium strips but one and then slowly pull out the final strip, the "control rod," just enough to reach criticality. If the reaction looked like it would grow out of control, all that was needed was to slip the control rod back in place. The reaction would die down almost immediately, as the cadmium absorbed neutrons from the chain reaction. As an added precaution there would be a small group of intrepid souls standing on top of the pile with buckets of a cadmium solution, prepared to douse the whole pile if for some reason SCRAM did not work. That would surely stop the reaction instantly but would also render the entire apparatus useless for further research.

By the end of summer 1942, the general plan was clear. The pile would be built out of town, in an area west of Chicago called Argonne Forest where Compton and his wife enjoyed horseback riding on weekends. Its distance from central Chicago made Argonne ideal. If an accident occurred, it would be far away from the densely populated urban area. Its distance also made it easy to isolate and maintain the kind of secrecy that was required. Work began on a facility for the pile using Compton's engineers of choice, Stone & Webster, a Massachusetts firm that worked under the auspices of the Army Corps of Engineers. In mid-September a colonel in the corps, Leslie Groves, was promoted to brigadier general and assigned to oversee the entire Manhattan Project, effectively putting Compton, Fermi, Lawrence, Oppenheimer, and the hundreds of other physicists who were drawn into the project over the previous year under military authority. The plan was to finish the facility by October 20, 1942, at which point the entire Met Lab would move there and complete the pile.

Beginning in September 1942, Fermi gave a series of lectures to the Met Lab scientists regarding the theory behind the pile, focusing on calculations of the reproduction factor. Notes of these lectures were taken by Anderson, Libby, and others, and they include some of Fermi's more endearing uses of American slang. In describing what to do if the reproduction factor ended up much greater than one—in other words, if the chain reaction got out of control—he said, "run quick-like behind a hill many miles away." The lectures were an important part of the program, designed to ensure that those working so hard on the pile maintained confidence in the science underlying it. They also allowed Fermi to indulge yet again in one of his favorite pastimes, teaching.

Groves, a bull of a man with a gruff manner and an ability to get things done, had just finished successfully overseeing the construction of the Pentagon, the world's largest office building. He wanted his next assignment to be overseas and only agreed to the Manhattan

Project assignment reluctantly in exchange for two assurances: first, that he would be promoted to general officer rank, and second, that he would have first priority for men and materiel, without restriction. He got both.

Groves visited Chicago in early October 1942 and met with the senior scientists on the project. He was impressed by the "crackpots," as he liked to call them, who were making progress toward the first chain reaction, and the meeting was productive. Owing to the new authority that Groves had extracted from his masters, graphite and uranium of increasingly higher quality now arrived in Chicago in truly massive quantities. Everyone concerned expected to make the move to Argonne and start work on what would become the first working nuclear reactor.

With total control of the program, Groves imposed military secrecy and ordered that key personnel—Fermi included—could no longer travel by air. The risks of losing essential assets in an air accident were simply too great. He also insisted that a handful of scientists—once again, including Fermi—travel at all times with bodyguards. The bodyguard assigned to Fermi was a suitably large former Chicago policeman named John Baudino. The two eventually became good friends, and Fermi joked that Baudino grew into a decent physicist in the process of sticking by Fermi's side for the duration of the war.

Finally, in the name of military security, key scientists were not to travel under their own names. They were to use code names suitably chosen so that the scientists would have no trouble remembering them. Fermi liked his new name, Henry Farmer. It sounded very American, even though his pronunciation of his new name sounded distinctly Italian.

Fermi had a sense of humor about his code name. All the senior scientists had one—Eugene Wigner was "Gene Wagner," Niels Bohr was "Nicholas Baker." One evening at Los Alamos, after a screening of a forgettable 1943 movie about the life of Madame Curie, Fermi could not resist approaching Bohr: "Mr. Baker, I've just seen a grand movie, *Madam Cooper*."

On Columbus Day 1942, Roosevelt repealed the enemy alien status of Italians in the United States, even though the country remained at war with Italy. Ironically, just as Fermi became free to travel as he wished, he was restricted by Groves to traveling only by train or car and with Baudino following him everywhere. The bodyguard was supposed to chauffeur his ward whenever the need arose, but on this point Fermi stood his ground. No one would be driving him. Baudino might accompany him, but the bodyguard would be in the passenger seat.

The preparation for Fermi's pile experiment in Chicago was not the only, nor even the main, focus of activity for the Manhattan Project during the summer and fall of 1942. For many participants the experiment Fermi was preparing in Chicago was a foregone conclusion. Planning moved ahead under the presumption that the pile would work according to expectations. Such was the priority of getting a workable bomb in the shortest conceivable amount of time that a number of tracks that depended crucially on each other's success were moving forward simultaneously. One track was theoretical work on fast-neutron fission, conducted by Oppenheimer and a team based in Berkeley.

Related to this was crucial work on ever more accurate initial calculations of what the critical mass of the uranium bomb would be. More work was also being done on plutonium as a possible fission material. Studies at Berkeley and Chicago indicated plutonium could be used to fuel a fission chain reaction, but the more the new element was studied, the more concern there was about its stability in the quantity necessary for a working bomb.

Groves also directed work to begin on a variety of schemes to separate U-235 from U-238. He chose a location fifteen miles west of Knoxville, Tennessee, a site later known as Oak Ridge. Vast isotope separation plants rose on this site as the Manhattan Project progressed. Oak Ridge was also the site of the first small plutonium production reactor, built once the Chicago experiment proved the concept. Oak Ridge was primarily a research reactor to produce just enough plutonium to begin a more serious set of experiments to determine the new element's physical properties.

As 1942 drew to a close, Groves selected a large, desolate desert area of southeastern Washington State, eventually known as Hanford, for the top secret location of giant plutonium production reactors. The Columbia River would provide cold fresh water for cooling purposes and the area was easily secured because it was so remote from any urban centers. It was by far the largest facility in the Manhattan Project, some 586 square miles in area.

The work at Berkeley and the selection of large sites in Tennessee and Washington proceeded under the assumption that Fermi's pile would succeed. Though few physicists who knew about the project doubted that in principle it could work, Fermi and the team were aware that unforeseen difficulties might arise, including issues of safety.

Soon after Groves's visit to the Met Lab in early October 1942, a labor dispute arose at the new lab facilities at Argonne, and by mid-October construction work stopped. Compton had a schedule to meet. Fermi had a pile to build. The two discussed the situation and Fermi suggested finding space on the campus in which to build the pile. Compton thought about it and decided on his own authority—without consulting the university's president, who almost certainly would have vetoed the idea on safety grounds—to authorize a change in plans. They would build the pile in a squash court under the west stands of Stagg Field, the abandoned football stadium.

In retrospect it was a remarkable decision, reflecting both Compton's sense of urgency and the trust and confidence he had in his extraordinary Italian colleague. Fermi had persuaded Compton not only by outlining all the safety features he had envisioned but also by referring to the oddly comforting fact that some small percentage of the neutrons released in uranium fission would be emitted moments later than the initial prompt neutrons, giving Fermi additional time to put the control rods in place if the reaction looked like it might run out of control.

Compton was convinced. He understood the physics. He believed in Fermi. Now the work began in earnest.

CHAPTER SEVENTEEN

"WE'RE COOKIN'!"

By October 1942, shipments of graphite bars and uranium in the form of uranium oxide powder and uranium metal eggs were arriving at a furious pace and activity was intense. Fermi placed Zinn in charge of what was effectively a high-pressure construction job without blueprints. Zinn and Anderson managed a team of young physicists and thirty-odd day laborers, drop-outs from the local high school—what young physicist Albert Wattenberg referred to as "Back-of-the-Yards" kids—to machine the graphite into proper shape and to bore holes for the insertion of the uranium slugs. The team also began the process of sintering the uranium powder, using the dilapidated press that Zinn and Marshall had used at Columbia. Working in three shifts of eight hours, the sintering team could produce some twelve hundred lumps a day, aiming for a total of twenty-two thousand in total. The team worked fast and made few mistakes. Volney Wilson's instrumentation team also shifted into high gear.

As a first step in the construction, Fermi had Anderson approach the Goodyear Rubber Company for a heavy rubber "balloon," shaped in a cube, large enough to surround a squash court. History does not record what the executives thought of Anderson's request, although they were almost certainly assured that it was for the war effort. The balloon would, if necessary, play the role of the tin can that surrounded the last Columbia experiment, allowing Fermi to pump air out of the pile to enhance the chain reaction. Goodyear delivered a cubic balloon that would do the job if needed.

The balloon's arrival on November 16, 1942, permitted final construction to begin. Because the pile would be built inside the balloon, the first task was to hang the balloon from the ceiling so the work could take place inside it. That done, the team worked in twelve-hour shifts, Zinn in charge of the day shift, Anderson managing the night shift. Teams drilled blocks of graphite to accommodate the uranium slugs and laid them layer by layer according to plans drawn up by Fermi. The wood frame rose alongside the graphite pile. After the completion of each layer Zinn and Anderson met at Eckhart Hall with Fermi, who then made a rough sketch of how the next layer should look.

The pile rose, two layers of uranium-embedded graphite interposed with a layer of pure graphite, resting on a layer of pure graphite set at the base. Fermi calculated that the internal uranium lattice would result in a fully operational exponential reactor when the pile rose to seventy-six layers, just below twenty-seven feet high.

The uneven quality of the graphite and uranium posed considerable challenges. To address these, Fermi decided to allocate the highest-quality material—the uranium metal and the purist graphite—to the center of the pile. This was to ensure that the highest reproduction factor would be deep within the pile, offering the best hope for achieving an exponential chain reaction.

The nonstop construction took its toll. Graphite dust filled the enclosed space of the squash court and the noise was incessant. When layer fifteen was completed, Fermi asked Wilson to start measuring neutron production within the pile. Every three layers, the team repeated the measurements and dutifully recorded the increase in reactivity. As the pile rose, they placed channels for the control rods and the instrumentation running deep into the pile. The horizontal control rods were managed by hand. A vertical control rod, the "zip" rod, ran right through the center of the pile, to be lifted out by a rope and tied off when the reactor was set to go critical. The zip rod was attached electronically to instrumentation that would reinsert it back into the reactor if the reactivity level rose above a certain point. The rope on which the rod was suspended could also be cut manually with an axe should the need arise to shut the pile down instantaneously.

During the last two weeks of November, Compton, who was monitoring the progress of the pile with great interest, was deep in negotiations with DuPont executives to handle the construction of all the plutonium-producing reactors planned for the project. Fermi, Szilard, and Wigner were already scoping out the design for the initial reactor in Oak Ridge. Seaborg had agreed to a series of experiments designed to separate the plutonium from the spent reactor fuel. This experience would guide larger processing plants to be built alongside the major plutonium production reactors at Hanford. DuPont executives were hesitant to commit to the project. The company had never worked in conjunction with the US military, had no knowledge of nuclear physics, and worried about the difficulties of coordinating with Groves's Army engineers. To bring DuPont along, Compton convened a review committee, including the young, dynamic son-in-law of DuPont's president, Crawford Greenewalt, to persuade the executives. Compton wanted Greenewalt to be present when the Chicago pile went critical. He hoped that Fermi's performance that day would be sufficiently exciting to persuade the up-and-coming executive to commit the company to the project.

By late November 1942, Fermi had enough data to recalculate when the pile would go critical and determined the fifty-sixth layer of the pile would be the last one needed. He gave instructions to build the pile to the fifty-seventh layer as an insurance policy. So promising were the data that he decided not to use the giant cubic rubber balloon hanging from the ceiling. On the evening of December 1, 1942, layer fifty-seven was complete. With the last of some forty thousand graphite bricks set and with about nineteen thousand slugs of uranium sitting snugly in place, Anderson, on night watch, locked all the control rods into the pile and sat guard, waiting for dawn. Fermi had extracted a promise from him not to bring the pile to criticality by himself overnight. After almost four years of work on the pile concept, after countless experiments and a beryllium powder accident that was destined to shorten his life, Anderson would not betray Fermi, tempting though it might have been to make history himself.

On the morning of Wednesday, December 2, 1942, Chicago was in the grip of a cold snap. The previous day the high was thirty-two degrees Fahrenheit, but when Fermi awoke the next morning the temperature had dropped to zero. Leona Libby accompanied him to the pile, where they took some measurements of reactivity to compare with the measurements taken the night before. Anderson, who had been up late, arrived next, and the three made the short walk to Libby's apartment, where she cooked pancakes. Then they returned to the squash court to begin the day's historic work.

The process began about midmorning. The crowd overlooking from the balcony grew as the morning progressed and eventually included Zinn, Anderson, Szilard, Wigner, and several dozen other physicists who played a role in the pile's construction. At 9:45 a.m., Fermi instructed three of the safety rods to be withdrawn. Immediately, the counters started clicking in response to neutron production, and Fermi watched as the production leveled off. Shortly after 10:00 a.m., having satisfied himself that his predictions to this point were correct, Fermi called out "Zip!" Zinn, who was responsible for the zip rod, now withdrew it completely and set it above the pile, hanging on its rope. Again the clicking of the counters began to race. Again the clicking leveled off.

Fermi instructed George Weil, who was manning the last control rod in the pile, to pull it "to thirteen feet," halfway out of the pile. The rod had been marked carefully to allow its operator to know exactly how much of it remained inside the reactor. The counters rose dramatically in activity. Fermi not only knew that the pile was subcritical but also was able to point to the spot on the graph where the pen would begin to level off. Level off it did. After a few minutes of calculation, Fermi instructed Weil to withdraw the rod another foot. The counters picked up, but then leveled off again. Fermi fiddled with his slide rule, doing some quick calculations, and according to Wattenberg, "seemed pleased" that the neutron production was developing in the way Fermi predicted it would. Weil and Fermi repeated this process, six inches at a time. "Every time the intensity leveled off, it was at the values [Fermi] had anticipated for that position of the control rod," Wattenberg later recalled. At 11:25 a.m., the intensity of the neutron production increased to the point at which an adjustment of the instrumentation scale was required, an adjustment Fermi oversaw with Wilson. As a test, Fermi asked for the safety rods to be reinserted in the pile, and the intensity dropped dramatically. He then asked Zinn to remove all the safety rods, and the reactor started up again, the counters ticking wildly for a moment before rather suddenly, at 11:35 a.m., a loud crash startled those watching. The instrumentation had recorded a level of intensity that tripped the mechanism holding a safety rod in place; the rod had come crashing down into the pile, bringing the reactivity to a complete halt. It was, however, a level of intensity that was still below criticality.

When Fermi understood the cause of the crash, he smiled with relief and announced to the group, "I'm hungry, it's time for lunch." All the control rods were reinserted in the pile, locked in, and the group braved the freezing cold to walk to the main university dining room at Hutchinson Commons. In the splendor of a glorious Gothic replica of the Great Hall at Oxford's Christ Church College, they had a quiet lunch and spoke of anything except what they had just witnessed.

They returned to the squash court at about two o'clock, and Fermi asked the team to return the safety rods up to their positions prior to lunch. Over the next hour or so, Weil withdrew the control rod gradually, according to Fermi's instructions. Each time, the instruments would chatter away and then level off. At about 3:25 p.m. Fermi ordered a full foot of additional withdrawal. As Weil followed Fermi's instructions, Fermi turned to Compton. "This is going to do it," he assured Compton, who joined the group after lunch, with a wide-eyed Crawford Greenewalt in tow. For Greenewalt, this was a moment he would remember for the rest of his life. "Now it will become self-sustaining," Fermi explained. "The trace will climb and continue to climb," he said, referring to the line being drawn across the graph drum attached to the counters. "It will not level off."

He was right. The counters picked up speed and this time did not level off. The clicking became a high-pitched whine. The line traced on the graph paper moved ever upward. Fermi took some measurements, fiddled with his slide rule again.

"I couldn't see the instruments," Weil later said. "I had to watch Fermi every second, waiting for orders. His face was motionless. His eyes darted from one dial to another. His expression was so calm it was hard. But suddenly, his whole face broke into a broad smile."

"The reaction is self-sustaining," Fermi announced. "The curve is exponential." Still he did not order the reactor shut down. Not yet. He continued to study the graph and the instruments, monitoring the exponential production of neutrons. He gave no indication of next steps.

FIGURE 17.1. CP-1 goes critical, December 2, 1942. Fermi is on the balcony overlooking the pile; George Weil is below, operating the control rod. *Painting by Gary Sheehan. Courtesy of the Chicago History Museum.*

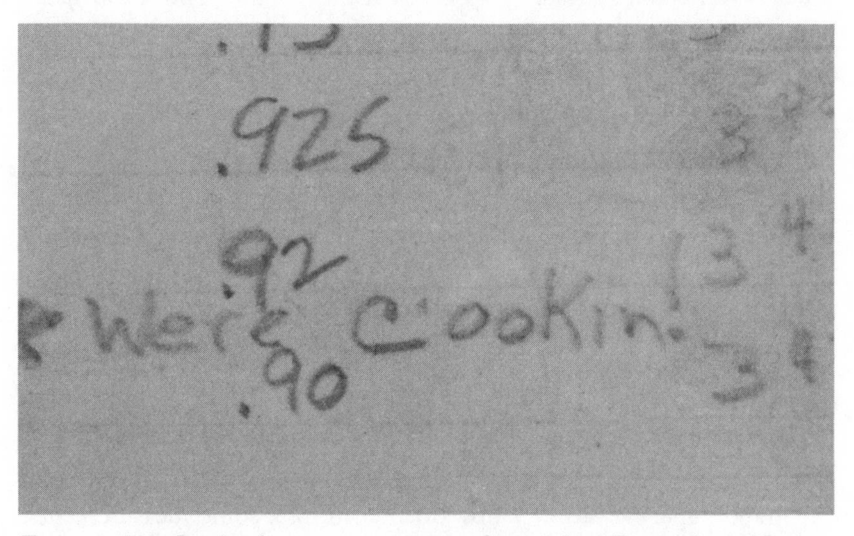

FIGURE 17.2. Logbook note at moment of criticality. Facsimile in Nuclear Science Museum, Argonne Laboratory. *Photo by Susan Schwartz. Courtesy of Argonne National Laboratory.*

Richard Watts, a member of Wilson's instrumentation team, memorialized the moment in the log book: "We're cookin'!"

The several dozen witnesses grew increasingly tense, but Fermi was calm. The team atop the reactor, led by Sam Allison, was alert to any sign of danger, ready at a moment's notice to flood the pile with the cadmium solution. At one point Leona Libby approached Fermi and whispered, "When do we become scared?" Fermi didn't answer. His attention was entirely on the instruments.

Twenty-eight minutes into criticality, he decided he had witnessed enough. "Zip!" he called out to Zinn, and Zinn dutifully released the safety rods into the pile. At 3:53 p.m., the world's first controlled fission chain reaction came to a complete halt.

The room was quiet. Fermi was elated, but said little. He was silent even as Leona Libby accompanied him home at the end of the day. Wigner had brought a bottle of chianti for the occasion. Those in attendance shared the wine out of paper cups, Fermi first. Later most of those present signed the straw cover surrounding the bottle. Perhaps the most famous chianti in history, it now resides in the archives at Argonne Lab. No toasts were made, no dramatic speeches offered. History had been made, but the future looked grim. Everyone understood that this was a major step toward the development of a fission weapon. Szilard

FIGURE 17.3. Fourth anniversary reunion of some CP-1 participants, December 1946, at the University of Chicago. *Back row, from left:* Norman Hilberry, Samuel Allison, Thomas Brill, Robert Nobles, Warren Nyer, and Marvin Wilkening. *Middle row:* Harold Agnew, William Sturm, Harold Lichtenberger, Leona Libby, and Leo Szilard. *Front row:* Enrico Fermi, Walter Zinn, Albert Wattenberg, and Herbert L. Anderson. *Courtesy of Argonne National Laboratory.*

recounts that he shook Fermi's hand in congratulations but warned that this would go down as a black day in human history.

Compton took Greenewalt with him back to his offices. The DuPont executive was sufficiently impressed by Fermi's performance—according to Compton "his eyes were aglow"—that he swept away the remaining obstacles and the company promptly concluded a contract with the government to build all the project's reactors. Compton wanted to let Conant and the rest of the Manhattan Project leadership know of the pile's success, but the two had not agreed on a secure means of communicating the news by phone. Compton rang up Conant and, in a burst of uncharacteristic lyricism, reported, "The Italian navigator has just landed in the new world. The earth was not as large as he had estimated," Compton continued, "and he arrived at the new world sooner than he had expected." Conant picked up the reference immediately. "Is that so? And were the natives friendly?" he asked. "Very friendly," replied Compton. The message had been passed.

LAURA FERMI HAD BEEN PLANNING A COCKTAIL PARTY FOR MEMBERS of the Met Lab that evening at her home. She had no idea what her husband was working on, and when he left for work that morning he gave no indication that anything momentous would be occurring. After he returned for dinner, she asked him to buy some cigarettes for the guests. Enrico—who hated cigarettes—refused, claiming rather absurdly that he did not know how to buy cigarettes and that they left a foul stench after a party that took days to air out.

As soon as the party began, however, Laura noticed something unusual. The Zinns were the first to arrive, and Walter made a point of shaking Enrico's hand, offering congratulations. His was only the first of many congratulations offered as party guests arrived. Mystified, she knew better than to ask Enrico, who would have put her off with a silent shrug. Instead, she asked Leona Libby, the only woman on the team. Libby, bound by official secrecy, was embarrassed, because she felt close to Laura and did not want to dissemble. Without thinking it through, she blurted out that Laura's husband helped sink a Japanese admiral—meaning, she later wrote, that it was *as if* Enrico had done just that. Laura was naturally quite astonished. The Met Lab was a long way from the war in the Pacific. "Are you making fun of me?" she asked, with justifiable irritation. Anderson, sensing perhaps that Leona was digging herself into a hole, joined the conversation. "Do you think that anything is impossible for Enrico?" Leona later wrote that this was perhaps the most embarrassing moment of her life, having been forced to lie to a lovely, intelligent woman, effectively a second mother to her, in order to maintain confidentiality.

After the party, Laura grilled Enrico about the Japanese admiral, and her husband, characteristically, was playfully evasive. Only after the war did she learn of the significance of December 2, 1942, and the part her husband played.

THERE IS SOMETHING A BIT CONTRIVED ABOUT THE EVENTS surrounding December 2, 1942. Many have waxed lyrical about the importance of the moment. It was the first time humans tricked nature into releasing, in a sustained, controlled way, the energy embedded in the nucleus of the atom. Using knowledge painstakingly derived from experiment after exhaustive experiment to achieve this goal, Fermi showed that a controlled chain reaction was possible and helped clear the path to a fission weapon. He also led the way to the exploitation of uranium for peaceful purposes. Compton summarized the impressions of many when he wrote in 1956: "The first self-sustained atomic nuclear chain reaction, achieved on December 2, 1942, did indeed usher in a new age. Henceforth, the vast reserves of energy held in the nucleus of the atom were at the disposal of man." On the twentieth anniversary of the event, Wigner, always more circumspect, would write:

> Do we then exaggerate the importance of Fermi's famous experiment? I may have thought so at sometime in the past but do not believe it now. The experiment was the culmination of the efforts to prove the chain reaction. The elimination of the last doubts in the information on which our further work had to depend had a decisive influence on our effectiveness in tackling the second problem of the Chicago project: the design and realization of a large scale reactor to produce the nuclear explosive, plutonium. . . . Even though our hearts were by no means light when we sipped the wine around Fermi's pile, our fears were undefined, like the vague apprehensions of a man who has done something bigger than he ever expected to. Our forebodings did

> not concern concrete events. In fact, our hopes, some of them very far-reaching, preponderated.

One is left, however, with the impression that much of Fermi's performance that day was a show put on by a master to impress and inspire an eager, receptive audience. Consider that he had accumulated enough data, through operating the numerous noncritical piles constructed at Columbia and during the summer months in Chicago, to enable him to predict how neutrons would be produced in a pile with an ellipsoid structure. He had systematically taken data on virtually every step of the final pile's construction, allowing him to predict the moment the pile would become critical "almost to the exact brick," in the words of official historians Corbin Allardice and Edward R. Trapnell. At each step of the way on the fateful day he was able to predict what would happen at the next withdrawal of the control rod and then, at a moment of high drama, the unexpected release of the zip rods at 11:35 a.m. that morning, he suggested that the team break for lunch, just as he had in October 1934. When the pile went critical at the moment he predicted it would, he decided to let it run for almost half an hour without a word to those around him, all of whom anxiously awaited his order to shut it down. One is struck by the impression that he could simply have asked Weil to move the control rod to the point at which he predicted criticality, measured it, and called it a day. He could have chosen to go critical with Anderson, Zinn, and Allison in attendance the previous evening, with no audience at all, because the pile was ready to go the night of December 1, 1942. Yet he did none of these things. He may have had some underlying concerns over the safety of the experiment and proceeded slowly to ensure that it would not run out of control. But Fermi also clearly understood the drama of the occasion and rose to it. In this instance he chose to play the role not of the disinterested experimental genius but of the showman his peers wanted him to play. He played with his slide rule, although one wonders if he even really needed to do that. He made sure that he looked the part of the physics genius that day. This is not to disparage him in the slightest. He worked for almost four years for this moment, in what he and others considered a desperate race against time. He was absolutely confident in his ability to make this happen, a confidence he shared with others, notably with Teller early in 1942 when he ventured his idea for a fusion weapon. For Fermi, the results must have been a foregone conclusion, but he had the insight to know that the moment required a great show, and he was happy to oblige.

It was, nevertheless, the culmination of years of hard work and enormous dedication by a large team of talented individuals, a team that found exceptional, inspirational leadership in a brilliant émigré physicist, one who had, in his youth, predicted the possibility of releasing enormous amounts of energy from within the atom. It is instructive to consider that the German effort was at this point already years behind Fermi's effort. The Germans never constructed a real, working reactor. Part of this difference results from the early decision, by Fermi and Szilard, to use graphite rather than heavy water and to press for greater purification of the graphite when impurities impeded the efficiency of the chain reaction. Graphite was plentiful, cheap, and easy to mill with woodworking tools. When the Germans tried working with graphite, the material they used was far too impure to do the job, but the physicist responsible for much of this work, Walter Bothe, did not believe that the impurities could be removed. Bothe and Heisenberg decided to use heavy water, extremely difficult to make in bulk. This effectively doomed the German project.

Part of the difference, surely, was Szilard's uncanny ability to squeeze the best-quality uranium and graphite from commercial suppliers, helped along the way by Walter Zinn and Sam Allison. Part of it was Fermi's extraordinary, intuitive understanding of how neutrons would be produced in various configurations of uranium, an understanding that had as its foundation the two years of solid, lonely work he did in Rome with Amaldi after the discovery of slow neutrons in October 1934, on which they reported in a long and exhaustive 1936 paper. In the words of a Chicago colleague, Fermi learned how to "think like a neutron." No German had done the exhaustive, painstaking work required to develop this intuition. German physicists spent much of the 1930s either fleeing from Hitler or pondering the meaning of quantum theory, but they simply had not completed the grinding experimental work on neutron physics that Fermi had.

He was also, in his own unassuming, casual way, a brilliant and inspirational leader. He led from the front, easily accessible to senior and junior scientists alike, able to cut through thorny puzzles quickly and decisively, confident in his ability to understand any problem thrown at him, simplifying experimental setups wherever possible. He was utterly unafraid of the "quick and dirty" solution that worked. His confidence was contagious and his example led them to achieve well beyond what they thought they were capable of. He succeeded in the extraordinary challenge of scaling up the kind of team he developed in Rome to a much larger American organization under much greater pressure. It worked and produced an historic success. A direct line can be drawn from those early neutron experiments in Rome to the experiments by Fermi and Szilard at Columbia in early 1939, through the endless subsequent experimentation with different piles, to the events at Stagg Field that cold winter day in December 1942. The Stagg Field experiment capped off a major phase of research, but it was also the beginning of another chapter in the story of the Manhattan Project, a chapter in which Fermi was to continue to play a pivotal role. Plutonium production reactors now needed to be built and the "Italian navigator" would also play a leading role in this new phase of the work.

CHAPTER EIGHTEEN

XENON-135

FROM JANUARY 1943 THROUGH THE SUMMER OF 1944, FERMI divided his time between three places: the newly relocated Met Lab in Argonne Forest, just west of Chicago; Oak Ridge, where the vast isotope separation plants were taking shape and where Fermi and a team from DuPont built the first plutonium research reactor; and the eastern desert of Washington State near the village of Richland, where the large-scale plutonium production reactors began their deadly work. Though he made the occasional trip to other locations—to Berkeley, and to the new village of Los Alamos, New Mexico, soon to become the epicenter of bomb design and construction—the majority of his time was spent with Met Lab colleagues who had worked with him in the squash court under the stands of Stagg Field.

IN FEBRUARY 1943, AFTER THE LABOR PROBLEMS THAT BEDEVILED the construction of the lab facilities were resolved, the entire Met Lab moved to Argonne. In the short period between December 2, 1942, and the February 1943 move, Fermi and the team continued to experiment with the pile, at one point driving the pile up to two hundred watts.

In February, the pile, originally dubbed CP-1 (Chicago Pile 1), was reconstructed at Argonne, in a new and more flexible configuration that became known as CP-2. Fermi reconfigured it so it had a central core that could be removed to test out different lattice structures and different quality materials. When it went critical in May 1943, its reproduction factor was much higher than CP-1. It took a little over one minute for CP-1 to double in power. CP-2 took about five seconds to do the same.

In CP-2, not only had Fermi created a prototype for the eventual production of plutonium, he also created a veritable neutron factory. He would never again have to fuss with delicate neutron sources in glass bulbs. All the neutrons he needed for research existed in the heart of CP-2. Beginning almost immediately, with the enthusiasm of a child with a new toy, he ran systematic experiments to test neutron reactivity on a variety of targets. He found a way to create beams of ultraslow neutrons and studied neutron diffraction and refraction in great detail. He also developed ways to test the purity of graphite and uranium now being produced in unprecedented quantities for the huge plutonium production reactors.

Working closely with Anderson, Zinn, Marshall, Libby, and the rest of the team, Fermi developed new detectors that measured the precise number of neutrons produced in the reactor. These detectors enabled the team to carry out new studies with unprecedented precision.

Work at Argonne required new routines for everyone. Fermi could no longer walk the half mile or so to his lab. Argonne was some twenty miles southwest of Hyde Park, along fabled US Route 66, and the drive, with Baudino at his side, took the better part of an hour. The hour-long drive drew the two men—opposites in so many ways—close together. Fermi might have initially resisted the idea of a bodyguard as unnecessary, but they soon became friends. Occasionally, Baudino lent a helping hand, stacking graphite bricks or moving blocks of paraffin. Libby, Anderson, and Marshall, who normally slept in a primitive dorm on the lab site, sometimes rode with their boss when they needed to get into town, and they discussed the day's objectives as they drove through the cornfields and prairie land between the city and their new workplace. Fermi got to know the vast fertile plains that make up the American heartland. He was fascinated by the landscape, which was so very different from what he was used to in Italy.

Laura adjusted well to the new environs. Chicago was different from New York, and she liked her new home. She enjoyed socializing with the growing group of physicists and their spouses at the Met Lab—a very social group, even under the constraints of official secrecy. The party she hosted on the evening of December 2, 1942, was but one of many she hosted during this period. The Fermis did not drink very much at all—when they had wine, it was always with a bit of water added—but they served alcohol to their guests. They enjoyed playing party games like charades, which Enrico took very seriously. Old friends like the Ureys and Mayers made Laura feel even more at home. Laura also got along well with Leona Libby, although one senses Laura was a bit jealous Enrico was spending more time with this attractive twenty-three-year-old than he was with his own wife. During a snowstorm, Leona was a passenger in Fermi's car as he drove home from the lab late at night. When Leona suggested they might have to stop the car and sleep there overnight, Fermi objected, saying that it might harm his reputation. The young woman, by now married to John Marshall and pregnant with his child, suggested that perhaps *she* should be worried about *her* reputation and because she wasn't, why should he be? He asserted with characteristic confidence that his reputation was more important and they drove on. Unspoken, perhaps, was his concern over what Laura might have said the next morning.

Leona's pregnancy posed some problems at Argonne. The health and safety officers assigned there would certainly have prohibited her from working at the new pile if they had known. She was, however, able to persuade those around her to keep the secret and wore baggy clothes to hide the pregnancy. Zinn apparently never knew or if he knew he never let on. Fermi became convinced he might have to perform midwife duty at the lab some nine months on, given the twenty-seven miles between Argonne and the Chicago hospital where Leona and John planned to have their baby (why exactly he didn't expect John, who worked there every day, to perform these duties if necessary is not clear). Fermi even researched what he would be required to do. Fortunately for everyone involved, including Laura, it never came to that and the Marshalls had their first child at the hospital.

As she settled into life in Chicago, Laura also devoted time to the children. By this time, Nella was twelve years old and Giulio was almost seven. They were both enrolled in the famous Lab School established in 1896 by American educator John Dewey. The school, founded on progressive principles of child development and education, was somewhat less permissive than Nella's New York school. Giulio was unhappy, but

Nella adjusted well—perhaps because she was older and her temperament was better suited to the more structured program. Also, Giulio was upset that he was taken away from his friends in New Jersey without sufficient explanation. It was a resentment he carried with him throughout his life.

Some of the new members of the Argonne team had heard of Fermi's extraordinary abilities but were now witnessing them for the first time and were suitably impressed. One of the most important of these was an experimental physicist named Luis Alvarez, a student of Compton at Chicago in the early 1930s. Alvarez's work in those days was the stuff of Chicago legend by the time Fermi arrived. After getting his degree, Alvarez left Chicago to join Lawrence's team at Berkeley, doing a series of important experiments at the cyclotron. At the beginning of the war, Alvarez worked in England and at MIT on the development of radar, and in the summer of 1943 he arrived at Argonne, climbing on top of CP-2 and working with Libby to design and build new instruments, something at which he was particularly adept.

Alvarez was never a modest man, so his report of an early encounter with Fermi is particularly instructive. He joined a conversation in the Argonne cafeteria with Fermi, the Marshalls, and Herb Anderson, who were discussing how neutrons might obey a law of refraction similar to those of X-rays. Fermi commented that he could not remember the exact formula for X-ray refraction. Alvarez pointed out that it was contained in the classic textbook on X-ray diffraction written by Compton and Allison. Alvarez had seen a copy of the textbook on a desk next door and offered to get it. Fermi told him not to bother—he would derive it.

Alvarez goes on to describe Fermi's performance:

> As a student of Compton's I had thought long and deeply about X-rays, but I had never seen the refractive-index formula derived from basic principles. Enrico wrote James Clerk Maxwell's classic electromagnetic field equations on the blackboard and then in six separate steps derived the formula. The most remarkable aspect of this tour de force was that Enrico worked through his derivation line by line at a constant rate, as if he were copying it out of a book. That night at home I reproduced it and was quite pleased with myself. If one step was easy

> enough to allow me to go faster than he did, the next was so difficult that I could never have managed it alone. But Enrico worked the difficult steps at exactly the same rate he worked the easy ones.

In time, the two of them became good friends. Alvarez, back at Berkeley after the war, would occasionally call Fermi to ask for recent graduate students for post-doc positions at Berkeley and Fermi would happily oblige.

Increasingly, young American physicists who encountered Fermi for the first time were impressed by his ability to work his way through thorny problems. Many of the brilliant physicists of Alvarez's generation—about ten years younger than Fermi—had never worked closely with people of Fermi's European peer group, the Heisenbergs and Diracs of the physics world. The closest many of them had come to this group was Oppenheimer at Berkeley, who in the late 1920s brought European-style theoretical physics back to the United States. To see one of the quantum pioneers at work must have been quite inspirational. Increasingly, young American physicists were exposed directly to Fermi and shared Alvarez's sense of awe.

That awe was not confined to the younger generation. At Fermi's memorial service in 1954, Sam Allison recalled that on a train ride to Hanford, Compton decided to make small talk with Fermi. He remarked to Fermi that during his time in the Andes pursuing cosmic-ray studies his watch did not keep good time. "I thought about this considerably," Compton explained, "and finally came to an explanation that satisfied me. Let's hear your discourse on the subject." Fermi immediately pulled pencil, paper, and slide rule from his pocket and after a few minutes came up with an answer that explained the phenomenon and even predicted the amount of the watch's inaccuracy. Compton's look of wonder was something that Allison, Compton's long-time collaborator, would never forget.

INCREASINGLY, FERMI'S ATTENTION WAS DRAWN TO PROBLEMS ARISING in the planning and construction of the reactors. The first major plutonium reactor, known as X-10, was rising at Oak Ridge, with three objectives: to determine how fast plutonium could be produced, to develop the chemistry to separate plutonium from the other by-products of controlled uranium fission, and to provide small samples of plutonium to determine its suitability for use in a fission weapon. Fermi's old friend Emilio Segrè, now at Berkeley, but soon to relocate to Los Alamos, would be one of the key scientists researching this latter issue.

The construction was a helter-skelter affair, with engineers working round the clock from the sketchiest of drawings. Based at Argonne, Fermi provided guidance and advice on scientific and technical questions. Eventually, the enormous cube took shape atop a small hill at Oak Ridge. It looked more like CP-2 than CP-1, but there was a difference. It was built so that, after the reactor had been running for an appropriate length of time, the uranium rods could be easily removed and placed in acid baths to initiate the extraction of plutonium. A large cube of graphite, twenty-four feet on each side, it had 1,260 long channels bored into one face. Into these channels rods of uranium metal would be inserted. Along a perpendicular face, channels were drilled to accept cadmium safety rods, which were attached to a mechanized system to pull them

FIGURE 18.1. The author at the control panel of the X-10 reactor at Oak Ridge. *Photo by Susan Schwartz.*

out or reinsert them as conditions required. Designed to run hot, at one thousand kilowatts (one megawatt), to produce enough plutonium to study in a relatively short amount of time, it required a cooling system that neither CP-1 nor CP-2 had. The system chosen was air cooling. Air flow supplied by motorized fans circulated air in and around the channels for the uranium rods. At the intended power level, radiation from the pile became an important safety issue, so engineers encased the pile in concrete shielding seven feet thick, creatively formulated to retain water even after curing so that it would be more effective at slowing down any excess neutrons escaping from the pile.

The Oak Ridge pile went operational on November 4, 1943, less than a year after the first pile in Chicago. It was a tremendous achievement and, in recognition of this, Compton and Fermi traveled to Oak Ridge to witness the event. They were sleeping in when they were awakened before dawn—apparently with some glee—by the crews loading the uranium fuel, who were ahead of schedule. The pile went critical at five o'clock that morning. Soon it was producing plutonium, which was transported by courier—on a commercial flight—to Los Alamos for study by Segrè. Segrè's discoveries regarding these samples, which were often just a few milligrams in size, would alter the entire course of the Manhattan Project.

For the time being, however, the success of the Oak Ridge pile must have given great satisfaction to the Chicago physicists and to DuPont's leadership, especially Crawford Greenewalt, who had come to worship Fermi. Seaborg's chemical separation unit next door would soon give Groves and the civilian leadership confidence to start an all-out effort to build industrial-scale plutonium production reactors at Hanford.

DURING LATE 1942 AND EARLY 1943, WORK UNDER OPPENHEIMER at Berkeley continued to focus on issues relating to the use of U-235 for a fission bomb; all signs pointed to the conclusion that fast neutrons would cause that particular isotope of uranium to split. This was progress for the project, because fast neutrons would be the only way the energy locked in the uranium nucleus could be released in the short time required to create an explosion. Gifted with an ability to read people and situations quite accurately, Oppenheimer charmed the blunt anti-intellectual Army general. He did not spend a lot of time regaling

Groves with meditations on high culture. Instead, he evinced a complete confidence that, with the scientific team at his disposal, he could deliver the goods. Groves agreed and selected Oppenheimer as the scientific director of the effort to design and build the actual weapons once sufficient materials were produced.

The story of how Oppenheimer persuaded Groves to designate a mesa about twenty miles northwest of Santa Fe, New Mexico, the site of a boys' boarding school called Los Alamos, as the site—code-named Site Y—for the technical effort to design and build the bombs is told elsewhere. Fermi attended the first organizational meetings Oppenheimer held there in April 1943, along with a stellar cast of scientists, including Rabi, Bethe, Segrè, and many other old friends. Fermi's presence was eagerly anticipated. A senior group of scientists were lunching at Fuller Lodge, the canteen for the high-level scientists, and Teller assured them that Fermi would be arriving within the week. Stanislaw Ulam, a Polish mathematician who in later years was credited with a major breakthrough in the development of the hydrogen bomb, intoned "*Annuncio vobis gaudium maximum, papam habemus.*" To the bewildered group, John von Neumann, Ulam's brilliant Hungarian colleague who played a central role at Los Alamos over the next few years, translated the Latin used by the church to announce the election of a new pope: "I announce to you with greatest joy that we have a new Pope." The group knew of Fermi's nickname in Rome and burst into applause. They had been looking forward to Fermi's arrival and loved Ulam's allusion.

When the entire group was assembled at Los Alamos, they discussed a "primer" on bomb design prepared by Oppenheimer student and collaborator Rob Serber, the result of the Berkeley theoretical group's months of study and debate. Oppenheimer began to plan a division of labor, which he was able to sustain until late summer 1944, when events forced reorganization. This was Fermi's first exposure to the magnificent wilderness of the New Mexican portion of the Rio Grande valley. Over countless millennia, the river ground out the valleys the scientists now hiked in and explored. The wilderness around Los Alamos was like nothing Fermi had ever experienced, and when Norris Bradbury, the first postwar director of Los Alamos, invited Fermi back for a number of summers after the war, it was not a hard sell for the inveterate outdoorsman.

The April 1943 meetings at Los Alamos were the first time all the senior scientists came together to discuss the design and construction of the bomb. In later years Oppenheimer recalled that Fermi seemed taken aback by the enthusiasm of many in the group. "I believe your people actually *want* to make a bomb," Oppenheimer reported Fermi commenting to him. "I remember his voice sounded surprised," Oppenheimer added. Fermi, like Szilard, understood the gravity of the challenge they were facing and, though he understood it might be necessary, especially because the Germans were known to be pursuing nuclear weapons on their own, he did not share the majority's enthusiasm.

He was, however, capable of coming up with rather brutal and ruthless ideas for winning the war. A letter from Oppenheimer to Fermi dated mid-1943 survives, commenting on an idea being considered by the two physicists for massive radiation poisoning of the German civilian population. How the idea arose is not clear, because Fermi's side of the correspondence has never been found, but it is clear that both Oppenheimer and Fermi were open to considering a range of ideas to end the war quickly. Fermi might not have been enthusiastic about these ideas, but he probably considered it an obligation to think them through anyway.

Fermi returned to Chicago after the April 1943 meetings and continued his CP-2 experiments and his consultations with DuPont over the construction of the plutonium production reactors until August 1944, when he finally moved to Los Alamos. In the interim, Oppenheimer and others at Los Alamos continued to consult with him on a variety of problems, and he traveled to Los Alamos on several occasions, but his home base remained Chicago.

WHAT FERMI MADE OF OPPENHEIMER, DURING THOSE INITIAL April 1943 meetings or later, is hard to divine. He made only the occasional, brief comment to close colleagues. Yet even the casual observer would have noted the differences between the two physicists. Oppenheimer was the product of a highly cultured upbringing on Manhattan's Upper West Side. The son of assimilated and wealthy German-Jewish parents, he attended the exclusive Ethical Culture School and then Harvard, where he distinguished himself in a wide range of disciplines. He then studied at Cambridge University under J. J. Thomson—the man who discovered the electron—and took his doctoral degree with Max Born at Göttingen. Oppenheimer was a man of the world, well read, conversant on art, history, and philosophy. He felt right at home at Göttingen, where a few years previously young Fermi felt an outcast. He was a prodigious theoretician who contributed to the quantum theory of his generation. At Berkeley he developed a reputation as a stern, even cruel, taskmaster, someone who occasionally took delight in ridiculing hapless graduate students or post-docs who were guilty of an error in calculation. One might accept C. P. Snow's judgment that Oppenheimer would have traded his entire career to have made just one discovery of the magnitude of Fermi's three great contributions and yet still respect Oppenheimer's achievements as a physicist of high caliber. Certainly, he had the respect of his colleagues at Berkeley and, increasingly, those of the growing Manhattan Project for which he now found himself in a leadership role.

The contrast with Fermi was striking. Fermi was neither well-rounded nor interested in high culture. He came from a distinctly middle-class background and never felt the need to rise above that. He could be socially charming, but in a more informal way. Everyone who knew him commented on the surprise they first felt when they met the legendary genius and discovered he was extraordinarily approachable, even diffident.

Fermi also found Oppenheimer's style of physics more than a bit foreign. In 1940, he had traveled to Berkeley to give the Hitchcock Lectures, and in his spare time sat in on a lecture by an Oppenheimer acolyte. Later he confided to Segrè:

> Emilio, I am getting rusty and old. I cannot follow the highbrow theory developed by Oppenheimer's pupils anymore. I went to their seminar and was depressed by my inability to understand them. Only the last sentence cheered me up; it was, "and this is Fermi's theory of beta decay."

By 1943 Fermi was a legend, even before his colleagues began to appreciate the beta decay paper of 1934. Many things contributed to the legend: his landmark discoveries, his single-minded commitment to physics at the highest level, his astonishing ability to work his way through complex problems, his ability to simplify those problems in ways that facilitated understanding by those less gifted than he. He could be blunt, even dismissive if he believed someone was wrong, and his fondness for teasing those closest to him could be irritating, but he was never deliberately cruel. Physicists presenting new research knew they were in trouble if Fermi interrupted and said, "Excuse me, but there is something here I do not understand." That was enough to convey his belief that the speaker was wrong. He had already won a Nobel Prize, something about which Oppenheimer may have had a bit of envy. He had also been in on the very start of the project that was now under Oppenheimer's guidance.

The April 1943 meetings were not the first time the two physicists met. They encountered each other during Fermi's occasional visits to Berkeley in the 1930s, particularly in 1937 when he spent time learning about Lawrence's cyclotron and in 1940 when Fermi delivered the Hitchcock Lectures there. Now, though, they would come into increasingly regular contact. In spite of their differences, they developed a deep mutual respect. Oppenheimer respected Fermi's way of thinking, his way of solving problems, his confidence and the way that confidence inspired others, and of course his enormous contributions to physics prior to the war. Fermi respected Oppenheimer less as a physicist than as the person in whom the US government, to which he now owed total allegiance, had vested authority over the entire Manhattan Project. If he ever disagreed with Oppenheimer, it would never be in public.

In later years, Oppenheimer was subtly dismissive about Fermi: "Not a philosopher. Passion for clarity. He was simply unable to let things be foggy. Since they always are, this kept him pretty active." And yet he seems to have been profoundly influenced by, and perhaps even obsessed with, his wartime colleague's intellect and the scale of his contributions to physics. Leona Libby recalls a dinner party in Pasadena after Fermi's death at which Oppenheimer suggested a parlor game he called "Who do you want to be on your day off?" One would venture a name and the others would try to analyze what the choice signified, in a sort of amateur psychoanalytic fashion. Astonishingly, Oppenheimer chose Fermi. There was silence in the room. No one knew what to ask him. It is significant that Oppenheimer himself chose the game and then chose Fermi, reflecting an inscrutable, deep-seated need to identify with his Manhattan Project colleague.

Once Oak Ridge was up and running, attention turned to the construction of the first major plutonium production reactor at Hanford. Construction at the vast high-security site began in March 1943.

The project was vast, involving the construction of hundreds of buildings, the recruitment of many thousands of laborers, the creation of a virtual state-within-a-state—many times larger in area than Oak Ridge. The centerpiece was three plutonium production reactors, built alongside the river, and a huge chemical separation plant at some distance from the reactors, where spent fuel rods were transported by truck for disintegration and chemical treatment for plutonium extraction. The first reactor to go live was the "B" reactor, followed by the "D" and "F" reactors.

The work was a triumph of organizational collaboration between Groves's Army Corps of Engineers and Greenewalt's DuPont engineers. The decision to go ahead and build these behemoths was made after the success of CP-1, but before CP-2 had been completed and well before the Oak Ridge facility went critical in November 1943. Groves was rolling some very heavy dice here, backed by the complete confidence he earned from the civilian leadership of the project and his own faith in his "crackpots" at the Met Lab.

From Chicago, Anderson, the Marshalls, and Wheeler would eventually relocate to Hanford to contribute to the design and construction of the reactors.

Construction on the B reactor began in October 1943, based on a scaled-up version of the Oak Ridge X-10 reactor. The core was significantly larger than the X-10 core, consisting of a graphite block twenty-eight feet wide, thirty-six feet deep, and thirty-six feet high. The face into which fuel rods were fed had channels for 2,004 fuel rods, compared to the X-10's 1,260 channels. Cadmium and boron safety rods could be controlled automatically in channels running perpendicular to the fuel rods. Fermi and Wigner calculated that in order to produce the volume of plutonium required at the pace demanded by Groves and the team's leadership, the reactor would have to run very hot indeed—at some 250 megawatts of power, more than some million times the top power achieved by CP-1. This required special treatment of the graphite to cure it against expansion from the ambient heat of the reactions, as well as a more aggressive cooling system to prevent the aluminum cladding of the uranium fuel rods from melting.

The Columbia River provided plentiful fresh water to be pumped through the core's channels, in the space between each rod and the wall of its channel, and then returned to the river when deemed safe. The creation of 2,004 channels for fuel was an example of overengineering. Neither Fermi nor Wigner thought that so many channels would be necessary, but Greenewalt's senior engineer, George Graves, in consultation with Wheeler, decided to pack more capacity into the core in case they needed more power. This, it turned out, was perhaps the most fortunate decision made by the DuPont team.

On the occasion of Fermi's first visit to Hanford, with Wigner in tow, the two were stopped at the gate to the high-security B reactor site. Wigner momentarily forgot that his code name was Gene Wagner and when the guard asked him to identify himself, Wigner said "Wigner—oh, excuse me please, Wagner!" The guard was immediately suspicious. He began to grill the timid, quiet man with the strong Hungarian accent. "Is your name really Wagner?" Fermi immediately intervened. "If his name is not Wagner, my name is not Farmer." Wigner continues: "And the guard let us pass. That quick self-assurance was so typical of Fermi."

Fermi arrived at Hanford a week before the B reactor was scheduled to go critical, in mid-September 1944. Greenewalt and Leona Libby escorted him around the vast facility, including the assembly line for the fuel rods that encased the uranium in aluminum/boron sleeves, where technical problems had delayed the start-up of the reactor. Greenewalt never quite got over the excitement of watching Enrico Fermi bring CP-1 to criticality, and Fermi respected the man whose team of engineers could make a large-scale plutonium production plant—reactor, separation facilities, and all the facilities needed to support these—rise from the desert floor in just over a year. They made an incongruous pair, yet the two of them worked well together—so well, in fact, that their partnership lasted long after the war was over.

The problems related to the canning of the uranium rods were finally solved, and by September 26, 1943, the fuel rods were being loaded into the face of the hulking core. Fermi helped in the operation, perched on a scaffold high above the floor, gently easing a rod or two into its channel. The reactor went critical that afternoon and everyone went home to relax as the pile started to generate plutonium. According to Wheeler, the plan was to power the pile up to nine megawatts and then maintain that power level for a while. Then they would run it hotter and hotter until it reached the 250-megawatt level for which it had been designed. Sometime the next morning, however, the reactor began inexplicably to lose power. The operators were puzzled and decided to pull the safety rods further out of the pile to keep the power at a steady nine megawatts, but the power continued to drop. By midafternoon the control rods were almost all the way out of the pile and still the operators were having trouble maintaining nine megawatts. At about four o'clock, Fermi suggested bringing the power down to 400 kilowatts to

FIGURE 18.2. The author at the control panel of B reactor at Hanford. *Photo by Susan Schwartz.*

see if they could hold the reactor there. They couldn't. By the end of the day, the B reactor—over a thousand tons of graphite and uranium rising thirty-six feet off the ground—was, for all intents and purposes, dead.

Panic now gripped everyone in the reactor building. Over $7 million ($95 million in today's dollars) had been spent on the reactor alone, not including the staggering cost of the fuel rods. The US government was in the midst of spending what would be $350 million in total, including two other production reactors, separation facilities for the fuel rods, and all the associated facilities for a small city of forty-five thousand people—nearly $5 billion in today's terms, an unprecedented expenditure by the US government at that time. If this reactor failed to do its job, though, it would not simply be money and time wasted, it would be a crucial logjam in the timeline for production of one of the main materials for the bomb itself. There was enormous pressure to figure out what was going wrong and to do it fast.

Wheeler had spent time over the past year worrying about reactor "poisoning." By his calculations, the fission by-products of the controlled chain reaction, particularly at high power, might prove troublesome if allowed to build up. Those by-products could absorb neutrons out of the chain reaction, slowing or even stopping the reaction from working—"poisoning" the reaction. This is why he and DuPont engineer George Graves had decided to "overengineer" the reactor by installing one-third more fuel rod channels than Wigner thought were required for the reactor to work at 250 megawatts.

In any case, it was not immediately clear just what was causing the loss of power. Fermi and Leona Libby suspected that water had leaked into the fuel rods from the cooling system, but upon inspection the water cooling system was intact. There was also concern about a possible leak in pressurized helium being pumped into the graphite core to replace air that might reduce the reproduction factor, but they could find no such leak.

The reactor itself gave the team an important hint at the cause of the problem. Spontaneously, it started up again. Throughout Wednesday afternoon the power returned to the reactor and by Thursday afternoon, September 28, 1943, the reactor reached nine megawatts again, only to falter once more a few hours later.

Wheeler suspected that some radioactive element had been produced as a by-product of the chain reaction, with an enormous ability to absorb neutrons and a fairly brief half-life, on the order of around eleven hours, after which the chain reaction could revive again at full force. He also suspected that because the drop-off occurred only after the reactor was able to reach the fairly high running power of nine megawatts, the real poison was a product of another radioactive by-product that itself did not absorb neutrons but decayed into something that did. Otherwise, the reactor would not have been able to reach nine megawatts in the first place. Wheeler checked a wall chart that listed isotopes that might be created in fission reactions, along with half-lives, looking for a likely suspect. The only one that really seemed to fit the profile was xenon-135, produced in the decay of iodine-135, a known first-generation by-product of uranium fission. Xenon-135 has a half-life of just over nine hours.

He made some rough calculations of the ability of xenon-135 to absorb neutrons. He found that the culprit was produced in about 6 percent of all fission events. He also discovered, to his and everyone else's astonishment, that its ability to absorb neutrons in its general vicinity was vastly higher than any element previously studied. When Fermi and his Columbia team studied a list of which elements were particularly good neutron absorbers, they had discovered that cadmium was one of the most potent. They had not tested xenon-135, because the isotope was extremely rare and quite unstable. But Wheeler's calculation suggested that this form of xenon was *one hundred thousand times* more potent than cadmium.

Fortunately, with xenon's half-life of only nine hours, the solution was clear. If the reactor was fully loaded with all 2,004 of the uranium fuel channels filled, the reactivity would swamp the effect of the xenon, and the reactor would be able to operate smoothly at its rated power level.

Wheeler and Graves deserve enormous credit for deciding to overengineer reactor B in the event that fission by-product poisoning required an increase in the power of the reactor. In so doing, they allowed the Manhattan Project to keep to its tight deadlines and salvaged the multi-billion-dollar engineering project on which so much depended. When Groves was apprised of the problem, he was furious and lashed out at Compton, whose somewhat feeble response was that the problem would be studied in greater depth at Argonne, where a new pile, CP-3, had been built to supplement CP-2. Groves did not reproach Fermi, though well he might have.

In Fermi's brilliant career, he demonstrated his fallibility on only a few occasions, none more dramatically perhaps than this one. His 1934 failure to realize that he had split the uranium atom led later to some embarrassment, but we can take some comfort that his failure deprived the fascist government of a four-year head start on nuclear weapons. The reactor B mistake was far more serious and might have led to the loss of the plutonium side of the project altogether. He never commented on his failure to anticipate xenon poisoning, but he must have been enormously embarrassed. Anticipating it would have saved much time, because the adjustments required to swamp the xenon poisoning delayed the running of the reactor at full power for some five months. The B reactor only achieved full power in February 1945, at which time two other reactors, D and F, were nearing completion. Fermi could have forecast this problem, and also could have Wheeler, who had been concerned about poisoning for some time prior to the completion of the reaction and had also not anticipated the xenon problem. At that moment, though, Fermi had other pressing things on his mind. During the summer of 1944, the work at Los Alamos was running into potentially catastrophic roadblocks. Oppenheimer decided that a total reorganization was required and Fermi wound up with another new role, one that would bring him into residence on the secret mesa northwest of Santa Fe.

CHAPTER NINETEEN

ON A MESA

LAURA AND THE CHILDREN WERE THE FIRST FERMIS TO ARRIVE ON the dusty, high-security mesa that was quickly becoming the focus of the Manhattan Project. Her husband was still shuttling between Chicago, Oak Ridge, and Hanford and would not arrive until early September 1944. In midsummer 1944, she took the train, as instructed, from Chicago to Lamy, New Mexico, a town fifteen miles south of Santa Fe. Like all those destined for Los Alamos, she was ignorant of her final destination. Arriving at Lamy, she almost missed her ride into town. A young Army officer was eagerly looking for "Mrs. Farmer." Laura was not aware of her husband's code name and at first did not respond. After a few moments of thought, though, she asked if he was looking for "Mrs. Fermi." The young man looked her over, figured she was indeed the person he was looking for, and brought Laura and the children to 109 East Palace Road in Santa Fe, where the cheerful Dorothy McKibbin dutifully checked her in and gave her and the children ID cards.

From Santa Fe, she was driven along the twenty miles of dirt road, narrow with frequent switchbacks, to the top of the mesa. She settled herself and the children into a modest apartment, on the second floor of a barracks built to Army specifications for standard issue housing. The Fermis could have insisted on more spacious housing along "Bathtub Row," a street with private houses that was built for project VIPs, where Oppenheimer and his wife Kitty lived, next to Berkeley physicist Edwin

McMillan and his wife, Elsie. Those houses were equipped with bathtubs rather than the showers that prevailed in more standard accommodations, hence the name of the street. In typically modest fashion, the Fermis decided not to pull rank and lived where the Army assigned them. Their downstairs neighbors were German theorist Rudolf Peierls and his wife Eugenia. Their old friends the Segrès were also nearby, having moved from Berkeley not long before.

This was the third major upheaval in five years for the Fermi family: first to New York, then to Chicago, and now to Los Alamos. Of the three, the move to Los Alamos was the most dramatic and the most disorienting for the upper-class woman from Rome and her children. They were enclosed in a compound where the highest security procedures prevailed, where the simple act of going into town (Santa Fe, in this case) to buy provisions was difficult and sometimes impossible, and where they knew almost nothing of what was going on around them. The children were not allowed to wander off-site and attended a small school on the grounds. There, Nella and Giulio joined other children of scientists and engineers in elementary school studies. Nella recalls these days as a great adventure; she and the Peierls' daughter Gaby would sneak out of the compound and then check back in at the main gate, causing considerable consternation. For Nella, it was all very exciting to be set down in the middle of the New Mexican wilderness, with all sorts of important things going on, none of which she understood.

Certain aspects made life at Los Alamos bearable for Laura and the children. Familiar furnishings from the Leonia house, left behind on the assumption that the Fermis' move to Chicago would be temporary, now arrived to fill the apartment and make it feel more like home. Many of the families on the mesa were of European origin and were old friends of the Fermis. It must have been pleasing to see the Bethes, and meeting up with the Segrès certainly reminded her of happier days in Rome. There were many others as well. Thrust together in the most extraordinary circumstances, they bonded and supported each other. Laura eventually found work helping the doctor in the Tech Area, the most sensitive section of the facility, where work on the bomb itself was being conducted. She was one of the first people to learn firsthand about the dangers of radiation poisoning. She socialized actively with other wives.

Being one of the older women in the group, she was a bit of a mother hen to younger wives whose husbands—straight out of undergraduate school, in many cases—were drafted into the project.

The call for the Fermis to relocate to Los Alamos was perhaps inevitable, but it came specifically in response to a series of major crises in the project. In response to these crises, Groves and Oppenheimer decided to reorganize the Los Alamos project, and Fermi played a key role in that reorganization.

THE MAIN PROBLEM WAS A MATTER OF PHYSICS. THE WORK AT Los Alamos had always assumed that the "gun" method of assembly—shooting one subcritical chunk of metal, either uranium or plutonium, into another subcritical chunk at high velocity so that they would together form a critical mass—would be the most reliable way of creating a fission explosion. All the knowledge developed about U-235 suggested that the gun method would work for the uranium isotope. The initial studies of plutonium, conducted by Seaborg, Segrè, and others at Berkeley with material created in Lawrence's cyclotron, suggested the same thing. Early on, physicists knew that "implosion"—that is, compressing a subcritical mass of either uranium or plutonium—would also achieve criticality and a fission explosion, but the challenge of actually executing an implosion was daunting.

However, as early as November 1942, doubts began to arise regarding the suitability of plutonium for use in a fission weapon. Those doubts only grew over time. Throughout the early part of 1944, Segrè had been working in a small shack located in a remote corner of the mesa, focused on analyzing the properties of the plutonium produced at Oak Ridge and Hanford. His research, which threatened to bring the entire plutonium project to a grinding halt, suggested that the plutonium produced in the heart of nuclear reactors was quite different from the plutonium produced in the cyclotron. Under the intense neutron bombardment over extended periods of time in the Oak Ridge reactor, some of the atoms of Pu-239 absorbed an additional neutron, creating the isotope Pu-240. That extra neutron threw the nucleus into turmoil, creating spontaneous fission. It was Fermi who, looking at Segrè's data, suggested that Pu-240 was the culprit, prefissioning Segrè's microsamples. At the power levels in the reactors at Oak Ridge and Hanford, Pu-240 could compose as much as 7 percent of the plutonium being produced. Spontaneous fission was going to be a major problem in using plutonium. As the news of Segrè's findings was absorbed, it became clear that the only method of building a plutonium bomb would be the implosion method. A sphere of plutonium, even with 7 percent Pu-240, could be made sufficiently subcritical that spontaneous fission would not be a problem. If the subcritical sphere were then explosively compressed into a sufficiently dense sphere, it would become critical and a plutonium bomb would work.

That was, however, a pretty big "if." In order to achieve an effective critical mass, the sphere would have to retain a perfectly spherical shape from the initiation of the implosion through to the moment of criticality. Otherwise, the fission reaction would pass through the sphere unevenly, resulting in a misfire. To maintain the plutonium in a perfect sphere throughout the implosion, the shock waves of the explosion would have to arrive across the entire surface of the subcritical sphere at exactly the same time. There was very little room for error.

To do this involved an unprecedented technical challenge. Imagine standing in a swimming pool, with one hand on a floating beach ball so that the ball is halfway submerged, the other holding a penny. When the water in the pool settles down and is perfectly calm, the penny is dropped. The penny creates a circular wave that travels on the surface of the water and strikes the beach ball. The front of the wave strikes the beach ball first, and then the rest of the wave breaks across the surface of the ball. A convex surface is striking a concave surface (or vice versa, depending on the point of view). The first moment they touch is at a single point on the surface of the ball closest to where the penny dropped. Blast waves behave like the waves created by the penny, emerging outward in circular fashion from the point of detonation. They begin to compress the subcritical sphere at the single point where the blast wave first touches the surface of the sphere, thereby deforming it almost immediately. Adding to this complexity is that the wave comes at the target sphere in three dimensions, unlike the wave in the pool. It is extremely difficult, but possible, to create simultaneous detonations all around the sphere at various equidistant points. However, the only way to make sure that the sphere retains its shape throughout the implosion is to shape the blast wave that emerges from each detonation so that it has the same shape as the sphere when it arrives there, microseconds after detonation. Scientists would need to *reverse* the shape of the wave and they would have to do it very quickly after the charge detonated. No one had ever done anything quite like this before. To do it at all, and with the requisite accuracy, would require an enormous amount of scientific and engineering brilliance.

The Los Alamos reorganization involved the redeployment of personnel from the plutonium gun project to the implosion project. Research into implosion, which was under the auspices of a low-priority group headed by Navy officer William S. "Deak" Parsons, was transferred to a high-priority group headed by a flamboyant, Russian-born physical chemist named George Kistiakowsky, on secondment from Harvard. As part of the shake-up Oppenheimer named Fermi associate director of Los Alamos with overall responsibility for research and theory and for all special problems related to nuclear physics. This was an honorific title that gave him little administrative authority but that allowed him to poke his nose into issues as needed and as might interest him. He was also given direct responsibility over a special division—F Division, *F* for Fermi—under which a variety of projects not subsumed in other divisions were grouped. These included theoretical and experimental work on the "water boiler" project and on the "Super" project.

The water boiler, a project particularly close to Fermi's heart, was a high-intensity reactor that used powdered uranium enriched to 14 percent U-235 and mixed into ordinary water. Its location at a remote site code-named Omega in a canyon off one side of the mesa was an ideal place for Fermi to keep experimentally active, particularly when so much of his time was spent helping with other scientists' projects. The water boiler ran at low power, but even so was sufficiently reactive, owing to the enriched uranium, that the water's tendency to absorb neutrons could be ignored. The liquid was contained in a sphere one foot in diameter, with instrumentation surrounding it to measure neutron production and absorption, as well as control and safety features. Later configurations of the water boiler ran with increasingly enriched uranium. The water boiler was useful in the study of the critical mass of uranium. It also produced refined studies of neutron production, one of which, conducted by Fermi's old friend Bruno Rossi, determined how quickly "prompt" neutrons emerge from fission reactions. Fermi's Omega site team included L. D. P. King, Herb Anderson, and a young woman named Joan Hinton. King ran the project for Fermi. A Purdue-trained physicist, he worked closely with Fermi during these next few years. A graduate of Bennington and the University of Wisconsin, Hinton also became a daily colleague of Fermi at the Omega site. Segrè describes Joan as "very athletic," perhaps because she could clamber down into the ravine that had been chosen to locate the water boiler. She was quite sympathetic to left-wing causes, and in 1948 as the revolution in China came to a close, she moved there and lived out the rest of her life under communist rule. For the time being, though, she was eager to serve as Fermi's assistant.

The Super was the fusion bomb (hydrogen bomb) project that had preoccupied Edward Teller practically every moment since early 1942 when Fermi first suggested the possibility. To his enormous frustration, Oppenheimer could not get Teller to work on anything else, either at Berkeley or at Los Alamos. Oppenheimer judged the likelihood of a real breakthrough on the Super too low to devote significant resources to it, but he wanted to find a way to keep the creative Teller happy and gainfully occupied, so he worked directly with Teller on the project. After the reorganization, Teller would become Fermi's problem.

A MAJOR PROBLEM PHYSICISTS HAD TO CRACK WAS THE MATTER OF critical mass. Given the extraordinary expense of producing U-235 and Pu-239, the project leadership required more than vague estimates. They required accurate calculations, based on theoretical considerations that were being explored for the first time. Fermi's old friend Hans Bethe, the head of the Theoretical Division since the outset of Los Alamos, had been thinking deeply about this problem, as had many others, including, for example, a youngster from Queens, New York, named Richard Feynman, who had already annoyed military security with his penchant for breaking into locked safes and leaving "guess who?" notes. Bethe and company had been helped enormously by the arrival of a team from Britain, including Rudolph Peierls and his young protégé Klaus Fuchs who had been part of a parallel project, run by the British government since the spring of 1940, to explore the possibility of fission weapons.

The story of the British project, known by the code name "Tube Alloys," is fascinating and in many ways mirrors the Manhattan Project, although it started earlier and made significant progress before the US and British governments revealed to each other what they had been working on. As part of this project, German refugees Rudolf Peierls and Otto Frisch did important theoretical calculations regarding the critical mass of uranium 235 and, although the approximately one kilogram mass they calculated was actually too small for a true critical mass, their work indicated that the problem was not intractable. Fermi's old student Bruno Pontecorvo was also involved, working on a plutonium production reactor project run by the British at Chalk River in Ontario, Canada. Pontecorvo had several meetings with Fermi in Chicago before Fermi arrived in Los Alamos, and they discussed various aspects of reactor design.

Peierls spent time in Rome before the war and came to know Fermi well at Los Alamos. They worked together and lived one below the other in the cramped apartments built by Groves's team. Their wives also hit it off and became fast friends. Peierls admired Fermi greatly, but he was also a subtle and observant critic. He noted that Fermi seemed deliberately to choose problems that could be radically simplified, that when he came to a stage in a problem where complex mathematics would be required to move forward, he "generally left them. He didn't choose to go beyond that." Peierls concedes that for Fermi "the range of things that seemed simple to him covered very many things which were complicated to all of us until he explained them," but when a problem seemed like it would involve more work than he felt it was worth, he lost interest. This critique rings true.

By the time Peierls came to Los Alamos under the cooperation agreement between the Manhattan Project and Tube Alloys, he had in tow a younger colleague named Klaus Fuchs. Fuchs was a member of the German communist party who left Germany in 1939 and moved to England. He and Peierls worked together, and Peierls brought the younger physicist with him to Columbia, and then to Los Alamos. The two of them worked in the Theoretical Division together under Peierls's old friend Hans Bethe. From this vantage point, Fuchs was ideally placed to pass vital intelligence to the Soviets, for whom he had begun to spy several years earlier.

The critical mass problem that preoccupied the Theoretical Division was amenable to brute force calculations. In an era prior to the ready availability of electronic computers, the most effective way of doing these calculations was to rely on slow, simple mechanical calculators, operated by teams of young women, called "computers," who sat at their desks for eight-hour shifts of mind-numbing work. They were overseen by the exuberant young Feynman. An undergraduate at MIT before he was chosen by Oppenheimer for Los Alamos, Feynman had never before met Fermi. Feynman was mightily impressed with the Italian émigré, not because of Fermi's reputation, which mattered little to him, but because of Fermi's ability to interpret the results of calculations. Many years later he remembered an early encounter with Fermi:

> We had a meeting with him, and I had been doing some calculations and gotten some results. The calculations were so elaborate it was very difficult. Now, usually I was the expert at this; I could always tell you what the answer was going to look like, or when I got it I could explain why. But this thing was so complicated I couldn't explain why it was like that. So I told Fermi I was doing this problem, and I started to describe the results. He said, "Wait, before you tell me the result, let me think. It's going to come out like this (he was right), and it's going to come out like this because of so and so. And there's a perfectly obvious explanation for this—" He was doing what I was supposed to be good at, ten times better. That was quite a lesson for me.

Feynman later engaged Fermi in an hour-long argument about a technical issue related to the water boiler, and when Fermi finally conceded that Feynman was right, the younger physicist regarded this as a sort of triumph.

The respect between the two physicists was mutual. As a mark of that respect, Fermi naturally felt at ease teasing him. At Los Alamos Feynman one day picked up the phone. It was Fermi at the other end. He had just read a report that Feynman had produced and explained to Feynman that he considered the research too trivial to merit publication. He claimed the results were obvious even to a child. Feynman countered, "Only if that child is Fermi." To which Fermi replied, "No, even an ordinary child."

THE CRITICAL MASS PROBLEM MAY HAVE BEEN DAUNTING, BUT the most challenging technical problem Fermi worked on involved calculations for the implosion device. Kistiakowsky's team understood that high-explosive "lensing" would be required. Lensing is a technique that changes the shape of a blast wave through high-explosive material in the same way that an optical lens changes the shape of a light wave, by slowing it down. High-explosive material of differing densities through which the blast wave would travel at differing speeds, resulting in the proper shape of the wave just as it reached the subcritical plutonium sphere, would push the entire sphere inward at exactly the same time.

Lensing required technical expertise in many areas. Kistiakowsky was perhaps the greatest expert in the world on high explosives—he loved blowing things up—but that was not sufficient. Other expertise was needed, particularly in the physics of optics. Luis Alvarez knew a great deal about optics from his work prior to the war and was drafted into the project. So did Ed Purcell from Harvard, another optics specialist. A Hungarian Jewish mathematician, however, would be the central figure to do the calculations required to structure the high-explosive charges around the plutonium sphere. His name was John von Neumann.

Von Neumann is regarded by many as one of the greatest mathematicians of the twentieth century. Born in Budapest, he went to high school with the other Los Alamos Hungarians, Szilard, Teller, and Wigner. They all considered him the smartest of the bunch. By the age of eight, young "Johnny" was able to multiply eight-digit numbers by eight-digit numbers in his head, far faster than anyone could do it on paper. He had an idiot savant's ability to calculate, but he was no idiot. He was highly social, at ease in groups, and a great storyteller. He was even shorter than Fermi and had an impish, mischievous look about him, which he reinforced with colorful but crude jokes. He also had an explosive temper and would erupt with anger frequently, certainly more frequently than the normally placid Fermi.

He emigrated in 1933 when the Institute for Advanced Studies at Princeton offered him a tenured position. By the time the war started, he had contributed to virtually every area of mathematics and had, while dabbling in physics, published major work giving a formal mathematical basis for the quantum work of Heisenberg and Dirac. He joined the war effort early on and worked on conventional explosive shock waves before arriving as a consultant at Los Alamos. Fermi knew of his work, but the two had never met. At Los Alamos they were thrown together frequently, for long stretches. Fermi quickly recognized the Hungarian's superior mathematical ability but always tried to outdo von Neumann when it came to calculating. He rarely succeeded. Bethe, Fermi, and von Neumann could often be found sitting together in a quiet room inside the throbbing heart of the Theoretical Division, challenging each other to solve complex integral equations related to pressure waves. Sometimes Oppenheimer would join them. Von Neumann usually left these other three brilliant physicists in the dust.

Von Neumann's mathematical abilities never ceased to amaze Fermi. Years later, returning from summer work at Los Alamos after the war, he regaled colleagues over lunch at the University of Chicago faculty club with a story of how von Neumann masterfully solved a particularly thorny mathematical problem. As Fermi's young physics department colleague Courtney Wright recalls, Fermi observed of his own role in solving the problem: "You know, I felt like the fly who sits on the plow and says, 'We're plowing.'"

Not every calculation was done on paper or in von Neumann's head. Like the work on critical mass, work on the implosion device required a variety of mechanical calculators operated by the "computers." Fermi enjoyed using them himself, so much so that one of the first things he did upon arriving at Los Alamos was write to Pegram at Columbia to send him the calculator he left behind when he moved to Chicago. Fermi also used the newest wave of IBM mechanical calculators that were driven by punch cards. This experience left a mark on him and inspired him after the war to become one of the first physicists to use computers to simulate physical interactions. For von Neumann, the IBM machines inspired him in another direction and led him after the war to design the first programmable, fully electronic computer.

FERMI SOON SETTLED INTO A FAIRLY INTENSE BUT REGULAR ROUTINE. After his traditional simple breakfast prepared by Laura, he would walk or bicycle to the highly secured Technical Area where the daily work on the bomb took place. The mornings were his alone, and he concentrated on any particular physics problems that were bothering him. He also tried to keep abreast of the myriad administrative duties involved in managing the wide range of scientific efforts under his supervision. The afternoons were for others and he opened his office door to all. It soon became clear that if a physicist or an engineer had a difficult problem to solve, approaching Fermi would almost always lead to a quick, clear solution. Segrè recounts a moment when there was a problem with a particular electric circuit. Fermi analyzed the problem, listed the characteristics of an electronics tube that would solve the problem, and a few hours later a tube with those characteristics had been found, inserted into the circuit, and the problem was solved. He was pulled into one meeting after another, to give advice and counsel.

He also began to give lectures in physics to anyone who cared to attend. These became a regular series. The younger staff members particularly appreciated having the opportunity to break away from their work to hear one of the greatest physicists in the world lecture on neutron physics.

Frequently he would climb down into the ravine to the Omega site, working with King, Anderson, Hinton, and others on the water boiler. He also now had responsibility for Teller and the Super. Teller had mastered the fusion equations that were required for the work, but no one could figure out how to keep the assembly of fission and fusion devices together long enough to produce a true fusion explosion. These and other less vital, but still pressing, technical issues preoccupied Fermi at work. But Los Alamos was not just about work.

Even with the punishing work schedule at Los Alamos, Sundays were for leisure, and Fermi took active advantage of them. The surrounding Jemez Mountains were ideal for long, strenuous hikes with his colleagues and friends. As in his youth, he would plan out the excursion, lead it, and walk ahead of most of the others. Geoffrey Chew, a twenty-year-old physicist from George Washington University who worked in the Theoretical Division, was a tall, athletic young man. He recalls that he was one of the few who could keep up with Fermi. He doesn't recall talking physics at all or in fact talking about much of anything except the beautiful austere Southwestern landscape. During the winter it snowed on the Jemez range, giving Fermi an opportunity to go skiing. He was joined by some of his former European colleagues who had learned to ski when they were young, notably Hans Bethe and Niels Bohr.

Fermi eagerly ventured off-site in other directions as well and explored the Bandelier forest, about ten miles southwest of the Los

FIGURE 19.1. Fermi skiing with colleagues in the mountains around Los Alamos. *Courtesy of Los Alamos National Laboratory.*

Alamos mesa, where old Pueblo Indian ruins could be found at the end of long, inspiring hikes. He also tried to pick up the art of trout fishing in the many streams at the base of the ravines surrounding the mesa. To the amusement of those who had labored to learn the difficult skill of fly fishing, he insisted on using live bait, usually worms, arguing that it was more humane to give fish live bait for their final meal. Segrè once pressed him on this and Fermi explained he could not see the point of fly fishing. Segrè patiently lectured Fermi on how it was harder to catch a trout with a fly lure, that it took real skill to put the fly down on the stream in such a way as to fool the fish into thinking it was an insect. Fermi grinned and said, with only mild irony, "I see, so it's a battle of wits!"

The social life at Los Alamos was also active. In general, those with white badges—the several hundred scientists who were cleared to know anything and everything about the project—socialized among themselves, and those with blue badges—everyone else—socialized among themselves. It was easier that way. The white badge parties were frequent and lively. Sometimes they were held at the cramped apartments of the scientists, although the Oppenheimers, who lived on Bathtub Row, also hosted parties in their more spacious digs. Sometimes there were grander affairs, held either at Fuller Lodge, a larger log cabin–style structure that served as the white badge mess hall, or even at the much larger general mess hall for the entire community. Neither Enrico nor Laura were big drinkers, but they enjoyed these parties and hosted many of their own, at which they led guests in the types of parlor games Enrico always tried to win.

Chew tells the story of one party he and his young wife hosted, which the Fermis attended. Chew suggested a party game involving the passing of a scissors around a circle; you passed it "closed" if your legs were crossed and "open" if your legs were not crossed. A few people knew the secret, while others had to guess as they watched. Round and round the scissors went and with each cycle Fermi grew more and more agitated because he could not guess the rule. Laura picked up the rule fairly early, and finally, sensing her husband's growing frustration, she leaned over and explained it to him. He was so upset that the two made excuses and left the party early. The Chews were mortified, but the incident seemed to have no lasting effect. After the war Chew went to Chicago and was one of Fermi's graduate students.

Some of the bigger parties involved square dancing, which was an entirely new experience for the Fermi family. Bernice Brode, the wife of physicist Robert Brode, who developed the fuse mechanisms for the first fission weapons, was one of the leaders of the square dance group, which met frequently throughout the period. In later years she recalled that when the Fermi family initially came to the sessions where she and others taught newcomers, they just sat and watched, presumably terrified at the thought they might have to execute these complex dance patterns. Eventually, Laura and Nella joined in, but Enrico still sat on the sidelines, studying the moves:

> He said in his mild and reasonable voice he would let me know when he was ready to join a square, and one could almost see his mind watching and remembering. Then one evening he came up to me and said, "Well, I think I am ready now, if you will be my partner." He offered to be head couple, which I thought most unwise for his first venture, but I could do nothing about it, and the music began. He led me out on the exact beat, knew exactly each move to make and when. He never made a mistake then or thereafter. I wouldn't say he enjoyed himself, for he was so intent on not making a mistake, which the best of us did all the time. Although I congratulated him, I also kidded him and admonished him to relax and have fun. He laughed tolerantly, but I knew he could continue to dance with his brains instead of his feet.

He eventually learned to enjoy square dancing, so much so that it became a feature of the many parties hosted by the Fermis in Chicago after the war.

By March 1945, the plutonium production reactors at Hanford were producing at full speed, the giant facilities at Oak Ridge were enriching uranium on an industrial scale, and shipments were arriving at Los Alamos with increasing frequency. Much work had been accomplished regarding critical mass in uranium, most notably the famous "tickling the dragon" experiments that took place alongside the water boiler in the Omega site. A subcritical slug of enriched uranium was dropped through another subcritical block of enriched uranium with a hole bored through it just large enough to accommodate the slug being dropped. For a brief instant the whole apparatus was supercritical, not quite long enough to explode, but long enough to give high confidence that estimates of U-235 critical mass were accurate.

The implosion device, however, required more work. One critical experiment consisted of carefully controlled implosions of aluminum spheres. The configuration of the high-explosive charges around the sphere, designed by Kistiakowsky's team guided by the calculations of Bethe's Theoretical Division, proved sufficiently promising to begin manufacturing a sphere of plutonium metal. To the consternation of

Robert Bacher, who headed the group responsible for weapons physics, Fermi's attention turned now to the issue of the "initiator," the device that would provide the fast neutrons at exactly the moment of criticality. Bacher was a respected experimentalist who would go on to a long and distinguished career after the war. Fermi and Bacher had enormous mutual respect, but Bacher felt that Fermi was becoming a bit of a nuisance. Fermi had been playing with a number of different initiator concepts, all of which Bacher considered hare-brained. In later years, Bacher, who remained friendly with Fermi in spite of his annoyance over the initiator issue, described it this way:

> I think Fermi began to be very worried about the fact that this terrific thing that he'd sort of been the father of was going to turn into a great big weapon. I think he was terribly worried about it. . . . I think he [Fermi] was worried about the whole project, not just the initiator. But focusing on the initiator was the one thing that he thought he could look at. The thing really might not work.
>
> And I think he also felt an obligation to take something that was as hare-brained as this was and try to find a way in which it really wouldn't work. So he did look into every sort of thing, and I think every second day or so for a period, I'd see him and he'd come up or he'd see Hans [Bethe] and come up with a new reason why the initiator wouldn't work.

Perhaps, as Bacher surmised, Fermi finally understood the enormity of the project and was looking for ways to demonstrate that the problem was impossible to solve. Or perhaps Fermi was struggling with the admittedly difficult technical problem of how to get enough neutrons out of the initiator to create a full explosive chain reaction before the initiator was destroyed. Regardless, Bacher and Oppenheimer decided to give the initiator assignment to Niels Bohr and his son Aage, who arrived as residents of the mesa during late 1943/early 1944. Bacher believed Fermi would accept a design approved by the Bohrs, and he was right. The Danish father and son team thought about the problem for a few days and came up with an elegant solution, dubbed "the urchin," a spherical version of the original neutron sources that Fermi had used in Rome that would sit inside the plutonium sphere and would be triggered under the enormous pressure of the implosion, releasing between ten and one hundred neutrons before being destroyed. Those neutrons would be sufficient to initiate a full explosive fission throughout the plutonium sphere. Fermi, confronted with this elegant solution, conceded that it would probably work.

Bacher was not a psychiatrist and it is difficult in any case to speculate on Fermi's inner life, given his profound reticence. Yet Fermi may well have been experiencing some deep level of psychic stress. He had climbed on board the Manhattan Project express train early on and had helped stoke the engine for more than five years. In fact, he was the best coal man the train had and that train was now bearing down on its destination. Perhaps somewhere deep inside, he was feeling a growing sense of panic that he was responsible for launching a physics project that would result in terrible consequences. Of course, it might also have reflected a concern that the project on which his advice and judgment and physics expertise largely relied might fail in the end, with unimaginable personal embarrassment for him. We have seen a certain reluctance all along, in his expressions of doubt that the project would work, in his lukewarm presentation to the military in March 1939, and in his decision not to spend the summer of that year devoted to fission, that fateful summer when his erstwhile colleague, Leo Szilard, petitioned Einstein to write a letter to FDR urging work on fission weapons. Now, in early 1945, his mind latched on to the idea that the initiator could not in principle be made to work. If this were true, then the United States would not be able to make nuclear weapons, but then neither would the Germans. If this were true, a lot of money would have been spent demonstrating that the demonic device was technically impossible. The fact that Bohr and his son came up with an initiator that would work in some subtle way let Fermi off the hook, but it left the project intact. Fermi would have one last chance to try and stop the express train from reaching its destination, but for now, at least, he stopped fighting it and returned to supporting it.

The war with Germany effectively ended in late April 1945; on May 2, the Soviet flag flew over the Reichstag. The Allies' most important European enemy had been comprehensively defeated and with it, the fear that Heisenberg and his colleagues would beat the Allies to the atomic bomb.

The Germans had come nowhere near building an atomic weapon, though, as Jeremy Bernstein conclusively demonstrates, they tried hard to do it. Their decision initially to pursue a heavy water model doomed the project from the start. Heisenberg later encouraged suggestions that he had deliberately chosen that route knowing that it would delay the project, but the consensus today among historians is that this explanation was somewhat self-serving. In addition, when the Germans turned to graphite moderation, the graphite they used was filled with impurities that altered the material's ability to serve as a moderator. To cap it off, they grossly overestimated the critical mass of uranium. In the end, they were never even able to achieve a self-sustaining chain reaction.

THE END OF THE WAR IN EUROPE PROFOUNDLY CHANGED THE NATURE of the Manhattan Project. In the eyes of many of the scientists involved, including Fermi, the main justification for the project had evaporated. Germany had lost the race to build the first nuclear weapon. Looking to the future, no one took seriously the threat of a possible Japanese atomic bomb. Groups of scientists at the Met Lab, most importantly James Franck and Leo Szilard, now believed that the Manhattan Project should be slowed or altogether stopped, and that certainly the bomb, if developed, should not be dropped on Japan. The scientists at Los Alamos were aware of the growing discomfort of the Chicago scientists and some of them expressed misgivings as well.

Thinking within the political and military leadership, however, was not tending in the same direction. Groves wanted to push the weapon through to completion, believing it was a quick way to end the war against Japan and also wanting to measure the weapon's effectiveness. Those few around the new president, Harry Truman, who knew about the project—most notably, Secretary of War Henry Stimson—shared Groves's perspective. So did the president himself, who only learned of the project the day he assumed office.

Against this background, Secretary of War Henry Stimson called four key scientists to Washington in late May: Arthur Compton, Fermi,

Lawrence, and Oppenheimer. They were to advise the "Interim Committee," newly established to provide the president with high-level political and strategic advice on the future of the whole nuclear project. Members of the Interim Committee, chaired by Stimson, included a small group of the government's highest level civilian and military leadership.

The meeting, held at Stimson's Pentagon office, started around ten o'clock in the morning, continued over lunch, and ended late in the afternoon. Compton later recalled the meeting for his memoir. The conversation covered the current status of the project, estimates of the weapon's effectiveness, whether, and if so when, and how to let the Soviets know of the project, how to manage the still secret technology in the postwar world, and, most importantly, how to use the weapon to end the war with Japan.

Over lunch, Compton and Lawrence both advocated exploring a demonstration, inviting the Japanese political and military leadership to view an explosion. The thinking was that the experience would be so dramatic that the Japanese would quickly sue for peace. Oppenheimer strongly disagreed. He wanted the bombs to be used against targets in Japan. He could not imagine a demonstration that would be sufficiently dramatic to persuade the recalcitrant Japanese to surrender. Nor did he believe the weapon would be significantly less humane than conventional bombing, which had already leveled Japan's great cities, including Tokyo, and killed some two hundred thousand people.

Oppenheimer's opinion counted; however, the political leadership attending were already leaning heavily in favor of using the bomb against Japan, either to end the war quickly or to make subsequent invasion easier. The quickest way to do this, in their opinion, would be to use the bomb in a dramatic fashion against major cities involved in the Japanese military-industrial effort. It was also clear that only one or two bombs would be ready for use in the immediate future and they opposed using one simply for a demonstration. If the Japanese did not respond constructively to the demonstration, perhaps only one weapon would remain in the arsenal. If the demonstration proved to be a dud, that would only make matters worse.

Fermi limited his participation to an estimate of the amount of enriched uranium that would be required for future research after the war—about twenty pounds initially, and half a ton in the next phase of work.

What Stimson thought of Fermi, or Fermi of Stimson, is not recorded. Fermi had been working at the heart of the Manhattan Project since its inception but had never been this close to the highest levels of political power. He had encountered important Americans before, but none of Stimson's stature. Stimson, a product of Andover and Yale and the ultimate Establishment man, must have been bemused by this short, unprepossessing Italian immigrant with a thick accent who, by all accounts, was essential to the project's success. As the meeting adjourned, Stimson instructed the scientists to prepare a short report to the Interim Committee with advice on the use of the bomb against Japan. It is unclear why Stimson bothered to do this, because the very next day presidential adviser and future Secretary of State James Byrnes reported to the president the Interim Committee recommendation that the United States drop the bombs on Japanese targets. Stimson may have been concerned about the growing consensus among scientists in Chicago and Los Alamos that the bomb should not be dropped and ordered this report from the nation's top scientific leaders in order to undercut the developing consensus. As it happened, the four scientists, not knowing of Byrnes's report to the president, canvassed colleagues at their respective labs. Compton's job was perhaps the most difficult. He asked James Franck, whose moral authority was respected by most of the scientists at Chicago, to prepare a report. Franck was already concerned about how the bomb might be used in the aftermath of the German surrender. He drafted a letter, eventually signed by Szilard and Seaborg, among others, reflecting the views of many in the Met Lab that either a demonstration use or a decision to keep the very existence of the bomb secret would be preferable to using the bomb against Japan. While still leaning toward the option of a demonstration, Compton did not add his signature. Franck personally brought the letter to Washington and gave it to Stimson. It is not clear whether the four scientists who met with the Interim Committee ever saw a final copy.

Oppenheimer scheduled a meeting in mid-June 1945 at Los Alamos with the other three who had briefed the Interim Committee to continue discussions and to prepare the recommendation that Stimson had requested. The meeting began on June 15, 1945, and produced three reports. One relatively uncontroversial report recommended funding the postwar atomic research at a level of $1 billion annually. The third recommended that the Manhattan Project under Groves be extended for the duration of the war to continue oversight of nuclear technologies. It was the second report, however, that historians remember. It addressed the immediate issue of the bomb's use. Just 350 words in length and presented under Oppenheimer's signature on behalf of the four scientists, it recommended that the allies be apprised of US progress in atomic weaponry; noted that, though opinions differed among the scientific community, they themselves leaned toward immediate military use to end the war as quickly as possible; and observed that as scientists they recognized that they had no special competence in the political, social, or military aspects of the issue of atomic weapons.

How Oppenheimer crafted this "consensus" view is not exactly clear. Most accounts are based on Compton's recollection, published in his 1956 memoir. Both Compton and Lawrence advocated demonstration, in opposition to Oppenheimer, at the May 1945 meeting in Washington. Compton returned to Chicago, where he faced active and increasingly frustrated lobbying by Franck and others who opposed using the bomb against Japan. Even Lawrence, a perennial hawk on defense issues, was somewhat fraught and continued to prefer a demonstration, as he argued at the Washington meeting. Oppenheimer was a forceful advocate of using the bomb, though, and skillfully drafted language that Compton and Lawrence could sign on to in spite of their reservations. Compton reports that Lawrence was the final holdout, but reluctantly went along with the consensus.

Oppenheimer's secretary, Anne Wilson Marks, was interviewed in 1983 and tells an entirely different story. Soon after the June 15, 1945, meeting, Oppenheimer revealed to her that Fermi was the last holdout of the four. According to Marks, Fermi leaned heavily in the direction of the Franck Report, arguing not only against demonstration but also that the bomb should simply not be used, that it should be kept secret for as long as possible. This accords with views expressed later in his life, as he considered the question of whether to proceed with work on the hydrogen bomb. In his view of human nature, warfare was a permanent aspect of human life. Eventually another war would be fought and men would use these terrible weapons against each other. The entire project should therefore be kept secret as long as possible. Oppenheimer reported to his secretary that it had taken Oppenheimer till five o'clock in the morning of Sunday, June 17, 1945, to persuade Fermi to agree to the "consensus" recommendation on use. There is no particular reason to doubt her memory of the conversation.

Why did Fermi relent? We can only speculate. He maintained public solidarity with Oppenheimer and the others through the rest of his life. Oppenheimer was an effective, energetic spokesman for any position he promoted, and he was certainly in favor of military use of the weapon against the Japanese. It may be that he actually changed Fermi's mind or perhaps Fermi felt it was more important for him to show solidarity with the man charged with running the project than it was for him to press his own perspective on the matter. This is generally consistent with Fermi's view of himself as a scientist with only limited expertise in political matters. Finally, it may be that he was still sensitive to his position as a foreign-born national who only recently became a citizen of his adopted country. It might have been more important to underscore his loyalty to his country than to express any personal political or moral qualms.

In the end we cannot know his private motivations, only that he came around to the consensus view. He never spoke or wrote about this decision, never indicated anything but solidarity with Oppenheimer on this issue. We would not even know about his reservations were it not for journalist Peter Wyden's interview with Anne Wilson Marks some thirty-eight years after the event.

Szilard, increasingly alarmed that the government might use the bomb militarily against Japanese targets without an initial demonstration, organized his own petition, signed by some seventy Met Lab scientists, and sent it off to Groves. Groves stamped it "secret" and placed it in a drawer.

The irony in all of this, of course, is that the recommendations of the group led by Oppenheimer hardly mattered at all. Neither did the Franck Report. Nor did the Szilard petition. The president and those closest to him had already come to their decision, based on political and military considerations, to use the weapons against Japanese cities.

The only issue was the selection of which cities to bomb. Stimson, who had traveled in Japan as a youth, struck the magnificent cultural capital Kyoto off the list of potential targets, but all other cities were fair game.

Scientists have long fretted about their role in the decisions of July and August 1945. Could they have been more forceful? Could they have been more persuasive? They needn't have worried. The decision makers in Washington were not listening to them, one way or the other.

CHAPTER TWENTY

AN UNHOLY TRINITY

HAVING SIGNED ON TO THE OPPENHEIMER RECOMMENDATION, Fermi now returned to the task at hand. The preparations for a test shot of the weapon began in March 1945 with a decision to halt all other work at Los Alamos and direct all scientific resources to the test, at a site about two hundred miles south of Los Alamos, in a flat, barren desert now part of the White Sands missile testing range. Oppenheimer created a new division to organize and conduct the test, the TR (Trinity) Division, headed by Kenneth Bainbridge, a physicist who had been working on studies related to high-explosive charges. Bainbridge's new division eventually totaled some 250 men, with Fermi formally designated as a consultant.

Work at Los Alamos had provided high confidence in the uranium gun-type configuration. Trinity would test the plutonium device with its more complicated, more problematic implosion configuration.

Fermi arrived by car a few days before the scheduled test on July 16, 1945, joined by Allison and Baudino. The car they used was on its last legs. Fermi quipped to Allison that if they survived the drive to the site they would have survived a danger greater than anything the Trinity test might pose. As soon as he arrived, Fermi got himself into trouble. Several members of a military police unit assigned to Bainbridge at the test site heard Fermi taking odds on whether the earth's atmosphere would burn up as a result of the test. They became sufficiently alarmed to report their concerns to Bainbridge, who had better things to do than deal with an outbreak of panic among the support crews. Justifiably furious, Bainbridge pulled Fermi into his makeshift office and delivered a chastening rebuke.

Fermi and Anderson commandeered a couple of Army tanks and, working with welding crews and engineers, modified the tanks with lead linings to shield them from the radiation and outfitted them with a remote-controlled scooper to collect material from the test site for study at the lab after the blast. The activity grew increasingly frantic and stressful as young men clambered over the iron tower erected to hold the device, putting the final touches on it and wiring it up to the instruments that would record the event in fine detail.

The VIPs at the test were given "front row" seats at a location on a small rise called Compania Hill. Oppenheimer, Groves, Rabi, and other notables viewed the test from atop the hill. Fermi, however, wanted to be out in the open, with as little as possible between him and the detonation. He chose a viewing site in the open desert some ten miles from the tower. Several other physicists joined him there, including two MIT-based physicists, Victor Weisskopf and Phillip Morrison, and Fermi's water boiler colleague, L. D. P. King. Early on the morning of July 16, 1945, the weather was not cooperating. It rained on and off all night and serious consideration was given to postponing the test for a day. However, by four in the morning the rain had stopped and Bainbridge made the decision to go ahead with the test.

The various observation sites were fitted out with intercoms, which Allison used for the final countdown as 5:30 a.m. approached. Music by Tchaikovsky serenaded those waiting during the hour or so before the detonation. Then Allison, watching the clock carefully, counted down to zero, at which point an electronically controlled process took over. An electric signal from the control unit passed down a cable across the desert and up the tower, where it was split into thirty-two pulses that arrived at the high-explosive fuses at exactly the same time. The high-explosive lenses detonated simultaneously, pushing the subcritical sphere of plutonium in on itself and smashing the urchin initiator nestled in the center of the device. The urchin emitted the requisite neutrons, leading to an explosive chain reaction of plutonium fission in less than a millisecond.

FIGURE 20.1. Trinity test moments after the initiation of the explosion. *Courtesy of Los Alamos National Laboratory.*

Fermi, King, Weisskopf, Morrison, and a few others had positioned themselves lying face down on the ground, heads away from the detonation as they had been instructed to do, with arc welding dark glass in front of their eyes. For a period of a few seconds, the entire desert and mountains around them flashed bright white, brighter than anyone had ever seen, brighter than any light ever created on the planet. Fermi and his colleagues stood and turned toward the detonation, and as the white light passed over them, they were able to see the fireball of the bomb rise slowly off the ground, shifting in color from purple to pink and varying hues in between. Fermi began counting by seconds as soon as he saw the light and pulled some strips of paper out of his pocket. This was why he wanted to be out in the open, not in a bunker. He held the shreds of paper high above his head and when he counted off the seconds for the blast wave to arrive he began to drop the strips of paper. As the roar of the blast wave swept over them at a speed of about eleven hundred feet per second, he dropped the strips of paper into the wave and watched them blow along with it. They traveled about eight feet (2.5 meters). Fermi announced to those within earshot that the blast was the equivalent of about ten kilotons of TNT.

He had prepared this little experiment the evening before. He had, after all, spent months thinking about pressure waves with Bethe and von Neumann, so the experiment almost suggested itself. It was classic Fermi—a simple experiment based on some back-of-the-envelope calculations that arrived at an answer that was impressively close. Instrumentation specialists later measured the blast at 18 KT, but Fermi got a reasonable answer faster than anyone else.

Within an hour of the detonation, Fermi and Anderson jumped into their respective tanks and drove out to the explosion site. (Baudino presumably did not offer to drive Fermi on this occasion.) The crater was some five feet deep and thirty feet wide. The tower on which the device was perched had actually evaporated. The sand had been melted into radioactive glass. Sand in the engine block caused Fermi's tank to break down before he could get close enough to the site, so it was Anderson who in the end scooped up the radioactive glass left behind by the intense heat of the blast, a substance we now call trinitite, which was taken to Los Alamos for complete analysis.

The Trinity test left a lasting impression on everyone who witnessed it and many recounted their experience for the history books. Fermi's water boiler assistant Joan Hinton had a particularly compelling account:

> It was like being at the bottom of an ocean of light. We were bathed in it from all directions. The light withdrew into the bomb as if the bomb sucked it up. Then it turned purple and blue and went up and up and up. We were still talking in whispers when the cloud reached the level where it was struck by the rising sunlight so it cleared out the natural clouds. We saw a cloud that was dark and red at the bottom and daylight at the top. Then suddenly the sound reached us. It was very sharp and rumbled and all the mountains were rumbling with it. We suddenly started talking out loud and felt exposed to the whole world.

The light from the detonation could be seen several hundred miles away. At Los Alamos, some people had heard a rumor that a test would be conducted at dawn and that they might be able to see it even at that distance. Elsie McMillan, wife of physicist Edwin McMillan, saw the flash from the window of her Bathtub Row cottage. In Santa Fe, Dorothy McKibbin climbed a hill and saw a flicker in the direction of the test. Other citizens of Santa Fe also recalled seeing a flickering light on the horizon, but assumed it was a distant lightning storm.

Fermi was so absorbed in his paper experiment that when Laura later asked him to describe the sound, he could not remember it at all.

The scientists and officials who witnessed the detonation experienced a range of strong emotions immediately after the fact. Fermi remained outwardly placid and calm, in spite of his problems with the tank, but the stress of the day finally caught up with him. When he drove home with Sam Allison that night he asked his friend, most uncharacteristically, to take the wheel. "It had seemed to him as if the car were jumping from curve to curve, skipping the straight stretches in between," recalled Laura years later. She also reported that he went to bed without a word.

Allison later described the trip home. The two of them had a flat tire and Allison hitched a ride to the nearest garage, leaving Fermi to stay with the car. Before he could get a lift back to rescue Fermi, Fermi drove up in the car and explained that a passing physicist was traveling by with a canister of argon gas, with which Fermi had filled the tire. It may be the first and only time that a vehicle has driven on a tire filled with the safe but expensive gas.

Many years later the Centers for Disease Control and Prevention looked into the issue of radiation exposure of the indigenous groups living in the general area of the Trinity test. Some sixty-three ranches and three camps were within thirty miles of ground zero. Since groundwater in that part of the New Mexico desert has always been brackish, local residents have long relied on rainwater for their drinking needs. At Trinity it rained the next day, contaminating rainwater cisterns with radioactive debris from the test. Ranchers reported that fallout "snowed down" for days afterward. Although health radiation professionals understood the risks of exposure to radiation, residents in the surrounding area were neither informed of the risks nor evacuated. To quote the study, "Different standards of safety were applied to informed project workers than to uninformed members of the public."

Activists such as Tina Cordova, founder of the Tularosa Basin Downwinders Consortium, have worked to hold the US government accountable for the radiation exposure and subsequent health problems suffered by those in the area and their descendants. Her efforts and those of others in the area resulted in a CDC study published in 2009 that detailed the exposure and the possible consequences. Studies continue, but compensation has yet to be determined.

On July 24, 1945, at a summit conference in Potsdam, Germany, Truman told Stalin about his new weapon. He did this on his own, with neither a US translator nor any other official to witness it. He reported that Stalin showed no special interest, but the Soviet dictator congratulated the president and hoped that it would be used effectively against the Japanese. Stalin probably did not know beforehand that Trinity had been a success, but he certainly knew a great deal about the Manhattan Project, via information provided by Klaus Fuchs, Peierls's colleague who had been keeping the Soviet Union abreast of the issues facing the project, providing Soviet scientists the news that plutonium, not uranium, was the material of choice in the construction of these new weapons. This information would save the Soviet Union years of work as they pursued their own project.

The declaration issued by the Potsdam Conference on July 26, 1945, warned of devastating consequences if Japan did not surrender immediately. It made no mention of the new weapon. Over the next few days, the US Air Force dropped hundreds of thousands of leaflets on cities throughout Japan warning citizens of imminent catastrophic destruction, but again, no mention was made of a revolutionary new weapon.

The partly assembled components of two bombs, one gun-type uranium bomb christened Little Boy and one implosion plutonium bomb known as Fat Man, were flown from Los Alamos to the South Pacific island of Tinian in the Northern Mariana island chain some 120 miles northeast of Guam, where a group from Los Alamos conducted the final assembly. On August 6, 1945, some ten days after the Potsdam declaration, a long-range bomber flew from the Tinian Island base and dropped Little Boy on the city of Hiroshima, killing some seventy thousand and leveling most of the city. Factions within the

Japanese government frantically tried to contact US officials to negotiate surrender, but the senior Japanese military council decided to fight on, betting that there would be only one more bomb in the US arsenal. On August 9, 1945, Fat Man fell on the port city of Nagasaki, a secondary target chosen because the primary target, Kokura, was obscured by cloud cover, and leveled it. Some eighty thousand people died that day. Between the blasts, the Soviet Union declared war on Japan and Stalin's troops crossed into Japanese-occupied Manchuria. It took another five days for the Japanese emperor, shaken by the calamity that had befallen his country, to broadcast an announcement that Japan would surrender without condition. It is clear that the Soviet attack played a role in the final Japanese deliberations over surrender, and perhaps even in the US decision to drop the bomb in the first place, but the bombs were seen by US leaders to have done what the key decision makers intended them to do: end the war quickly, thus avoiding a prolonged and bloody battle for the Japanese islands themselves.

The residents of Los Alamos initially received the news with jubilation. The wives of those with white badges now knew what their husbands had been working on and the blue badge employees now knew the ultimate purpose of their labors. The nation absorbed the news about a secret installation in the mountains above Santa Fe and about secret towns near Knoxville and Richland. The men and women of Los Alamos and the Manhattan Project became heroes.

On the day of the first bombing, Eugenia Peierls famously ran upstairs to the Fermis' landing and shouted to a somewhat bewildered Laura Fermi, "Our stuff was dropped on Japan!" Laura had no idea what Eugenia was talking about. She writes that the first time she understood what her husband actually had been doing for the previous six years was when, shortly after the bombings, he handed her a copy of a book by Henry DeWolfe Smyth, in preparation since 1944, which informed the public about the Manhattan Project. To her amazement, her husband had a starring role in the book, which sold over a hundred thousand copies in its first eight months of publication. That best-seller and the public fascination engendered by the book made him a national figure. Fermi endured the attention with good humor.

Laura herself was more circumspect than some about the use of the bomb, and we can assume her husband was, as well. For both of them, the merits of ending the war quickly outweighed the enormous loss of life involved. They also knew that the massive conventional fire bombings of major Japanese cities had led to similar, or even greater, fatalities than the atomic bombings of Hiroshima and Nagasaki. On the other hand, they understood that these weapons had their own unique psychological terror and that, if used by both sides in a major future conflict, no one would be safe.

Maria Fermi had strong views about the project in which her brother had a starring role. Soon after the news of the bombings, she wrote to Enrico:

> People of good judgment abstain from any technical comment, and realize that it would be vain to seek who is the first author in a work which is the result of a vast collaboration. All, however, are perplexed and appalled by its dreadful effects, and with time the bewilderment increases rather than diminishes. For my part I recommended you to God, Who alone can judge you morally.

Fermi and his family remained at Los Alamos until the end of 1945. Useful research remained to be done, on both the Super and the water boiler. Fermi wanted to finish his comprehensive lectures on nuclear physics, which were declassified in time for publication in the *Collected Papers* in 1965. As published, they were not written by him but rather put together as lecture notes by several of those who attended the lectures, reviewed and supplemented for publication by Fermi's young collaborator, Bernard Feld. The papers cover virtually every aspect of nuclear physics then known to Fermi and thus to the physics world more generally, in a straightforward manner.

Fermi also lectured on the then-current state of the Super project. Teller had not been able to solve some of the basic problems with the design of the weapon, and Fermi's lectures point out the flaws in the conception of the device as it stood in August 1945. These lectures remained classified until 2009 and were not included in the *Collected Papers*. Reviewing the progress Teller and Fermi made in thinking about a fusion device, they are highly technical, but the final paragraph indicates with typical Fermi humor the basic differences between the two physicists on the subject, at least during the final months of 1945:

> In concluding this series of lectures, it should be stated that they may represent a somewhat pessimistic view, in that Teller who has been in charge of most of the work reported is inclined to be more optimistic than is the lecturer. The procedure that has been adopted to try to resolve the question of practicability of the super [*sic*] is that Teller shall propose a tentative design which he considers somewhat overdesigned, and the lecturer will try to show that it is underdesigned. (This makes the pope the devil's advocate!)

By the end of December 1945, with his lectures completed, Fermi was ready to return to Chicago, to a position in a new institute at the University of Chicago organized just for him.

How did Fermi's former colleagues in Italy spend the war?

Rasetti spent the war at Laval University in Quebec. Fermi invited him to come to Los Alamos to work, but Rasetti rejected the offer on moral grounds. He wanted nothing to do with weapons and nuclear fission, considering them an abomination. When Rasetti moved to a position at Johns Hopkins University in Baltimore after the war, he resumed contact with Fermi and the two men worked closely together on getting compensation for the slow-neutron patent.

Soon after Italy entered the war, Amaldi was drafted and served for some time on the front lines. Prior to this, he and his fellow physicists followed the discovery of fission and discussed its implications. They decided not to pursue work on fission weapons, once again largely on moral grounds. When Amaldi returned from the front lines, he endeavored to preserve what he could of the records of the Rome physics department and brought as much as possible to store at his home at Via Parioli, convinced that these records would be safer in his apartment than they would be at the more centrally located Sapienza campus. He also stashed expensive and vital equipment there. Italians credit him with preserving Italian physics during the war against tremendous odds.

Others, including former Fermi student Oreste Piccioni, stayed behind as well. Piccioni, with two colleagues, Marcello Conversi and Ettore Pancini, conducted an ingenious experiment under extremely difficult conditions in a basement in Rome at the height of the fighting over the city in 1945, which would have an enormous impact on postwar particle physics when the results were published in 1947. Fermi remained concerned about his Italian colleagues throughout the war and made several attempts to locate them and assure himself of their safety. Indeed, in one of his 1938 letters to Pegram, after accepting the position at Columbia, Fermi urged Pegram to find roles for some of his colleagues most at risk. Many of those he left behind, however, never made it to the United States once the war started; colleagues like Oreste Piccioni would have to wait until the war ended.

ENRICO FERMI IS OFTEN CALLED THE "FATHER OF THE NUCLEAR AGE." Looking back over the period 1939 through 1945, a more nuanced evaluation emerges.

Several points are beyond dispute. Fermi and Szilard came up with the first crucial experiments in 1939 to explore how chain reactions might be created using uranium and they made the providential observation that graphite would be a more suitable moderator for the first chain reactions than heavy water. The German effort foundered largely because Heisenberg opted for a heavy water reactor. Could the atomic bomb have been developed without these experiments leading to the first Chicago pile? The pile itself was an important invention in the process of weapons development not simply because it demonstrated the physics of fission chain reactions. It also served as a model for the development of plutonium production reactors. No amount of cyclotron bombardment of uranium would have produced enough plutonium for a bomb and the enormous Hanford reactors would not have been possible without the work Fermi and his colleagues completed in Chicago in December 1942. Perhaps of equal importance, the project now had a device that could generate huge numbers of neutrons to be studied for their properties and their ability to create an explosive chain reaction in uranium and plutonium. CP-2 and CP-3 were also invaluable in assessing the level of impurities in the various components that were required in the manufacture of these first weapons. Finally, as the water boiler story illustrates, these reactors could be used to test estimates of critical mass. Fermi's work on these new reactors is a central part of the Manhattan Project story.

His work at Los Alamos is harder to evaluate. Although the water boiler project was clearly important, much of his role consisted of day-to-day advice and counsel across a wide range of technical issues, ranging from neutron diffusion studies to pressure wave calculations to the design of complicated electronic circuitry. He was the "go-to" physicist when other physicists ran into problems that stumped them. He usually saw quick, practical solutions to the problems others brought to him and became a highly valued resource for everyone at Los Alamos. Oppenheimer came to rely extensively on his scientific judgment for a wide variety of issues, as did Bethe and other division heads.

And yet Fermi did not design the bombs themselves. Others did that work, although Fermi was consulted on a variety of matters relating to weapons design and weapons physics. Nor was he central to the work done by Oppenheimer and others at Berkeley from 1942 through 1943, on the physics of fast-neutron collisions with uranium. He contributed little to the work of Lawrence, Urey, and others on isotope separation and uranium enrichment, work that was every bit as important to the ultimate success of the project as was the Chicago pile. Nor was he a chemist. It was the methodical, dedicated Glenn Seaborg and his team who figured out how to extract plutonium from the spent uranium fuel rods once they were removed from the plutonium reactors. Von Neumann was the key man on the pressure wave calculations required for the implosion device, with Fermi contributing where he could. Kistiakowsky and a crack team of ordnance experts engineered the high explosives required for both gun-type and implosion devices.

In missing the possibility of xenon poisoning during high-power reactor operations, Fermi risked bringing the plutonium project to a complete halt. Nature, along with John Wheeler and Charles Graves, came to his rescue and the B reactor went on to produce a major portion of the nation's plutonium reserve. That it worked in the end does not diminish the fact that Fermi missed a major technical issue. As we have seen before, in the case of the discovery of fission in December 1938, he may have been the Pope, but he wasn't infallible.

The Manhattan Project was far from being a one-man show. Hundreds of scientists at Los Alamos, Chicago, Oak Ridge, and Hanford contributed to the effort, as did thousands of engineers, construction workers, "computers," secretaries, and lower-level military personnel.

Many of these thousands learned the ultimate objective of their work only when the president announced the use of these bombs on Japanese cities in August 1945.

The decision to team Groves and Oppenheimer was perhaps the single most important decision necessary for the project to succeed. Groves was one of the very few individuals in the country who could run a vast, multibillion-dollar project under the requisite tight security and even tighter deadlines. Oppenheimer was ideally suited to work with Groves, on the one hand, and to work with a group of scientists on the other. It is hard to imagine that the Manhattan Project could have succeeded without either of these two giants.

Other scientists were also crucial to the effort: Bethe, von Neumann, Alvarez, Segrè, Serber, Bainbridge, Feynman, Neddermeyer, to name but a few of the hundreds of brilliant physicists who came to the mesa and devised the bomb.

Was Fermi's work central to their success? No doubt it was. Would the project have succeeded without him? Perhaps eventually, but it would almost certainly have taken far longer to complete. At the particular moment in early 1939 when Szilard collared him at the King's Crown Hotel on 116th Street in Manhattan, Fermi was the most knowledgeable person on the planet regarding neutron physics, a result of the grueling, intensive work he did with Amaldi from 1935 to 1938. That Szilard latched on to him to push the experimental agenda that resulted in a working nuclear reactor is one of history's great coincidences. It is not at all clear that Fermi would have chosen to develop a controlled chain reaction without the brilliant, persistent, and occasionally annoying Szilard prodding him along. He may well have decided to pursue other types of technical studies. He may even have moved away from neutron physics completely, toward his next great adventure, particle physics. Could Szilard have approached other talented physicists? Of course he could have, but we will never know how these potential collaborators would have fared in the process. What we do know is that Fermi agreed to explore the chain reaction with Szilard and that exploration made history. It made the atom bomb possible and ushered humankind into the nuclear era.

In the process, Fermi and his colleagues became a species that had never really existed before the war: public scientists. Prior to the war, basic research was conducted on shoestring budgets at universities throughout the world and at a very few institutions like the Bureau of Standards, established to conduct research for the public good. It was conducted in small groups, often just a handful of scientists working together on tabletop experiments, freely communicating their results via professional journals with their colleagues around the world. Fermi became a government scientist, as did all of his colleagues during the Manhattan Project. Much of their work was conducted under government secrecy and continued under such constraints into the postwar era. The US government soon established great national laboratories throughout the country, legacies of the Manhattan Project, that continue to operate today at Los Alamos, Hanford, Oak Ridge, and elsewhere.

The Manhattan Project was not only the beginning of classified science. It was the birth of "big science," funded by governments on an increasingly enormous scale. Gone forever were the early-twentieth-century tabletop experimental days of Rome, Cambridge, Paris, and Columbia.

The profound consequences of this revolution were only just being understood when Fermi returned to Chicago, but a man named Walter Bartky, the dean of the University of Chicago's division of physical sciences, was a keen observer of the revolution. Out of these observations he developed a plan, one in which he hoped Fermi would again play a central role.

PART FOUR

THE CHICAGO YEARS

CHAPTER TWENTY-ONE

RETURN TO CHICAGO

THE FERMIS ARRIVED BACK IN CHICAGO ON JANUARY 2, 1946. BY June they moved to a new house not far from the old one, at 5327 University Avenue. Like the one on Woodlawn Avenue, it was a grand, three-story turn-of-the-century brick edifice. This house served as the Fermi family home until 1956, when, after Enrico's death, Laura moved to an apartment. Nella and Giulio both returned to the Lab School. Laura picked up where she left off in the summer of 1944, managing a busy social schedule, hosting numerous parties for faculty and students, many of which involved square dancing, often with Harold Agnew calling the moves. She started a book group, called The Paperbacks, which focused on a range of classics.

After the war, Enrico Fermi gradually became a more well-rounded, if not exactly worldly, person. He occasionally joined Laura's reading group. Though the fact that he might not have read a book never prevented him from having strong views on it, the group did seem to broaden his philosophical and cultural interests. He commented to Segrè that he would occasionally persuade the group of his views by "using the old Italian method of shouting louder than his opponent." In the summer of 1953, he was invited to speak at a seminar in Aspen for twenty young businessmen that was designed to expose them to a broad range of readings in history, culture, and philosophy. Fermi enjoyed the week and was amused to be considered a "philosopher" by the conference organizers. Fermi confided to Segrè that he had been thinking about the philosophical aspects of quantum theory, aspects that earlier held no interest for him. He did not, however, bother to commit any of those thoughts to paper. Amazingly, Laura even persuaded Enrico to attend a performance of the highly popular musical *South Pacific,* for Enrico a major concession to the musical arts.

In their teenage years, Nella and Giulio experienced their father in different ways.

Nella admired her father and had the easier relationship with him of the siblings. At a conference at Cornell in 2003, she described her father trying to teach her algebra when she was eleven. He was not particularly successful, but laid the foundation, she felt, for a quicker understanding of the subject when it came time to study it at school. She also described his carpentry projects around the house. They were more functional than aesthetic. When Laura complained about the crudeness of one such project, Fermi observed that the work was hidden behind the couch, so no one would notice. As Laura stormed off, he turned to Nella and, in a highly revealing moment, advised his daughter, "Never make something more accurate than absolutely necessary." The two of them would sometimes cook together when Laura was out of town. Not surprisingly, he could be quite literal in his interpretation of recipes. Nella remembers fondly the time when her father came home from the lab with a new substance for which he had been asked to think of some uses. It had the consistency of putty when pulled apart slowly but broke like glass when pulled apart quickly. He showed it to Nella and Giulio, but none of them could figure out what it might be good for. It was "Silly Putty," which was a popular toy in the 1950s and 1960s. For a time they played with another new toy, the "dunking bird," a plastic bird that, properly set up, would repeatedly dip its beak into a glass of water and then straighten up. They had good fun, even though, as Nella was quick to note, her father was never particularly demonstrative.

Giulio had a more troubled relationship with his father. A very bright boy with a tendency toward depression, he never felt comfortable in his father's shadow. He rarely if ever spoke about his father in his later years, but he did occasionally confide in Robert Fuller, his lifelong friend from his undergraduate years. Giulio told Fuller of his frustration at not being able to build a working electric motor and his humiliation every time his father would step in to fix the problem at hand. He talked about his feelings of inadequacy when he compared himself to his father and of his father's relative insensitivity to this. The problems Giulio faced came to a head when he was sixteen, about ready to attend college. He tried to commit suicide by slitting his wrists. Fuller notes that Giulio first realized just how much his father really cared for him in the ambulance ride to the hospital that terrible day. He recovered and, after a short and miserable spell at the University of Chicago, found some distance and peace at Oberlin College, where few of his fellow students knew of his relationship with one of the towering scientific figures of the twentieth century. By the time Giulio arrived at Oberlin, he had changed his name to "Judd," a name he kept for the rest of his life. The effect was to create a distance from his famous father and his Italian heritage.

It wasn't that Fermi was a particularly bad parent. He may have been inattentive, but no more so than other career-driven fathers in America during the immediate postwar period. Giulio had the misfortune of inheriting a delicate psyche, perhaps directly from his grandmother, Ida. His sensitive makeup was simply ill-suited to life with Enrico.

More generally, the war changed everything. Enrico Fermi was now a national figure, even a bit of a celebrity. The best-selling official report on the Manhattan Project was partly responsible for this. Journalists' intense interest in the Manhattan Project made Oppenheimer and Groves superstars, but Fermi became famous as well, although to a lesser degree. He was inundated with invitations and requests for interviews for the remainder of his life. He participated in documentaries about the Manhattan Project, gamely re-creating for famed CBS broadcast journalist Edward R. Murrow the moment in December 1942 when he achieved a controlled chain reaction. He posed impishly in a famous photo session in front of a blackboard, which displays a formula for the fine structure constant that is clearly wrong. Because he knew as much about that constant as anyone alive, he almost certainly wrote it incorrectly to see whether anyone would notice. Along with four other University of Chicago alumni of the Manhattan Project, he was honored

Figure 21.1. Fermi at the blackboard during a publicity photo shoot. The equation for the fine structure constant alpha directly above his head is incorrect—most likely Fermi's idea of a joke. *Courtesy of Argonne National Laboratory.*

by the US Congress with the Medal of Merit. A plaque commemorating CP-1 was unveiled at a ceremony on the tenth anniversary of the event, presided over by University of Chicago president Hutchins.

Over time, however, as the invitations piled up, so did the graceful but firm letters declining them. Segrè suggests that Fermi might have begun to appreciate how little time he had left. Fermi may not have known that he would die so soon, but he certainly knew that, in general, physicists' achievements tend to come early in life. He wanted to do as much as he could as fast as he could, while he still had the energy and mental acuity to do so.

In the summer of 1945, Walter Bartky, a University of Chicago astronomer, replaced Arthur Compton as dean of physical sciences. Bartky had the idea for new, interdisciplinary research institutes modeled on the Met Lab and Los Alamos. Endorsed by University of Chicago president Hutchins, Bartky invited Fermi to become head of an institute devoted to nuclear science. Though Fermi was enthusiastic about the new institute, he refused Bartky's invitation to be the director,

FIGURE 21.2. Fermi at the unveiling of a plaque at the University of Chicago commemorating the tenth anniversary of CP-1. University president Robert Maynard Hutchins is unveiling the plaque. From right to left, Fermi, Walter Zinn, Farrington Daniels, Robert Bacher, and William W. Waymack. *Courtesy of Argonne National Laboratory.*

fearing that an administrative role would constrain his opportunities to pursue his own research. He suggested Allison for the role, and Allison accepted, seeing it as an opportunity to build a world-class scientific research institution, building on Fermi's enormous reputation. Many of Fermi's Manhattan Project colleagues eventually joined, and by 1950, the Institute for Nuclear Studies boasted some thirty-four senior scientists and a number of more junior staff, many of them either current or future Nobel laureates.

Fermi may not have wanted to direct the new institute, but in a December 1945 letter to Bartky he detailed his thoughts on its research agenda. He wanted to use high-energy particle beams to explore the force holding together the atomic nucleus. Yukawa's theory that the force was conveyed by a particle, the "mesotron," was his starting point.

Hitting a nucleus with enough energy might stimulate the creation of these particles, which would then be used to probe the nucleus.

Fermi urged Bartky to build a cyclotron capable of accelerating relatively heavy protons to a level of sixty million electron volts (MeV)—enough, Fermi believed, to create the conditions for mesotron production. Eventually, a souped-up cyclotron, capable of a then-astonishing 450 MeV, occupied most of Fermi's experimental time when it came on line in 1951.

Fermi noted that during the five years it would take to build the cyclotron, cosmic-ray research could provide clues as to the behavior of mesotrons. This research would not require a cyclotron. Some of these rays were extremely high energy. Studying them could be useful in exploring the nuclear force in the absence of the cyclotron.

At the time he wrote this letter, Fermi was unaware that his Italian colleagues Conversi, Pancini, and Piccioni, working with makeshift equipment in a dingy Rome basement, had demonstrated conclusively that the mesotron was not the particle predicted by Yukawa, the manifestation of the field that holds the nucleus together. Their work had not yet reached the United States. The particle that did seem to match Yukawa's particle was discovered in 1947, a bit more than a year after Fermi's letter. It was originally called the pi-meson, a clumsy name that Fermi shortened to pion. The particle that his colleagues in the Roman basement studied eventually became known as the mu-meson, or muon, again a shortened name proposed by Fermi. In the context of what was known at the time, however, Fermi's letter was a clarion call for research into the increasingly complex world of subatomic particles, a compelling agenda for the infant institute.

Life at the institute was collegial and informal, a far cry from the formality and hierarchy of the Met Lab under Compton. Office doors were typically open, and Fermi encouraged researchers to wander the halls and find out what their colleagues were doing. Fermi might have gone home for lunch, but just as often he would pick up colleagues and wander over to the faculty club, or he might have lunch with students at Hutchinson Commons. This was his favorite place to grab a hamburger and Coke during the Met Lab years, challenging his younger colleagues with thought-provoking questions like "How thick does the dirt on the windows of the Commons have to get before it begins to fall off the window?"

Wandering the halls was one of Fermi's ways of finding out what was going on in the world of physics. He had stopped reading professional journals sometime before the war, relying instead almost exclusively on gossip with colleagues and institute visitors about interesting results and discoveries, challenging himself to figure out how the results were obtained. He also used these discussions to interest his colleagues in his current enthusiasms. One such enthusiasm was a concept called "spin-orbital coupling," the notion that the spin of a particle or a set of particles within the nucleus could affect the way those particles orbited within the nucleus. In late 1949 and early 1950, he tried to get his young graduate student Richard Garwin interested in the concept, but Garwin showed no interest, so Fermi wandered down the hall to where Maria Mayer, his old friend from Columbia days, was working on her own project.

Mayer later vividly recalled the moment. She was working on why certain specific "magic" numbers of particles inside a nucleus resulted in particularly stable arrangements, using a model of the nucleus arranged in proton and neutron "shells," similar to the electron shells surrounding the nucleus that determine the chemical properties of the various elements. She was getting nowhere. Fermi and Mayer chatted for a while, Fermi politely tolerating Mayer's chain-smoking (he didn't smoke and generally frowned upon people smoking in his office). They chatted about physics and the shell problem for several hours and then they were interrupted by a knock on the door. Fermi had received a phone call in his office. As he got up to leave, he casually asked, "What about spin-orbit coupling?" Mayer was thunderstruck: "Yes, Enrico, that's the solution." Fermi, always cautious, replied, "How can you know?" However, she immediately saw that spin-orbit coupling explained the problems with which she was struggling and within two weeks she successfully incorporated it into one of the first comprehensive theories of the nuclear shell model. Years later she would describe it with an analogy to a floor full of couples arranged in concentric circles dancing a waltz. Some circles are moving around the room in a clockwise fashion, others, counterclockwise; the couples are also each spinning themselves, some clockwise and some counterclockwise. In her analogy, each particle in the nucleus is like one of those couples; the overall effect of the relationship between their orbit and their spin influences the stability of the nucleus.

In 1963, her work was rewarded with a Nobel Prize, shared with Wigner and the German physicist J. Hans D. Jensen, both of whom also elaborated the shell model of the nucleus. She always credited Fermi's generosity with helping her make the crucial breakthrough. Fermi had been generous indeed. Having seen the connection between the nuclear shell model and spin-orbit coupling, he could easily have done the work and published it himself. Instead, he offered the idea up and stood back, letting Mayer do the rest and take the credit. Mayer wanted to include him as a coauthor of the paper, but Fermi rejected that outright, explaining that because he was more famous, most people would assume he had done all the work, which was clearly not the case. It was a generosity that characterized his later years and reflected a more mature Fermi, comfortable in his stature as one of the world's preeminent physicists.

The institute was exciting not only because of the staff or the easy collegiality that Sam Allison and Fermi set by example but also because it was a magnet for visitors from around the world. Many of the most celebrated physicists would visit and some would be invited to present at the institute's biweekly theoretical physics seminar, organized by Maria Mayer's husband, Joseph. These seminars covered a huge range of topics, all of which were cutting-edge science. Hans Alfven elaborated his theory of the intergalactic magnetic fields that Fermi believed were responsible for the high energy of cosmic rays. Feynman spoke about his work on liquid helium and its strange properties. Other visitors from the University of Illinois and the Institute for Advanced Studies at Princeton presented on topics of their choice. Most of the seminars, however, featured institute staff. Murray Gell-Mann presented on numerous occasions, as did Valentine Telegdi, Gregor Wentzel, and many others. In 1952 Fermi presented work on pion-nucleon scattering experiments he had just begun conducting at the newly operational cyclotron. These seminars ensured a continuing intellectual ferment within the institute community and kept everyone, including Fermi, abreast of developments in the world of physics. The seminars always started at four thirty in the afternoon. Staff members knew and visiting speakers were informed that they needed to finish by six o'clock. That is when Fermi would excuse himself and, irrespective of where the discussion stood, return home for the evening.

Gell-Mann recalls with some frustration Fermi's tendency, if he disagreed with a speaker, to ask questions with the persistence of a terrier chewing on someone's ankle. If Fermi raised his hand and said, "There is something here I do not understand," the speaker was in for trouble and, if he had seen Fermi in action before, knew it. One senses that Fermi enjoyed disagreeing with Gell-Mann the same way he enjoyed disagreeing with his old friend Teller. It may have reflected an actual disagreement, but it was also a sign of respect.

It would take five years for Fermi to achieve his dream of a major particle accelerator. Fermi was a consultant to the project and introduced a simple invention to remotely control the placement of the accelerator's target without breaking the vacuum of the ring in which the protons traveled. Sam Allison oversaw the project, Herb Anderson took day-to-day responsibility for its development, and John Marshall assisted. The accelerator ended up growing from ninety-two inches to its eventual one-hundred-seventy-inch size, with a magnet weighing twenty-two hundred tons, and involved close, if sometimes fractious, collaboration among General Electric, Westinghouse, and the US Navy's Office of Naval Research. It also involved a new technology, varying the radio frequency of the cyclotron voltage to maintain strict synchronization with the beam itself, in what is now called a synchrotron. The new technology was challenging to the electrical engineers constructing the machine, and Fermi regularly advised them on methods to overcome these challenges.

Because Fermi had no operational responsibility for the cyclotron's construction, he faced a lengthy period during which he could not conduct high-energy particle experiments. What would Fermi do while he, and everyone else, waited for the new accelerator to be built? He wasn't exactly the kind of person to bide his time. Though none of his postwar work had quite the lasting significance of the work he did before the war, he kept himself quite busy, in both experimental work and in theory. His pre-cyclotron research focus was on exploiting the neutron sources provided by CP-3 at Argonne. He also began a five-year study of cosmic rays and their origins and in the process developed a close professional and personal connection with one of the most unusual and important astrophysicists in the world.

At Argonne, a new pile, CP-3, had superseded CP-2. It was a fine source of neutrons that could be exquisitely controlled. For the next few years, Fermi undertook a series of important studies of neutron collisions using CP-3, collaborating mainly with Leona Libby but also with Herbert Anderson, Albert Wattenberg, and a number of younger graduate students. With Fermi's old colleague, Walter Zinn, the new director of the lab, it was the old Fermi team working together again, as they had under the stands of Stagg Field in 1942, this time on projects selected solely on the basis of scientific curiosity.

Unburdened of his bodyguard-driver, Fermi enjoyed driving himself everywhere with characteristic, prewar gusto. Wattenberg was a frequent passenger on the forty-minute commute to Argonne. He recounts a harrowing moment when Fermi raced successfully to get across a railroad track before being blocked by an oncoming train. What Fermi did not realize was that there was a second track, obscured by the first train, and on that track was a second train going in the other direction. They missed the second train by a matter of a few feet. Pulling off to the side of the road so that the two of them could recover from what must have been a heart-stopping near miss, Fermi turned to his young colleague to reassure him. "This is why it is very important for you to be with me; my time may be up. But yours isn't."

At Argonne, Fermi and his colleagues bombarded some twenty-two elements with neutrons—slow, fast, and moderate in speed—to study how neutrons diffracted within solid material. They used a wide range of techniques invented by Fermi, including a "neutron mirror," to measure the angular reflection of neutrons at various energies. Wattenberg recounts another characteristic Fermi moment. A group was looking for Fermi at Ekhart Hall, where the physics department was located, and someone told them that Fermi was in the machine shop in Ryerson Hall next door. They found him in discussion with one of the machinists, a fine and reliable tool and die maker with whom Fermi worked closely. Fermi was explaining to him how to use grit to produce the right kind of polish on a neutron mirror. The mystified machinist asked Fermi how to know if he had done the job properly. "I'll hold the mirror up," Fermi replied, "and if I can see my eyelashes it will be okay." Not an especially rigorous standard for a high-precision instrument, particularly, coming from a Nobel Prize–winning experimental physicist, but it was characteristic of Fermi's rough-and-ready approach.

These experiments, which continued through much of 1947, established an important base of information about how neutrons diffract through a wide variety of substances and were quite influential in setting the stage for research reactors to be used as test beds in materials science industry as well as in biological and medical research.

In his 1945 letter to Walter Bartky, Fermi mentioned the importance of studying cosmic rays. The more these rays were studied, the more fascinating physicists found them. Some of these rays pounded the earth's atmosphere with almost unimaginable energy, colliding with atoms in the earth's atmosphere and creating a wide range of subatomic particles in the process. The universe, it turned out, was itself an enormous particle accelerator and, if a physicist was sufficiently diligent and armed with the right type of detectors, cosmic rays could be studied at very high energies, far higher than those achievable in man-made accelerators.

After the war, scientists debated the origins of these high-energy rays. Some, like Edward Teller, believed that they originated in the sun and that solar processes heretofore not understood were responsible for them. Others believed that they originated deeper in the universe, in interstellar space. There seemed to be no clear way to decide between these two propositions. These questions interested Fermi deeply, as did another question: How did they get their almost inconceivable energy? What could possibly accelerate these particles to such high speeds?

Fermi first began to think about these issues in 1947–1948, largely at the instigation of Teller. Fermi—perhaps, as he joked later, simply to find a way to disagree with his old friend—decided to argue that they come from deep space, well beyond the solar system. Thinking perhaps of how earth-bound accelerators push charged particles to ever higher energies, he constructed a theory to explain the high energies of cosmic rays. Fermi hypothesized that the presence of large-scale interstellar magnetic fields could explain the speeds with which cosmic-ray particles showered the earth. The Norwegian astrophysicist Hans Alfven visited Chicago in 1948 and gave a lecture arguing for the existence of these interstellar magnetic fields. Fermi liked the lecture enormously, because it added weight to his own theory of cosmic-ray acceleration, although Alfven always doubted the existence of fields strong enough to account for cosmic-ray energies.

Fermi soon found another foil for his musings on the origins of cosmic rays, with one of the most fascinating physicists of the twentieth century—Subrahmanyan Chandrasekhar. Born in India, "Chandra," as Fermi came to call him, wrote his undergraduate thesis on the relationship between "Compton scattering"—the photon experiments that won Arthur Compton his Nobel Prize in 1927—and Fermi-Dirac statistics. Chandrasekhar won a doctoral fellowship to Cambridge University and did post-doc work under Max Born at Göttingen and Niels Bohr at Copenhagen. In short, he had as strong an academic pedigree as any major physicist working in Europe at the time.

While at Cambridge, Chandrasekhar was the first to predict that, for a star over a certain mass that runs out of nuclear fuel and goes cold, the gravitational force would be sufficient to overwhelm the degeneracy pressure implicit in Fermi-Dirac statistics, and the star would have no choice but to collapse upon itself to a point-like "singularity." This mass came to be known as the Chandrasekhar limit, that is, the limit beyond which a white dwarf star would collapse into what John Wheeler later dubbed a "black hole."* However, Chandrasekhar's work was initially ridiculed publicly by the powerful astronomer Sir Arthur Eddington, who made a point of ruining the young scholar's career. Chandrasekhar had no choice but to leave England, and, in a move of striking prescience, the University of Chicago hired him in 1937. He won the Nobel Prize for his work in 1983 and remained in Chicago until his death in 1995.

---

* A white dwarf star is the core remnant of a dead star in which the electrons and the nuclei are so compressed the atomic structure disintegrates, leaving the electrons in a degenerate state, compressed to the limit allowed by the exclusion principle. It is extremely dense, about two hundred thousand times the average density of the earth. National Aeronautics and Space Administration, "White Dwarf Stars," *Imagine the Universe!*, revised December 2010, https://imagine.gsfc.nasa.gov/science/objects/dwarfs2.html.

Given the central importance of Fermi's 1925 work on statistical mechanics to Chandrasekhar's entire career, the only surprising aspect of the relationship between the two is that it seems to have started so late. Chandrasekhar arrived at the University of Chicago in 1937 but waited until many years later to send Fermi a rather formal letter inviting him to visit the observatory some one hundred miles northwest of Chicago on the shore of Lake Geneva, Wisconsin, in November 1947. Perhaps the delay reflected Fermi's hectic schedule during the war or perhaps it reflected Chandrasekhar's shyness. He might simply have been waiting respectfully until Fermi had settled into the new institute. Fermi responded positively and met with Chandrasekhar. The two of them hit it off well, the letters between them became increasingly less formal, and in the summer of 1949 Chandrasekhar invited both Enrico and Laura to spend time with him and his wife at their home near the observatory.

In the fall of 1953, the discussions between Fermi and Chandrasekhar became more regular. Chandrasekhar would spend two days a week on the Chicago campus, and the two would meet to discuss cosmic rays, galactic magnetic fields, and the like over lunch at the faculty club. In discussions with Chandrasekhar, Fermi extended the work he did in 1948, taking account of the spiral arms of the Milky Way galaxy and the magnetic fields created by them. The two men published several papers on the subject together during this period.

Fermi had no formal training as an astrophysicist, but Chandrasekhar likened him to a musician who, confronted with a new piece of sheet music, could perform it the first time through with the brilliance of an artist. For Chandrasekhar, the experience of working closely with Fermi was one of the highlights of his life. For his part, Fermi was willing to confide in his new friend, especially when it came to explaining how he thought about problems.

In retrospect, Fermi and Chandrasekhar were only partly correct about how cosmic rays develop such high energies. Astrophysicists still believe that magnetic fields are responsible for some of the high-energy particles that collide with Earth, but these are not necessarily free-floating interstellar fields. Rather they are associated with supernovae and with some highly unusual objects that were unknown in Fermi's day, objects such as quasars, pulsars, and other highly dense, rotating objects with strong magnetic fields. Also, the universe is far larger than they knew then, with far more potential sources of radiation. Yet with all that was unknown at the time, Fermi's method of studying the subject and coming to tentative insights demonstrates the power of his approach.

THOSE LIKE BRUNO ROSSI, WHO HAD BEEN STUDYING COSMIC RAYS, knew that particles like protons, pions, and muons were continuously colliding with the earth's atmosphere, producing showers of other subatomic particles. The collision of pions and their subsequent decay into other particles was a subject of some interest to Fermi in the first part of 1947. Fortunately, he had a new graduate student who was not doing particularly well in his theoretical physics thesis topic and who was willing, indeed eager, to follow Fermi's suggestion and conduct an experiment to study muon decay in the atmosphere. In the process, the student unwittingly demonstrated for the first time the existence of one of the fundamental forces of nature. His name was Jack Steinberger.

Steinberger, a German Jew who arrived in Chicago with his parents before the war, had not only been having trouble with his thesis but had trouble with the physics program from the beginning. He failed the basic exams required prior to beginning work on his doctoral thesis and was, thanks to the generosity of department chair Walter Zacharaisen, the only student ever to be given a second chance. Fermi clearly liked the young, strikingly handsome Steinberger. He asked Steinberger to be his teaching assistant for a course in elementary physics that Fermi taught in the fall of 1946. Fermi also agreed to be Steinberger's thesis adviser after his student finally passed his basic exams.

Fermi could see that Steinberger was getting discouraged with his theory dissertation and suggested gently that perhaps an experimental project would be more to his liking. Fermi's old friend Bruno Rossi and others had been studying the way cosmic muons, created by cosmic pion decay, themselves decayed into electrons and had discovered far fewer electrons than predicted. The rate was lower than expected by a factor of two. Sensing that it might be an interesting story, Fermi suggested that Steinberger study the spectrum of electron energies resulting from these decays.

It is a rare PhD dissertation that makes an important scientific contribution to the field, but largely because of Fermi's sixth sense for important questions and also because of the unusual results of Steinberger's experiment, this dissertation research was an exception to the rule.

Fermi lent Steinberger an assistant, and together they made some eighty Geiger counter detectors configured appropriately to capture electron tracks from cosmic muon decays. The first phase of the experiment was completed at sea level and later repeated in 1948 at the top of Mount Evans in Colorado, at an elevation of 14,271 feet, to enhance the statistical significance of the results. Steinberger discovered that the energy spectrum of electrons emitted from cosmic-ray muon decay was continuous and even at the highest energies was not sufficient to account for nearly half the energy that the muon's decay should produce. Steinberger analyzed the data carefully and came to the conclusion that two neutral particles, "probably neutrinos," accompany each electron emitted from the muon decay. These two particles carried off the missing energy.

Neither Steinberger nor Fermi realized the most important implication of this result. Steinberger says that he himself was not "clever enough." It is difficult to see why Fermi missed it. Perhaps, as Steinberger suggests, "new ideas are not always easy to accept, sometimes even by the brightest and most open of people such as Fermi." In any case, several others did, including three University of Chicago graduate students, Tsung-Dao Lee, Chen-Ning Yang, and Marshall Rosenbluth. They proposed that what Steinberger discovered was a more general extension of the "Fermi interaction" underlying beta decay. They suggested that there was a "universal" force—alongside gravity, electromagnetism, and the "strong" force holding the nucleus of an atom together—that produced changes in particles and that the neutrino always appeared in these processes as a way to account for the energy that was not imparted to the other particles created in these processes. This universal force came to be known as the "weak" force, and the creation and/or destruction of neutrinos came to be called the "weak" interaction, because the force responsible for this can only be felt at the closest of distances.

Steinberger recalls that Fermi was extraordinarily generous with his time, organizing the logistics and funding for Steinberger, but never interfered with the actual conduct of the experiment itself, allowing the young physicist to make his own mistakes along the way. Fermi explained to Steinberger, as he would to Maria Mayer at about the same time, that if Fermi added his own name to the paper reporting the results of the experiment, everyone would think that Fermi had done all the work, which would be bad for Steinberger.

CONFERENCES CONTINUED TO REVEAL AND DISCUSS IMPORTANT NEW developments in postwar physics, but the action shifted to the United States, where a series of major conferences organized under the auspices of the National Academy of Sciences effectively replicated the prewar Solvay conferences. The first was scheduled for early June 1947 at Shelter Island, situated between the north and south forks of Long Island, New York. Fermi received an invitation and was eager to attend. The agenda included discussion of some of the most interesting new developments in physics: Willis Lamb on a strange anomaly he had discovered while measuring the energy of the two possible quantum states of the hydrogen atom, an anomaly which would soon be known as the Lamb shift; Rabi on his precise experimental measurement of the magnetic moment of the electron, a value that Dirac's theory of quantum electrodynamics could not compute; Robert Marshak on the Yukawa meson, whose experimental observation a few months later would vindicate Marshak's speculations; and Feynman on an informal, preliminary presentation of his work in quantum electrodynamics and his use of a strange new analytical tool based on graphic diagrams of interactions. The stellar attendance list included among others Bethe, von Neumann, Oppenheimer, Rossi, Teller, Uhlenbeck, Weisskopf, and Wheeler. Fermi would have been in his element.

Fermi never made it to the conference. Passing through Baltimore, he noticed that his vision was blurred. Understandably alarmed, he decided to have a doctor examine him. The cause was a torn retina, which would take the better part of a year to heal completely. During this time, his friends would often notice him counting the ridges of his fingerprints or carefully holding a pencil in front of his eyes, trying to focus on the edge of the pencil to make sure his eye was returning to normal.

A second conference, at a resort in the Pocono Mountains of Pennsylvania, took place in late March 1948. By this time Fermi's eyesight had recovered sufficiently for him to attend. In addition to the participants who were at Shelter Island, several other old Fermi friends attended, including Niels and Aage Bohr, Dirac, Wigner, and Wentzel. The main focus of the conference was quantum electrodynamics (QED), and two presentations on the subject made history.

For decades the niggling problem inherent in Dirac's version of QED defied repeated attempts, including those by Fermi, to solve it. The completion of the magnetic moment of the election converged not on a finite value, as one would expect, but instead diverged to infinity. Between 1946 and 1948, three young physicists independently gave the problem another shot, and this time all of them succeeded. It was a great achievement, one of the greatest of twentieth-century physics, and put QED in an almost uniquely effective category of physical theory. Two of these physicists were invited to the Pocono conference—Feynman and a thirty-year-old Harvard professor named Julian Schwinger. (The third, a Japanese theorist named Sin-Itiro Tomonaga, independently developed a solution similar to Schwinger's and would certainly have been invited to present at the conference if the organizers had known about his work.)

Schwinger and Feynman, though the same age, could not have been more different, as individuals or as physicists. Feynman was a gregarious and irrepressible showman; Schwinger was a quiet introvert, with no interest at all in entertaining his audience. As a physicist, Feynman's presentation reflected an informal, intuitive approach that depended upon his diagrams, which eventually found universal acceptance. Schwinger was more of a formalist, relying on equation after equation, carefully sequenced, to get to his solution. His presentation was so long and tedious that in the end only two physicists lasted through the entire lecture: Fermi and Bethe. Fermi took it as a badge of honor that he stayed until the last.

Although he had great respect for Feynman throughout his life, Fermi clearly preferred the Schwinger approach, based squarely on the traditional quantum field theory Fermi knew so well. Feynman's approach was anything but traditional, and the graphic tools Feynman used did not appeal to Fermi, at least not immediately. In 1951, Fermi and Franck jointly nominated Schwinger for a Nobel Prize. They cited Feynman's and Tomonaga's QED work, but they believed that "the greatest contribution was made by Schwinger." Feynman never knew this. He continued to correspond with Fermi on a wide variety of physics problems and visited Fermi from time to time at Chicago. Feynman had enormous respect for Fermi, and part of his eagerness to bounce ideas off Fermi may have been an unconscious effort to gain the validation as a physicist that he sought but did not receive at the Pocono conference.

Why is it that Feynman, Schwinger, and Tomonaga succeeded in 1947 where Fermi (and others) failed in the early 1930s? Were they just better theorists? Given what we know about Fermi's formal abilities in mathematical physics and his ability to crack profound problems in quantum field theory, such a conclusion seems, on the surface at least, too facile. Perhaps a better explanation is rooted in Peierls's observation that Fermi was attracted to problems in which the mathematics was relatively simple. Peierls's implication is that Fermi was an impatient theorist, that when a problem did not quickly succumb to his intellect he quickly lost interest in it. Another possibility: Fermi was an intensely practical physicist, not inclined to put more effort into a problem than he felt it was worth. It is entirely possible that he knew that the first approximation of the magnetic moment of the electron provided by Dirac's theory was good enough for most practical purposes and that finding a way to calculate the value out to five or six decimal places was simply not worth his time and energy. As Fermi told his daughter Nella in another context, "never make something more accurate than absolutely necessary." After the war it was not a problem to which he devoted any time. This is not to deny the historic achievement of these three, who shared the Nobel Prize for their work in 1965. Yet for Fermi, it may not have been that interesting a problem or perhaps he knew that by 1947 he didn't have the youthful brilliance and intellectual stamina required to crack it and was happy to give others a chance. For whatever reason, he was clearly more concerned with other scientific matters.

Academic summers once again became periods of relative freedom, and Fermi took advantage of them as he had before the war. Most summers he spent six or eight weeks at Los Alamos, doing research on various projects, mostly classified, and spending time with friends, old and new. Soon after the war, the McMahon Act of 1946 placed the control of all atomic energy research under US civilian government monopoly, focused on a handful of national laboratories, including Los Alamos. Congress initially considered a more restrictive law that would have kept research under military control. Fermi, eager to present a united front with Oppenheimer, supported this along with Compton and Lawrence. However, other Manhattan Project scientists were outraged by continued military control of nuclear research and particularly unhappy with the four scientists who gave their support. Herb Anderson delivered a stinging critique of the group that included his old friend and mentor:

> I must confess my confidence in our own leaders Oppenheimer, Lawrence, Compton, and Fermi, all members of the Scientific Panel, . . . who enjoined us to have faith in them and not influence this legislation, is shaken. I believe that these worthy men were duped—that they never had a chance to see this bill. Let us beware of any breach of our rights as men and citizens. The war is won, let us be free again!

Widespread opposition to the bill ultimately led its sponsors to withdraw it and replace it with a 1946 bill, sponsored by Connecticut senator Brian McMahon, putting the nuclear program under civilian control. It created the Atomic Energy Commission (AEC), whose first head was

FIGURE 21.3. Summers at Los Alamos after the war were spent with old friends. Here is Fermi, third from left, on an afternoon jaunt with Hans Bethe, L. D. P. King's son Nick (behind the wheel), and Edward Teller's son Paul. *Courtesy of Los Alamos National Laboratory.*

former director of the Tennessee Valley Authority David Lilienthal. It established a General Advisory Committee (GAC) of science and technology experts to advise the AEC. Oppenheimer was named the GAC's chairman and Fermi was selected for a four-year term. Under the AEC, new labs at Los Alamos, Argonne, Oak Ridge, Hanford, and Berkeley continued working to refine the country's atomic arsenal and pursued basic research on a variety of scientific fronts. The new director of Los Alamos, Norris Bradbury, was eager to bring Fermi back to the mesa, and Fermi was delighted to return, family in tow.

Upgrading from wartime austerity, the Fermi family settled into one of the nicer homes on Bathtub Row and Enrico cycled to work, while Laura and the children socialized and pursued summer activities. He worked on a variety of classified projects, but two unclassified projects were of particular interest.

First, he became enamored with computers. Before electronic computers were readily available, he designed analog computers that could help him with his calculations. One such computer, built in Chicago with the help of graduate student Richard Garwin, was capable of solving Schrödinger equations. Another one, built at Los Alamos with the help of his old water boiler colleague L. D. P. King, used Monte Carlo methods to trace the path of a neutron through matter, simulating the paper and pen calculations Fermi did on neutron diffusion before the war. It was affectionately dubbed "Fermiac," after the early electronic computer ENIAC.

By 1947–1948, von Neumann developed one of the first programmable electronic computers and it became the focus of attention at Los Alamos. The father of computational physics, a young Manhattan Project veteran named Nicholas Metropolis, used von Neumann's machine to explore how physics equations could be programmed into a computer to simulate physical processes and predict the outcome of experiments. With von Neumann and Stanislaw Ulam, Metropolis invented the modern computerized "Monte Carlo" method of simulating stochastic physical processes. Fermi jumped at the chance to put his own equations into the computer, working with Metropolis on a wide variety of studies, including what would happen if a high-energy pion hit the nucleus of a relatively simple hydrogen atom. Several years later, these simulations allowed Fermi to compare his theoretical predictions with actual experimental results when the Chicago cyclotron went live. Later, in 1953, Fermi worked with Ulam and another computational physicist, John Pasta, to study the problem that absorbed him in his 1918 application to the Scuola Normale Superiore. They programmed the equations of a vibrating string into the Los Alamos computer and simulated its behavior. The resulting paper became an early and crucial contribution to chaos theory by showing that the string would return to a specified state at regular intervals, which was inconsistent with its expected ergodicity. In this respect it was a direct lineal ancestor of Fermi's very early concerns about ergodic systems and their behavior.

A second project that captured Fermi's imagination during summers at Los Alamos was "Taylor instability." An important concept in hydrodynamics, Taylor instability refers to what happens when the surface between two fluids of different density—oil and water, for example—is perturbed in some way. The complex interaction between the two fluids on a surface is extremely difficult to model mathematically, but the phenomenon becomes extremely important in certain types of events, including nuclear explosions. Fermi and Ulam published several papers on this crucial aspect of hydrodynamics.

The periods at Los Alamos gave Fermi time to stretch his mind in the company of a group of extraordinary physicists and mathematicians. It also provided time for relaxation and exercise. Fermi continued hiking and fishing around the Los Alamos area, habits acquired during the war. He also enjoyed playing tennis with anyone who was willing to accept his challenge, as Ulam did frequently.*

BY 1951, THE CYCLOTRON IN CHICAGO WAS UP AND RUNNING AT A then-impressive energy of 450 MeV, one of the most powerful cyclotrons in the world. Fermi had been preparing for this moment for some time. He summarized his preparations in lectures presented at Yale in

* Ulam once won a match with Fermi, 6–4. Fermi refused to concede defeat. He pointed out that the difference between the two scores was less than the square root of the sum of the scores, 3.17. This is a shorthand method used by statisticians to determine whether a result is significant or within the limits of measurement error. Ulam found Fermi's response at once ridiculous and adorable and continued to play tennis with his competitive friend. *Adventures of a Mathematician,* 164.

the Silliman Lectures of April 1950, subsequently published in a short volume called *Elementary Particles*. In the lectures, he introduced a "statistical theory of pion production" providing a "plausible approximation" of high-energy collisions. When it turned out not to be an exact predictor, he resented the criticism, correctly pointing out that he had never intended it to be so. He also prepared a paper for presentation at the opening of the cyclotron in September 1951, an event attended by more than two hundred distinguished physicists from around the world. He had already conducted a few preliminary experiments and reported on these. The celebration gave him a chance to catch up informally with old friends.

He had thought carefully about what experiments he wanted to conduct and what might be interesting to explore. His main interest was an exploration of the "strong force," the force that holds the nucleus of an atom together, by probing the nucleus with the pion, the particle Yukawa first suggested in 1935. In 1951, physicists believed the pion was responsible for the strong force and would interact in interesting ways with "nucleons"—that is, the protons and neutrons that dance together inside the nucleus. Importantly, the Chicago cyclotron was one of the only machines powerful enough to create pions with sufficient energy to probe the nucleus.

Fermi's experiments, conducted with Anderson and a group of younger physicists, created a beam of pions by accelerating protons in the cyclotron to very high energies and then "smashing" them into a target. The protons stimulated the nuclei of the target to emit pions, and the pions were then used to probe other nuclei—in this case, hydrogen nuclei that consisted either of a simple proton or of a "deuteron," a proton and a neutron. Pions hit these nuclei and bounced off—"scattering," in physics terminology—and this scattering revealed interesting things about nucleons and their relationship to pions. The three different types of pions—positive, negative, and neutral—had slightly different masses, and keeping track of the way these different pions interacted with nucleons in the hydrogen nuclei told even more about nucleons.

In the process of studying these scattering patterns, Fermi realized that he produced a new particle, created when the proton in the nucleus was hit by a pion in a range of energies centering on 180 MeV. This was the first time protons were struck with such energy by a particle capable of exploring the strong force inside the nucleus. The particle, an extremely short-lived one, was at first known by its somewhat exotic quantum state and is now called the "delta plus plus" and fits quite neatly into the group theory framework of heavy particles (baryons) proposed by Murray Gell-Mann a decade later.

Physicists call this type of particle a "resonance." Imagine that the nucleus is a radio station broadcasting at a number of different frequencies. The cyclotron is a radio receiver that can be tuned continuously to frequencies up and down the spectrum. When the listener starts tuning, only static might be audible, but then specific frequencies will come in and out of range, with clarity. In an analogous manner, the cyclotron produces pions that "tune in to" the nucleon at energies centering around 180 MeV, stimulating it to produce the new, somewhat exotic particle.

Analysis of these results suggested that protons were fairly complicated particles, which could be stimulated into a number of different states. Perhaps they were not "fundamental" in the sense of the muon or the electron. Perhaps they had an internal structure that could be further explored. It was an important insight that pointed the way to future developments in particle physics, developments that would entirely change the way these heavy particles were viewed. Fermi followed up these experiments with computer simulations at Los Alamos, aided by Metropolis, that were among the first particle simulations ever conducted.

Fermi wrote nine papers based on these experiments and computer studies. They were the last experimental papers he ever published. Herb Anderson writes that from mid-1953 onward, three Fermi graduate students—Jay Orear, Arthur Rosenfeld, and Horace Taft—ran cyclotron experiments that prevented Fermi from pursuing his own studies more thoroughly:

> Through my illness [berylliosis], he lost a major supporter who was willing to help smooth the way and cater to his way of doing things. His new students, Rosenfeld, Orear and Taft asked his guidance and advice but wanted the work to be their own. So Fermi changed his role; he spent more and more time helping others by discussion and by frequently lending a hand in the experiment, but never again to the extent that would allow him to admit that the work was his own.

He goes on to suggest that Fermi's decision to study the origins of cosmic rays with Chandrasekhar was a result of being sidelined by his graduate students.

Anderson was surely too critical of Fermi's students. Fermi certainly did not need Anderson's support to get time on the cyclotron. Fermi was the most prominent and powerful man at the institute and if he wanted to do an experiment on his own he certainly would have done so. He also had a long-standing fascination with the origins of cosmic rays dating back to 1947 and was happy to pursue conversations with Chandrasekhar without being forced to do so. What Anderson misses—perhaps through jealousy because Fermi was beginning to adopt a new, younger group of physicists to mentor—is that Fermi actually enjoyed working with his graduate students, encouraging their work, supervising their experiments on a big new instrument with such potential. Fermi justifiably had first crack at the cyclotron and by mid-1953, with two years of solid work under his belt, probably felt it was time to let his younger colleagues have a chance.

Anderson did have a jealous streak, as Richard Garwin found out. In late 1952, Garwin, who was as close to Fermi as anyone during this period, had a meeting with Anderson. The latter was rather direct: there was only room for one of them at the institute going forward. Garwin took that to mean Anderson and found a job at IBM's Watson Lab in Yorktown Heights, New York. Anderson was rightly proud of his long-time collaboration with Fermi. Of all of Fermi's collaborators in Rome and in the United States, Anderson probably worked with Fermi the longest. He was used to serving as a sort of gatekeeper, particularly for younger members of the team and resented it when physicists like Garwin or Rosenfeld developed their own special ties to the great man. Perhaps Anderson could be forgiven, though. He would eventually die of a disease contracted in a moment of heroism, when he dashed into a lab room ahead of Fermi and Zinn to put out a beryllium fire. In doing so, he spared both colleagues a similar fate.

Freeman Dyson, the young theorist who did so much to reconcile the work of Feynman, Schwinger, and Tomonaga, provides a revealing coda to Fermi's work on pion-proton scattering. Dyson was a junior professor at Cornell, responsible for supervising a small group of graduate students. They decided to tackle theoretical calculations of pion-proton interactions using the same technique successfully used to analyze quantum electrodynamics. The forces governing the pion-proton interaction are much stronger than those of electrodynamics, but Dyson and his students did not consider this a problem and got results that were fairly similar to those Fermi achieved using the Chicago cyclotron. This was the work of several years, and when completed in the spring of 1953, Dyson took a bus from Cornell to Chicago to show Fermi what they had done.

Dyson was eager to show Fermi his work. They chatted for a while about personal matters and then Fermi turned to Dyson's results. Dyson recalled Fermi's judgment on the work in 2004, some fifty years later: "There are two ways of doing calculations in theoretical physics," Fermi explained. "One way, and this is the way I prefer, is to have a clear physical picture of the process that you are calculating. The other way is to have a precise and self consistent mathematical formalism. You have neither."

Dyson was understandably stunned and asked Fermi to elaborate. Fermi explained that the mathematical technique Dyson used was inappropriate for the problem he was trying to solve. When Dyson objected that his results came very close to the numbers that Fermi himself had measured in his 1951–1952 experiments, Fermi pointed out that the arbitrary number of parameters undermined Dyson's calculations. "I remember," Fermi replied, "my friend Johnny von Neumann used to say, with four parameters I can fit an elephant, and with five I can make him wiggle his trunk." With that, Dyson made his way back to Cornell with the sad news that his work of several years had not passed Fermi's test.

In retrospect Dyson was not bitter, but rather grateful that Fermi spared him from spending even more time on what was, really, a dead end:

> Looking back after fifty years, we can clearly see that Fermi was right. The crucial discovery that made sense of the strong forces was the quark. Mesons and protons are little bags of quarks. Before Murray Gell-Mann discovered quarks, no theory of the strong forces could possibly have been adequate. Fermi knew nothing about quarks, and died before they were discovered. But somehow he knew that something

> essential was missing in the meson theories of the 1950s. . . . And so it was Fermi's intuition, and not any discrepancy between theory and experiment, that saved me and my students from getting stuck in a blind alley.

THOUGH MUCH OF FERMI'S POSTWAR RESEARCH WAS PUBLIC, A significant portion of the summer work he pursued at Los Alamos throughout the period from 1946 through 1952 was classified, none more so than work on the hydrogen bomb. In this work, Fermi combined the roles of scientist and public policy adviser, the latter of which he found somewhat uncomfortable. Public policy was to become a major headache for Fermi and led to difficult, somewhat contradictory decisions. It also led to one of the most dramatic moments in his life, the defense of his old colleague J. Robert Oppenheimer against charges of disloyalty to the US government.

CHAPTER TWENTY-TWO

IN THE PUBLIC EYE

As a central figure in the Manhattan Project, Fermi could hardly have been surprised when he was nominated to serve on the first General Advisory Commission (GAC) to the AEC. His Manhattan Project colleague Robert Bacher, a commissioner himself, and another commissioner, Carroll Wilson, developed a list of scientists and engineers who could be relied upon to give expert advice to the AEC as it took control over the US programs on nuclear weapons and nuclear energy. The list contained all the usual suspects. Conant, Oppenheimer, Rabi, Seaborg, and Cyril Smith were also on the list, as were former MIT Radiation Lab director and newly appointed president of Caltech Lee DuBridge; Hood Worthington, an official at DuPont who worked on the Hanford reactors; and Hartley Rowe, another Los Alamos consultant. The inaugural meeting of the GAC was held in Washington the first week of 1947 and was convened there every several months for two or three days at a time.

Fermi underwent a full background check for this assignment, now under the auspices of the Atomic Energy Act of 1946, for the new "Q" clearance. His FBI file makes for interesting reading and is more thorough than the one prepared for him by the FBI in 1940. Those interviewed this time overwhelmingly supported granting him clearance, vouching for his trustworthiness and his brilliance as a physicist. Zinn described Fermi politically as an "ultra-conservative." Norman Hilberry and Walter Zachariasen both considered him "the greatest living physicist," which would probably have come as a surprise to Einstein and Bohr, both of whom were still alive. Hilberry went so far as to state that Laura's family had been fervently anti-fascist, which was simply not true. These statements reflected less the mature judgment of those who were interviewed than an enthusiasm for ensuring that Fermi passed his background check.

Perhaps more fascinating, though, is the testimony of John Dunning, his former Columbia colleague and the builder of Columbia's first cyclotron. Dunning started off by saying that he had no doubt as to Fermi's personal integrity and loyalty to the United States, but then added that he was opposed to giving foreign nationals security clearances of any type. He pointed out that once someone had changed allegiance, that person might change it again. In this regard he pointed to Fermi's decision to move from Italy to the United States in 1939. Incredibly, he also pointed to Fermi's move from Columbia to Chicago as an example of changing allegiance, moving from one university to another. Surely, he knew that Compton ordered Fermi to move to Chicago, along with the entire pile project. He must have felt left behind as a result and whatever resentment he felt toward Fermi bubbled up during his interview with the FBI.

The FBI wisely ignored Dunning's remarks, much as they had LaMer's in 1940. That Fermi's brain was the source of much of what was classified in atomic research did not seem to occur either to LaMer or to Dunning, or if it had, it was not enough to stifle their resentment. Fermi was granted a Q clearance, which he used both for GAC work and for summer work at Los Alamos.

No minutes were taken during the free-wheeling, scientific and technical discussions of the GAC, chaired by Oppenheimer. The topics were all highly classified, covering a range of issues relating to the development and refinement of the "conventional" nuclear arsenal. The hydrogen bomb remained a topic of intense interest, but so were the program on fission weapons and the continued development of nuclear reactors at Hanford and later at Savannah River, South Carolina, operated jointly by the AEC and by DuPont.

David Lilienthal, who headed the AEC from 1946 through 1950, was no stranger to the corridors of political power in Washington. Sophisticated though he was, the GAC scientists impressed him mightily, none more so than Fermi. In the first week of the GAC's deliberations, Lilienthal had lunch "in a terrible little cafeteria in the War Department building" with Rabi and Fermi. "To have spent the day with Fermi," he wrote that evening in his diary "is like saying that one spent the day with Copernicus or Galileo or the primitive who discovered fire."

Fermi attended almost all of the meetings during his tenure on the GAC and participated when he had technical expertise or might have had strong feelings. For example, in policy deliberations exploring the choice of expanding the national laboratory program or strengthening existing programs, he leaned toward the latter option. He was outspoken in his support for strengthening Los Alamos and, according to Segrè, thought that in a period of scarce resources the work at Los Alamos should have priority over the development of civilian nuclear power reactors. He also supported the building of a high-flux reactor for continued advanced research. Among the issues discussed at the GAC, the only official written record of his contributions relates to a series of critical meetings in late October 1949.

Overshadowing these meetings was the surprise detonation of the first Soviet fission test in August 1949. Like the launch of *Sputnik* eight years later, the first Soviet nuclear test raised alarm throughout the American national security establishment. There could be little doubt that brilliant Soviet physicists were also at work on a fusion weapon. President Truman and his national security advisers wanted advice on whether to launch an all-out effort to create a fusion weapon.

The technical problems were still as daunting as they were during the Manhattan Project: how to keep the device held together long enough to fuse a meaningful amount of hydrogen. These problems were discussed at length during a major classified symposium during the summer of 1946 at Los Alamos, attended by Fermi and many of his Manhattan Project colleagues, but the challenges remained. Teller and Ulam cracked the puzzle in 1951, but that was two years in the future, and in October 1949 many scientists, including Fermi, were still skeptical that a practical weapon was feasible.

Under significant political pressure to act, the AEC asked the GAC for its views. The question came to discussion and vote on October 29, 1949. The entire GAC voted unanimously against moving forward. The main report focused on three points: the hope that "the development of these weapons can be avoided"; a reluctance to see the United States take the initiative in this matter; and, finally, "that it would be wrong at the present moment to commit ourselves to an all-out effort toward its development." They were also concerned that such an effort would divert resources and energy away from what was seen to be an equally important national security priority, the continued development of weapons based on nuclear fission. Given the uncertainties surrounding the feasibility of the hydrogen bomb, it seemed more responsible to focus on improving the existing fission stockpile.

Two addenda accompanied the main report. One was written by Harvard president Conant and signed by seven of the nine GAC members. They argued that such a weapon should never be built; that it was far from certain that such a weapon would be technically feasible; that if the Russians were to develop such a weapon our atomic arsenal would provide an adequate deterrent; and that the United States had an opportunity to provide "by example some limitations on the totality of war." Having let the nuclear genie out of the bottle, these scientists now had second thoughts and were looking for ways to persuade him to return.

Rabi and Fermi offered a far more forceful "minority report," which focused explicitly on the moral component of the question. It was an extraordinary statement, probably drafted by Rabi, but with Fermi's collaboration. Rabi had long supported aggressive efforts to find a way to establish international control over these weapons and their fundamental technology, and Fermi was sympathetic to such an objective, although probably more skeptical than Rabi of its potential for success. For Fermi, opposition to moving ahead with a program for the hydrogen bomb may have reflected the same inner conflicts he experienced periodically during the Manhattan Project. The two friends began their report with the observation that the Super would only provide an advantage over conventional nuclear weapons if the destructive force were a hundred to a thousand times that of "ordinary atomic bombs," which could destroy some 150 to 1,000 square miles of territory. Such a weapon, they argued, went far beyond any military applications; indeed, it approached the level of great natural catastrophes. "It is clear that the use of such a weapon cannot be justified on any ethical ground which gives a human being certain individuality and dignity even if he happens to be a resident of an enemy country." Such use would place the United States in a questionable moral position in the eyes of "the peoples of the world." The radiation released by these weapons would make large areas of the planet uninhabitable for "a very long time."

The two scientists ended their minority report with a call to invite other nations to join the United States in the pledge:

> The fact that no limits exist to the destructiveness of this weapon makes its very existence and the knowledge of its construction a danger to humanity as a whole. It is necessarily an evil thing considered in any light. For these reasons, we believe it important for the President of the United States to tell the American people and the world that we think it is wrong on fundamental ethical principles to initiate the development of such a weapon.

The GAC report with addenda went to the AEC for consideration. After much debate, the commission voted to support the specific policy recommendations of the GAC, with predictable responses from Lawrence and Teller, both of whom frantically opposed ending work on the Super. Other physicists lined up on either side of the debate. For example, Leona Libby confronted Fermi sharply when she learned of his opposition and reports that, uncharacteristically, he responded with a volcanic outburst, defending himself against her attack. Debate within the scientific community raged over the next few months, but once again, as in the decision to use the atomic bomb against Japanese targets, President Truman paid little attention to the scientists. He believed, as did several members of the AEC itself, including most notably future AEC chairman Lewis Strauss, the United States should develop the weapons before the Russians did and more broadly had little faith in efforts to exert international control over nuclear weapons. The president also had to deal with the political fallout of the January 1950 revelation that Klaus Fuchs had been spying for the Soviet Union since the beginning of the Manhattan Project. Weighing both political and military considerations, President Truman rejected the AEC's advice and ordered a high-priority effort to develop the hydrogen bomb.

Truman's directive notwithstanding, the scientific obstacles to actually developing the fusion weapon continued to challenge scientists. During the summer of 1950, Fermi returned to Los Alamos and continued working on how to solve this major technical puzzle. He brought Richard Garwin with him and they worked closely that summer.

Fermi often expressed skepticism about the practicality of Teller's concept. By the summer of 1950 he was still a skeptic, as was Ulam. They both worried that with Teller's design, the hydrogen fuel would only partially fuse before the reaction burned out. The calculations to determine the nature of this problem were complex, and Fermi and Ulam decided on a good-natured race to see who could come up with the answer first. Fermi wanted to use slide rules only, but Ulam's technical department insisted on using the services of a group of young women "computers" using mechanical calculators. Ulam and Fermi did some preliminary calculations to set up a spreadsheet for the women to complete and the women came back with the results, which the scientists incorporated and used to direct additional rounds of calculations. One particularly attractive and buxom young woman, Miriam Planck (no relation to the quantum pioneer Max Planck), spent a fair amount of time with Fermi, which Ulam and John Wheeler both noted with amusement.*

Throughout the summer, Ulam and Fermi, and others, notably a physicist named C. J. Everett, repeatedly pointed out fatal flaws in Teller's Super concept. Teller would return time and again with ideas to overcome the concept's deficiencies, but each time someone would find a new fatal flaw. It took a flash of insight by Ulam in early 1951, who realized that compressing the entire fusion assembly during the fission trigger would create enough pressure to hold the configuration together just long enough for the device to work. Teller's contribution was to think of reflecting the high-energy X-rays emanating from the fission

---

* According to Ulam, Ms. Planck would bring the completed daily spreadsheet to the two of them, place it on the desk, lean over, and ask, "How do they look?" Ulam, gazing up at her bosom, would reply, to Fermi's great amusement, "They're marvelous!" Ulam, *Adventures of a Mathematician,* 218.

trigger to create that compression. Ulam and Teller worked out the main details of this solution in a paper written in March 1951. By November 1952, the Ivy Mike test, conducted on the South Pacific island of Eniwetok, confirmed the feasibility of a fusion weapon based on the Teller-Ulam concept.

That Fermi continued to work on the project after having helped to draft such a strong statement against the weapon's development is perhaps the single most puzzling aspect of his postwar public service. He spent the summer of 1950 at Los Alamos, immersed in the fusion project along with many old friends such as Ulam, Teller, Wheeler, and Bethe. Bethe, too, was initially opposed but came on board after the Korean War broke out in June 1950. It remains a mystery what drew Fermi to join the project when eight months before he labeled this weapon "necessarily an evil thing considered in any light."

Fermi never wrote about his decision to participate. Perhaps he wanted to be viewed as a patriot and, given the president's decision on behalf of the country to move ahead, he felt working on the project was his duty. The physics issues were as compelling as they had been at Los Alamos six years previously. This may well have added to the allure of working on the fusion weapon. Perhaps because the project was going ahead anyway, Fermi felt he should participate from inside rather than look in from the outside. On the inside, he could continue to play the role of the "pope who plays the devil's advocate," a role we know he enjoyed.

On the surface it is surprising that Fermi did not come up with the Ulam-Teller idea first, because in retrospect it seems so obvious and should have been to someone of Fermi's enormous intellect. Both Fermi and Garwin worked extensively on X-ray pressure issues during that summer, yet neither of them thought of using that pressure to keep the assembly together for the required amount of time. It fell to Ulam to come up with the original insight of compression and to Teller to extend the compression to X-ray pressure. Fermi's reasons for participating remain a mystery to this day. In the end, he was not the one who invented the hydrogen bomb.

When Fermi's four-year term on the GAC was up, in early 1951, Oppenheimer asked him to stay on. Fermi gave a number of reasons for refusing but ended by saying that he had come to distrust his own judgment. The timing, of course, was bad. The Chicago cyclotron was about to go live and over the previous five years he had been planning a number of important experiments that would now take priority. One also senses that this expression of distrust in his own judgment was not false modesty, that important public policy debates made him deeply uncomfortable. The further away discussion veered from physics itself, the more he felt that he had little of value to contribute, especially because within his own field of expertise he had so very much to contribute and had the confidence in his ability to make a difference. Public policy issues were, in the end, governed not by equations on a page or experimental results one could publish in journals but rather by an uneasy combination of technical fact and value-based opinion. When Fermi friends like Teller and Rabi could disagree among themselves so vehemently, there were manifestly no "correct" answers. Fermi must have felt himself at sea. From this point onward he would do what he could to avoid public policy debates and stick to his physics, with only partial success.

In 1952, public policy issues again appeared on Fermi's radar screen, when he received a letter from New York attorney Emanuel Bloch, who was representing the most notorious couple in the United States at the time, Julius and Ethel Rosenberg. Charged with providing atomic secrets to the Soviet Union, they had been convicted more than a year earlier and now Bloch was desperate to find some way to spare them from the death penalty imposed at sentencing. He wrote to Fermi asking him to opine on whether certain information was generally available in the public domain.

Fermi clearly wanted nothing to do with the case. He referred Bloch to Gabriello Giannini, his former student who, in his current role as the manager of the patent litigation proceeding on behalf of the Via Panisperna boys, would be in a better position than anyone to determine what knowledge was in the public domain and what was not. It was just about this time that many of the patents Fermi and Szilard had applied for were coming up for declassification review. Perhaps Fermi hoped that Giannini might know of information that would serve to lessen the severity of the sentence. Fermi himself chose to be of no direct help and, when Bloch wrote again in late 1952 asking for intervention on behalf of his clients, the letter appears to have gone unanswered. Fermi must have found this correspondence upsetting. He firmly believed that no scientific discoveries could remain secret forever, and so probably viewed Soviet acquisition of nuclear weapons as a foregone conclusion, especially given the considerable talents of Soviet physicists. Nevertheless he was intensely loyal to the country he now called home and must have despised the idea of a native-born couple who could betray their nation.

In 1947 Karl Darrow, the secretary of the American Physical Society (APS), the main professional organization of physicists in the country, invited Fermi to be nominated as vice president of the APS. The vice president position was a relatively undemanding role, typically given to someone who would agree, after a year, to become president of the APS. Fermi politely but firmly refused, citing professional commitments that would prevent him from performing either role. He had already accepted the GAC appointment and had a full teaching and research schedule at Chicago. He clearly had no intention of distracting himself further with a largely administrative role, much less the higher profile role of president the following year.

Darrow must have approached Fermi several times in the following years, because the message sent to Fermi in 1951 was more than a bit aggressive. When Fermi tried to decline, Darrow informed him that he was the nominating committee's only choice, that there was no way within the constitution of the organization to take him off the ballot, and that Darrow "dared to hope that you [Fermi] will not wish to join the Italian gentleman that Dante placed in the Inferno because he made 'Il gran rifiuto' [the Grand Refusal]." Under such intense pressure, Fermi reluctantly agreed to serve as vice president for the 1952 term and in 1953 gracefully became president. Fermi made sure that his old friend Hans Bethe came along for the ride as vice president, positioning him to take the reins in 1954. Bethe had been a friend from the early days, and the two had worked closely together at Los Alamos. Bethe also had a reputation as an easygoing, likeable personality with the scientific gravitas to step into Fermi's shoes when Fermi's term was over.

Fermi's vice presidency was uneventful and distracted him little from his experiments at Chicago. However, his tenure as president was an entirely different matter. Almost at once he found himself embroiled in a purely political controversy that, almost seventy years later, seems faintly ludicrous but that, at the time, preoccupied Fermi, the APS, and the physics community in general for several months.

Allen Astin, the director of the National Bureau of Standards (NBS), was suspended by the Secretary of Commerce to whom he reported, because the bureau had informed the US Postal Service that a certain battery additive did not measure up to the claims of its manufacturers, thus making it impossible for the company to use the US mail to distribute the product. The new Secretary of Commerce, a Republican businessman and politician named Sinclair Weeks, declaring that the marketplace should be free to decide these matters, directed the postmaster general to ship the offending battery additive, and summarily suspended Dr. Astin for his temerity. Legitimately fearing the chilling effect that such an action would have on advice given to government agencies, the APS, under Fermi's leadership, issued a strong statement in support of Dr. Astin's scientific integrity, urging the government to adopt policies designed to maintain the independence and integrity of scientific advice. In August 1953, Secretary Weeks reversed Dr. Astin's suspension but directed the postmaster general to lift the ban on shipping the product. No one at the APS was particularly satisfied with the outcome, but the organization could say with some satisfaction that it had defended one of its own.

Perhaps of greater significance in retrospect was a little-known controversy that Fermi presided over in relation to the summer 1953 meeting of the APS, scheduled to take place at Duke University in Durham, North Carolina. During the planning for the meeting, APS Secretary Darrow pointed out to Fermi that the hotels in Durham were segregated. Though some accepted African American guests, some were "whites only." There were very few African American members of the APS, but the question nevertheless arose: Should the APS take some sort of stand on the matter? When Fermi was apprised of the situation, he was upset. His preference was for the APS to send out a full list of hotels in Durham, not indicating which ones were segregated. Perhaps he believed that African American members of the APS armed with full knowledge of the situation would force a showdown with "whites only" hotels. Others on the APS executive committee disagreed and pressed for a letter from Secretary Darrow to all African American members providing a list of Durham hotels that would welcome them so as to avoid confrontation. Ultimately, the executive committee prevailed over Fermi's wishes and the list of hotels that would welcome African American guests was sent to all members.

His problems were not yet over. In late 1953, the AEC electrified the physics world with the announcement that it would review J. Robert Oppenheimer's security clearances in light of suspicion that he might have been a Soviet spy during the Manhattan Project.

THE OPPENHEIMER CASE WAS NOT THE FIRST TIME THE LOYALTY OF US government scientists had come under scrutiny. In 1949–1950, the University of California required all professors to sign an anti-communist "loyalty oath." Highly controversial and divisive, the requirement affected professors and staff across all disciplines. Jack Steinberger, who was at Berkeley at the time, had not hidden his left-leaning political orientation and refused to sign the oath. Steinberger had never gotten along with Luis Alvarez, a major figure at Berkeley and a fairly conservative Republican, and his refusal only made things worse. In the end, Steinberger was forced to leave Berkeley and moved to Columbia. Considering that Steinberger would go on to do Nobel Prize–winning work at Columbia, this was clearly Columbia's gain.

In one of the few comments we have from Fermi on political issues during this period, he wrote a letter to his former student Geoffrey Chew in which he comments on the Berkeley situation. Fermi's attitude in the letter is surprising only if one fails to consider that he signed an oath of loyalty to the Fascist Party as a requirement for retaining his professorship in Rome. Fermi indicated to Chew that he didn't understand what the fuss was about, because the Berkeley oath had been so "watered down" as to be, in Fermi's mind, essentially meaningless. Clearly, Fermi would have signed and was fairly relaxed about such oaths if they were the price to be paid for freedom in the lab.

Fermi was also aware of the case of Edward Condon, the director of the National Bureau of Standards in 1948, who came under fire from House Un-American Activities Committee (HUAC) chairman J. Parnell Thomas for supposed ties to the Soviet Union. The charges were baseless and the APS came to Condon's defense. President Truman used the opportunity to show his defiance of the HUAC's agenda and publicly endorsed Condon. Eventually, the case lost momentum, although the HUAC occasionally resurfaced the charges, with little substantive effect.

The Berkeley loyalty oath controversy and the Condon case were both high profile, but the case against Oppenheimer was in a different league altogether. Americans were astonished that the charismatic leader of the Manhattan Project could be suspected of disloyalty to the US government. It would become a defining moment in the relationship of science and government and an iconic drama in American Cold War politics.

There were many factors weighing against Oppenheimer. Most of these would have been set aside in a different political environment and some of the allegations against him were simply untrue, but in 1953–1954, with fear of Communist infiltration weighing on the political consciousness, these factors led almost inexorably to a confrontation. Most of those who knew Oppenheimer were at least vaguely aware of his left-wing political leanings, justification enough for a full investigation at the height of the McCarthy period. He had also mishandled a potentially disastrous situation during the war, when an old friend at Berkeley, Haakon Chevalier, approached him with an offer to pass along atomic secrets to the Soviet Union. He never acted on the suggestion, but he delayed reporting the overture to authorities and initially dissembled when doing so.

In this context, Oppenheimer's opposition to proceeding with the Super at the GAC session in October 1949 also counted against him. It was easy to paint him as the person who led the GAC to advise against proceeding with the Super, hence someone whose commitment to the nation's security was suspect.

Another factor weighing against Oppenheimer was Edward Teller's role in stirring up resentment against Oppenheimer. Teller bitterly resented the way that Oppenheimer sidelined him from 1942 onward and was furious over the 1949 GAC recommendation against the Super. The feeling, apparently, was mutual. Oppenheimer gladly took the opportunity of the hearings to roundly criticize Teller's behavior during the Manhattan Project, explaining that Teller consistently refused to do the work that he was assigned to do and focused full time on the Super. Oppenheimer was correct in deciding that the technical challenges to the Super would not be solved before the war's end, but Teller never forgave him. Teller's resentment of his treatment continued well beyond the Manhattan Project and led to his successful campaign to create a rival weapons laboratory in 1952 under the University of California's auspices in Livermore, California.

The charges that Los Alamos dragged its feet in the development of the hydrogen bomb continued to circulate. Fermi went public in October 1954 in defense of the Los Alamos leadership, saying that he was "deeply perturbed" at the implication in the book by Shepley and Blair, just published, that Los Alamos stalled the H-bomb project and praising the leadership of Norris Bradbury and the entire Los Alamos team. Fermi's graduate student Arthur Rosenfeld helped him draft a press release declaring, in part:

> Statements of this kind are bound to produce dissension and to set back the atomic program. It is true, of course, that Edward Teller is the hero of the H-production. But it is equally true that a single person cannot carry alone a job of that kind. A genius needs the support of many other men and organizations. The Los Alamos Laboratory developed and added to his ideas and brought them into practice.

In late 1953, with this debate over Los Alamos at the boiling point, Teller weighed in against his old Manhattan Project boss.

A letter in November 1953 from William Borden, a staff member of Congress's Joint Committee on Atomic Energy, to FBI director J. Edgar Hoover ultimately triggered the AEC action against Oppenheimer. Borden expressed his conviction that Oppenheimer was "in all likelihood" a Soviet spy. President Eisenhower ordered that Oppenheimer be completely cut off from the AEC and its work until the matter could be formally addressed. The FBI stepped up surveillance and tapped Oppenheimer's phone, recording privileged conversations between Oppenheimer and his lawyers that were then presented confidentially to the AEC. In December 1953, AEC chairman Strauss informed Oppenheimer that his clearances had been suspended and that the commission would hold hearings to allow all the facts to emerge before making a final decision. The hearings were held over several weeks, starting April 12, 1954.

Oppenheimer hired famed civil liberties lawyer Lloyd Garrison to represent him at the hearings. Garrison reached out to many physicists, including Fermi, encouraging them to support his client at the hearings. The vast majority of those who had worked with Oppenheimer during the war came to his defense and offered to testify on his behalf.

Fermi did not need persuading. On a personal level he certainly had reservations about Oppenheimer. The two were never particularly close. Hungarian refugee and physicist Valentine Telegdi was with Fermi when the news of the accusations made its way to Chicago. He remembers Fermi remarking over lunch at the faculty club, "What a pity that they took him and not some nice guy, like Bethe. Now we all have to be on Oppenheimer's side!" Fermi knew that Oppenheimer could be nasty, a trait that Fermi detested. He also knew his duty, believing the charges to be without foundation, and was prepared to act. To the extent that the charges involved Oppenheimer's opposition to the Super in October 1949, Fermi shared Oppenheimer's views and expressed them even more strongly. He knew Teller well, liked him personally, and enjoyed being his intellectual sparring partner, but he was also irritated at the anti–Los Alamos campaign Teller had promoted. He agreed to serve as a witness on Oppenheimer's behalf, graciously refusing Oppenheimer's offer, made through Garrison, to defray Fermi's travel expenses.

FERMI HAD ANOTHER, MORE PLEASANT TASK TO ACCOMPLISH BEFORE heading to Washington in defense of Oppenheimer. His term as APS president was coming to an end and he had the pleasure of handing the baton to his old friend, Hans Bethe.

The APS chose to hold Fermi's final annual meeting at Columbia University, because the meeting coincided with the university's two-hundredth anniversary celebration. Fermi was delighted to return to his old haunts at Pupin Hall. As outgoing president, he gave a lecture on the future of particle physics, during which he spoke at length about his work on pion scattering and about the future of accelerators. He predicted ever-increasing accelerator energies and, in a tongue-in-cheek moment, displayed a slide predicting the creation of a huge accelerator circling the globe from outer space. He also gave an informal talk about the beginning of the Manhattan Project at Columbia, regaling the audience with memories of the first experiments with Szilard and the rest of the team. He drew laughter from the audience at several points in this latter talk, including his description of Szilard as "a very peculiar man, extremely intelligent" and his description of climbing on top of the high piles of graphite, observing "I am not a tall man." The audience loved the talk and for a moment, at least, they could all forget about the crisis looming ahead for the entire physics community.

In the weeks leading up to the hearings, Garrison again reached out to Fermi. His client's spirits were flagging and Garrison hoped a call from Fermi might cheer up Oppenheimer. No record exists of whether Fermi called, but it is hard to imagine his refusing.

The hearings took place in closed sessions before the "Personnel Security Panel" of the AEC. The transcript of the hearings, famously entitled *In the Matter of J. Robert Oppenheimer,* was initially not made public, but a redacted version was released in response to public pressure later that summer. It comes to almost a thousand densely printed and occasionally gripping pages. Fermi's testimony is neither the longest nor the most important. Scientists like Bethe and Rabi gave spirited testimony in support of Oppenheimer. Rabi famously testified against the entire proceeding, pointing out, "[His behavior] didn't seem to me the sort of thing that called for this kind of proceeding at all against a man who had accomplished what Dr. Oppenheimer has accomplished. There is a real positive record, the way I expressed it to a friend of mine. We have an A-bomb and a whole series of it, [redacted] and what more do you want, mermaids?" A few, notably Edward Teller, testified against Oppenheimer. When pressed, Teller said that "I think I would like to see the vital interests of this country in hands that I understand better, and therefore trust more." This short statement, coming as it did in a situation fraught with significance for Oppenheimer, was enough to ensure that Teller lived out his career isolated from the mainstream physics community.*

* Teller led Livermore Labs for several decades and continued to advise US administrations on nuclear weapons policy. In the 1980s, he was a strong advocate of the Reagan administration's Star Wars program. Though Teller and Oppenheimer reconciled in 1963, Teller remained a pariah to many in the physics community who would neither forget nor forgive his performance at the Oppenheimer hearings.

Fermi's appearance was squeezed in between that of two other witnesses on April 20, 1954. His session was shorter than it might otherwise have been because a previous witness, former AEC chairman David Lilienthal, had to excuse himself midway through his testimony owing to another commitment. The chairman of the hearing agreed to put Lilienthal back on the witness stand as soon as he returned. A short session with Harvard's president James Bryant Conant followed, and then the panel turned to Fermi. Fermi spoke for ten to fifteen minutes before Lilienthal returned, at which point Fermi's testimony came to an abrupt halt.

In those ten to fifteen minutes, however, Fermi rejected the notion that Oppenheimer somehow influenced the members of the GAC to vote against the hydrogen bomb program in October 1949; rejected the implication that Oppenheimer stifled an open and honest exchange of views; testified that Oppenheimer aggressively pressed for continued work to enhance America's nuclear arsenal; and talked openly about his own reservations regarding the hydrogen bomb project: "My opinion at that time was that one should try to outlaw the thing before it was born. I sort of had the view at that time that perhaps it would be easier to outlaw by some kind of international agreement something that did not exist. My opinion was that one should try to do that, and failing that, one should with considerable regret go ahead." After some follow-up questions, Fermi was asked whether the members of the GAC had already made up their minds before the October 29, 1949, meeting. Fermi's response revealed the general ambivalence he felt about becoming involved in public policy decisions, an ambivalence that led to his 1950 decision not to renew his membership in the GAC:

> I would not know. I had and I imagine that many other people had sort of grave doubts. It was a difficult decision. Even now with the benefit of 5 years of hindsight, I still have doubts as to what really would have been wise. So I remember that I had in my own mind definite doubts. And I presume my ideas and I imagine those of other people, too, must have gradually been crystallizing as the discussions went on. However, I have no way of judging.

Another panel member finished the questioning by focusing on Fermi's attitude about scientific secrecy. Fermi had been thinking about secrecy and science for fifteen years, ever since his initial disagreements with Szilard. Fermi explained that in ordinary times scientific work should not be secret but that he joined with Szilard and others in the specific situation of impending war to voluntarily censor his research. Asked whether it is possible to conceal scientific information, Fermi replied that for a short period it could be, but not forever. Finally, Fermi was asked whether he might have "guessed" some of the scientific secrets behind the fission weapon if he had remained in Rome, and Fermi replied, "I think I might possibly have guessed some things, at least."

With that, Lilienthal reappeared at the hearing and Fermi was quickly excused. Fermi was never recalled for further testimony. The panel knew it was not going to get Fermi to denounce his old colleague and friend.

Fermi did what he believed was right. He stood with Oppenheimer and defended him to the extent he could, given the narrow nature of the questions. He was also frank about his own reservations about the hydrogen bomb and revealed his own unresolved concerns some five years after the event. He did not describe the weapon as "evil," a term he and Rabi used in their minority report in October 1949, but he left the panel with no doubt as to his own views at the time. He also gave the panel a good understanding of the way in which Oppenheimer ran the GAC meetings of that period.

In the end, however, neither Fermi, nor Bethe, nor Rabi, nor Groves, nor anyone else could save Oppenheimer from his fate. The panel voted 2–1 to revoke his security clearances, with Evans in opposition. The full commission approved the panel's recommendation by a 5–1 vote. Oppenheimer faced public humiliation and disgrace and retreated to the institute he directed, a broken and dispirited man. In 1963, in a belated effort to atone for its unjust treatment of one of the century's living legends, the AEC awarded Oppenheimer its highest honor, presented by President Johnson. It was an award named for his late Manhattan Project colleague, Enrico Fermi. Teller attended the ceremony and the two arch-rivals arrived at a measured reconciliation.

CHAPTER TWENTY-THREE

A PATENT FIGHT

Soon after the war, Fermi's personal finances clashed with national security policy. The story is a fascinating one, and though it ended reasonably well, it could have soured Fermi's relationship with the US government. It is perhaps the best example of how the demands of the new national security state came into conflict with traditional ways of doing things. That Fermi, who assiduously avoided any such conflicts throughout his lifetime, found himself in this situation is one of his life's greater ironies. It can only be appreciated in the context of Fermi's overall attitude toward financial matters.

Fermi's intense concentration on physics never distracted him from the state of his bank account. When he was in the process of deciding to move to the United States, he took a careful inventory of his and Laura's assets and noted them in the back of a ledger he later used for physics. Of the two, Laura—he referred to her as "Lalla," the affectionate nickname her closest friends used—was by far the wealthier. She had some 120,000 lire worth of bonds and about 1,000 shares of stock in various companies. These holdings alone would have made her a very comfortable, upper-middle-class woman. She also held title to the apartment at Via Belluno and one-quarter of the villa in Tuscany they visited every summer before they moved to the United States. Enrico owned the new apartment at Via Magalotti and the garage in the building next to their home at Via Belluno. He also seems to have owed 10,000 lire to Laura's father, Augusto Capon. When they left Rome in 1938, the family entrusted all their Italian assets to Enrico's sister Maria, who carefully protected them and sold them off after the war.

When he came to the United States, he meticulously recorded his monthly paychecks in his pocket diaries, which increased from an annual rate of $8,000 per year to almost double that by the time he passed away in 1954. At $15,000 per year, this made him one of the highest paid professors at Chicago. He also noted carefully all out-of-pocket expenses for work-related travel, to Los Alamos, to Hanford, to Oak Ridge, and to various universities and summer schools. He noted consulting fees from his association with DuPont and other US companies eager to have his advice on technical matters. He also recorded in detail the various stock and bond transactions he entered into throughout the postwar years. In another life he would have made a world-class accountant.

Giulio recalls his father as somewhat stingy and gives as an example Enrico's preference in the winter to keep the heat in the Chicago houses set at a chilly sixty degrees. Whether Enrico was genuinely miserly is debatable. For example, he lent Schrödinger money before the war and graciously wrote off the debt when Schrödinger tried to repay. However, he was certainly no spendthrift. He was always looking for ways to spend less money while, at the same time, always looking for ways to increase his income. Several of those postwar efforts to increase his revenues are noteworthy. One was an ill-fated attempt to replicate in the United States the financial success he achieved with the publication of his Italian physics textbook for high school students. Another was an eight-year effort to get financial compensation from the US government for the use of the slow-neutron patent granted to him and his fellow inventors in July 1940.

In November 1946 Fermi approached Macmillan Publishers with a proposal for a translation of his Italian textbook *Fisica,* designed for *liceo* students. For the head of Macmillan's education division, R. L. Knowlton, the idea of a high school textbook authored by the famous Manhattan Project veteran must have seemed too good to be true. Macmillan sent a senior editor, Martin Robertson, on December 9, 1946, and by December 16 the two men were outlining terms for the project. Laura was responsible for the translation into English. She was deeply involved in the Italian edition, which took shape at the Tuscan villa owned by her uncle during summers and holidays in the late 1920s. By March 1947, a "competent" high school teacher, Warren Davis, principal and physics teacher at Alliance High School in Alliance, Ohio, was selected to join the project. The choice of Davis is a bit of a mystery, because he had not yet published anything at this point and indeed never published anything during his career. Fermi had certainly never heard of him. However, after meeting him, Fermi agreed to the arrangement and on March 24, 1947, made a note of the division of royalties from the project. Royalties on the book would be split 60/40, with Fermi getting the larger percentage. Royalties for a laboratory workbook written by Davis would be split 60/40 the other way. Macmillan followed up with a detailed letter outlining the terms of the arrangement, including a review by Davis of all existing high school texts and a proposed completion date of November 15, 1948. This would allow for publication in early 1950.

From the start, the experience was not a happy one. By early October 1947 Laura was translating the second volume, but Fermi had heard nothing from Davis. A letter he sent to Davis at that time must have prompted a package of material, because the next letter in the file is from Fermi to Davis in late January 1948, in which Fermi—clearly unhappy with what he had read—suggests that he mark up the sections on fluid dynamics with a series of editing "codes" for Davis to consider. These codes give a flavor of the difficulties Fermi encountered in Davis's manuscript. He would mark "1" for grammatical errors, "2" for incorrect physics, "3" for material that was too obscure, "4" for an unnecessary deviation from Laura's translation, "5" for material that should be deleted, "6" for material that required additional illustrations, "7" to indicate where the use of letters would simplify the text, and "8" for questions in the text that should be changed or deleted. Fermi was not impressed with what he had read, yet he still had the patience to see if the collaboration would work.

In early August 1948, Fermi wrote another letter to Davis. The high school teacher revised the opening chapters, but they still were unsatisfactory to Fermi, who marked them up for further revision. He also felt that Davis had added enormously to the word count, perhaps doubling Laura's translation. By the end of August 1948, Fermi took the time to read the drafts more carefully and his verdict was even more negative: "Unfortunately, I find that the manuscript is still in a very unsatisfactory shape, and I am afraid that it will be so even after the corrections I am suggesting are incorporated into it. I am afraid that the trouble is much more serious than I realized."

Realizing the seriousness of the problem on their hands, the editors at Macmillan decided to step in and review the manuscript that Davis had presented to Fermi. They made some minor changes to one chapter, which Fermi seems to have approved, although he must have cringed at the editors' decision to replace the term *velocity*—which has a very precise meaning in physics—with the term *speed,* which has a precise, but different, meaning. By this time, the process had become quite confusing, with Davis, aware of the looming deadline, sending a spate of draft chapters to Fermi and to Macmillan for review and editing. It took Fermi a few months to review and revise chapter four, and he submitted it to Macmillan in mid-March 1949, now some five months past the deadline. Macmillan agreed it was a major improvement over Davis's draft.

The next letter in the file is a draft of correspondence from Fermi to Davis, which must have been sent some time later. It is a disavowal of the entire project. Fermi criticizes Davis for not following the original book more closely. He observes that the language in places seems more suited for "younger children." The draft of the letter ends with Fermi's concern that he no longer could afford to continue the collaboration, but this sentence was crossed out. Because the actual letter he sent does not exist in the file, it is difficult to know what the final version said.

Some four years after the original deadline, in November 1952, Macmillan wrote a letter to Fermi acknowledging his decision to end the project. The long-suffering editor, Martin Robinson, put it this way:

> I am sincerely sorry that the plan did not work out to everyone's satisfaction, and personally regret any misunderstandings that may have developed. Needless to say, I still hold you and your work in the highest esteem and hope that if you find the time to prepare a college text or a reference work, that you will give us an opportunity to show you that we can really make good books without so much struggling.

The project's disintegration must have been agonizing for everyone involved and its end, some relief, albeit tinged with regret, for Macmillan.

For a host of reasons it is a great pity that the project failed. Obviously, it would have made a significant amount of money for all parties concerned—Macmillan, Davis, and the Fermis. Sticking to the translation provided by Laura might well have proved the best course for everyone. Back in Rome, Amaldi took over the project of updating and revising the Italian textbook, which continues to be a best-seller. When Edoardo passed away, the publishing project passed to his son, Ugo, named after Edoardo's father, who continues to update it. It has sold more than three million copies to date and is still the basic textbook for high school physics in Italy.

Fermi was interested in the money, but he was more interested in getting the textbook right. He had very high standards and could not bear the thought of a slipshod textbook being published in his name, irrespective of the financial opportunity involved. Davis seems to have been out of his depth and certainly not on the same wavelength as his coauthor. High school students to this day are the poorer for Davis's failure.

FERMI'S 1940 POCKET DIARY CONTAINS A SEVEN-DIGIT ENTRY FOR July 3—"2,206,634." It represented a milestone for Fermi. The day before, the US Patent Office issued Fermi and his Rome colleagues a patent for the slow-neutron technique. Gabriello Giannini had filed the patent on their behalf on October 3, 1935, almost a year after the patent was filed in Italy. In the interim, the application had worked its way through the bureaucracy, emerging on July 2, 1940, as a fully registered US patent. Fermi learned of his good fortune the next day and scribbled down the number of the patent for his records.

Thus began a thirteen-year saga, which eventually embroiled the five named inventors—Fermi, Amaldi, Pontecorvo, Rasetti, and Segrè, as well as D'Agostino and Trabacchi, both of whom were promised shares in any royalties—in extensive and complex litigation against the US government.

IT MAY HAVE BEEN CORBINO'S IDEA TO APPLY FOR A PATENT ON THE slow-neutron idea, but Fermi and his colleagues eagerly adopted it as their own. Corbino saw the commercial potential for a technique that could be used to develop new radioactive isotopes for medical use. He could hardly have anticipated that slowing down neutrons would enable physicists to split the uranium atom, much less create a chain reaction. Though the fission weapon did not rely on slow neutrons, slow neutrons created the world's first controlled chain reaction and were an essential part of the vast reactor facility at Hanford that churned out plutonium for America's nuclear arsenal. The slow-neutron technique had enormous potential civilian power generation applications as well. To understate the case, the slow-neutron patent was valuable intellectual property.

As the Manhattan Project developed, Fermi, Segrè, and Pontecorvo saw the growing value of their patent, but seeking compensation for its use by the US government was problematic. First, Fermi and Segrè were overwhelmed with work on the bomb and had little time for the administrative or legal effort required to pursue the matter. Second, the true value of the patent was apparent only to those who were cleared to know what was happening within the Manhattan Project. Giannini, even though a US citizen, was not cleared, and of course Amaldi, D'Agostino, and Trabacchi were citizens and residents of Italy, at that point an enemy nation. A third complicating factor was the policy developed haphazardly during the project that all intellectual property arising from it would be signed over to the US government. Fermi and his Manhattan Project colleagues filed applications for at least twelve patents between 1944 and the end of the war related to their work, the most famous being patent 2,708,656, filed jointly by Fermi and Szilard, for the "neutronic reactor." All of these were effectively the property of the US government and remained classified until the 1950s. It would be another ten years, in the 1960s, before the patents were granted. The slow-neutron patent was filed before the United States entered the war, but pursuing compensation while the war was on would, in the inventors' view, run afoul of the way the US government was handling intellectual property related to nuclear weapons. A fourth issue was the uncomfortable fact that Amaldi, D'Agostino, and Trabacchi were all Italian citizens living in Rome and thus enemies of the US government.

Nevertheless, Fermi and Segrè made several ineffective attempts to discuss compensation for the slow-neutron patent with the relevant Los Alamos lawyers, Navy captain Robert Lavender and his assistant, Ralph

Carlisle Smith. After the war, and as the most business-minded of the group, Segrè was chosen to liaise with Giannini and Lawrence Bernard, the Washington, DC, lawyer Giannini chose to represent the team, but the end of the war only further complicated matters.

The Atomic Energy Act of 1946 rendered the entire nuclear industry a monopoly of the US government. Previously, Giannini and Bernard negotiated with Captain Lavender in Vannevar Bush's Office of Scientific Research and Development (OSRD). OSRD proposed a variety of compensation schemes, initially offering a lump sum settlement of $900,000. However, with the 1946 act's creation of the AEC, the jurisdiction of the dispute shifted to the commission's office responsible for patent work.

The change of jurisdiction introduced a myriad of new delays. The AEC would not engage in serious negotiations before establishing its own set of policy guidelines, which took several years to complete. The slow-neutron patent was but one of many the AEC had to consider. When Bernard finally filed for compensation in October 1948, Giannini told him to suggest $1.9 million: $1 million as a lump sum payment and nine annual payments of $100,000. Bernard did as he was told but also suggested that the inventors would settle for the original $900,000 offered by Bush in 1946. Bennett Boskey, a lawyer in the Office of the General Counsel of the AEC, was assigned to examine the claim.

An Ivy League lawyer, Boskey vigorously defended the AEC's monopoly on nuclear science and wrote his report accordingly. It was, as scholar Simon Turchetti puts it, "not good news" for the inventors. Boskey questioned whether the slow-neutron method was essential for the production of fissionable material. Undaunted by the facts and not a scientist, he rejected the claim that the basic R&D for the Manhattan Project relied on slow neutrons. He also painted a demonstrably false picture suggesting that the work of Rutherford and Chadwick held priority over the work of Fermi's team in 1934. He also claimed the production of radioactive isotopes did not rely on the slow-neutron method. In short, he systematically rejected every one of the patent's claims, an astonishing performance given that a trained US patent examiner had actually granted the Italians a patent almost two decades previously.

To muddy the waters even further, he darkly noted the 1947 treaty ending hostilities between the United States and Italy entitled the

United States to ignore any property or commercial claims made by Italian nationals like Amaldi, D'Agostino, and Trabacchi. To top it off, he argued Fermi's membership in the GAC made him an employee of the US government and therefore it was a crime for Fermi to litigate against the US government. Irritated, Fermi contemplated resigning from the GAC to clear up the conflict but was able to persuade the AEC that his impending departure from the GAC in January 1951 would solve the problem.

Boskey was smart enough to realize that this first, highly aggressive salvo would shake the confidence of the inventors and he was right. However, their confidence took a greater hit when one of the inventors, Bruno Pontecorvo, took the occasion of a summer holiday in Italy to disappear with his wife and reappear in Moscow. Intelligence services quickly discovered his defection, which was publicly reported in October 1950 by Reuters.

The process of litigation against the AEC stimulated the FBI to review the backgrounds of all the inventors. Their attention came to rest on the left-leaning Pontecorvo, all of whose siblings were members of the Communist Party in Italy. Bruno was known to be sympathetic to Communist causes, as well. Reports on the FBI research were passed along to British authorities at MI5 and MI6. British security officials determined that although there was nothing definitive to cause concern it would be wise to place Pontecorvo in a less sensitive position. The unmasking of Klaus Fuchs in 1950 as a Soviet agent landed Pontecorvo under even greater scrutiny. The exact circumstances that caused him to defect when he did are not known, but it is possible that he was tipped off by the British liaison to US intelligence in Washington, Kim Philby, who had been working as a double agent for the Soviet Union for decades.*

The news of his defection came as a shock, especially to the other litigants, who now felt that their position had been terribly compromised.

* It remains unclear whether Pontecorvo was an active Soviet agent. Frank Close studied the issue carefully and discovered that Philby knew of the FBI's interest in Pontecorvo. Within weeks, the physicist defected. See Close, *Half Life*. Others, particularly in Italy, believe he was not working for the Soviets.

Segrè, perhaps the closest to Pontecorvo, decided it would be best for everyone if he resigned as the liaison with the legal team, and Rasetti, who had found a position at Johns Hopkins in Baltimore, took his place. Negotiations dragged on until, eventually, the sides agreed on a disappointing $300,000 lump sum payment, far less than they expected and far less than they felt they deserved. After paying out the approximately $180,000 of legal fees, each litigant was left with a little less than $30,000. Pontecorvo obviously never got his share.

At that time, $30,000 was a significant sum, enough money to buy a grand house in the Hyde Park section of Chicago. The inventors had, however, been expecting millions. For Fermi, the whole misadventure was an embarrassment. He indicated privately that he would probably never have pursued litigation at all were it not for the loyalty he felt for the other members of the team. The litigation put him in an extremely awkward position with the AEC, which was paying him a consulting fee as a GAC member and as a Los Alamos consultant.

Nevertheless the outcome seems unjust. The patent was granted prior to US entry into the war, well before the Manhattan Project was organized. The enormous power wielded by the AEC during the immediate postwar years was sufficient to determine the outcome irrespective of the merits of Fermi's position. The odds were stacked against the Italians and at some deep level they must have known it. Along with Szilard's chain reaction patent and the Fermi-Szilard patent on the nuclear reactor, the slow-neutron patent remains one of the most important patents in the history of atomic energy. The commercial value of these three patents is quite literally incalculable. The Italians settled not because they wanted to but because, given the circumstances, they had to.

When Fermi received his payment, he made sure to invest immediately in securities, under the heading "The Patent Fund." As with all of his investments, he tended to the Patent Fund carefully.

CHAPTER TWENTY-FOUR

BRILLIANT TEACHER, BELOVED MENTOR

NOT ALL GREAT RESEARCH PROFESSORS ARE GREAT LECTURERS. Rabi and Teller were both world-class research scientists, but by general consensus they were both dismal in the classroom.

In contrast, Fermi was an outstanding lecturer. His teaching in Rome was legendary, and students flocked to summer sessions in the United States to hear him lecture. At Los Alamos, physicists clamored to attend his lectures, particularly after the end of the war. Now, back in Chicago, he threw himself into teaching with renewed vigor. The results were spectacular.

ONE OF THE VERY FEW FERMI LECTURES THAT WE CAN LISTEN TO today is the speech he gave before the American Physical Society in January 1954, when he told the story of his fission work at Columbia. He did not speak from a fully prepared text. After the war he would routinely accept lecture invitations and inform his hosts that he would not be preparing a text of the talk. Instead, he worked from fragmentary notes that gave him topics on which he would elaborate to create an overarching structure to the talk. The Columbia talk exists in print only as a transcription of the recorded version.

He spoke slowly. His voice was deep and heavily accented even in 1954, some fifteen years after he moved to the United States. He could read his audience and when he felt a digression was appropriate—as was the case when he started to describe his work with Szilard—he would take the opportunity to enjoy a good laugh along with the audience.

In contrast to other lecturers who, like Fermi, deliver from fragmentary notes, Fermi knew exactly what he was going to say and delivered each thought in full, complete, grammatically correct sentences. In the frequent long pauses during the talk, one can almost hear him think through each sentence before speaking it. The transcription needs little editing, because he delivered it perfectly.

This speech provides just a hint of why his students, as well as his colleagues, found his lectures so illuminating. No matter how complicated the subject matter, he moved through it slowly, at a pace allowing less gifted students to keep up as well as giving those more gifted students a chance to appreciate Fermi's specific approach to problem solving: eliminate the extraneous considerations, strip the problem down to its essential elements, and proceed step by step toward solution.

In his brilliant essay on Fermi's Chicago years, colleague and collaborator Valentine Telegdi notes that Fermi was obsessive about his preparations for course lectures, writing out notes on large sheets of paper to which he would refer during the course of any given lecture. He never showed annoyance with students who did not understand a point the first time around. "On the contrary," Telegdi writes, "if Fermi had to repeat an explanation he seemed to derive twice the pleasure."

Fermi's enthusiasm for teaching is evident in the course load he managed throughout his postwar Chicago career. A man of his stature could easily have found ways to do the bare minimum of teaching required, but he consistently taught two or three courses each term. His course load for the academic year 1946–1947 included introductory lectures on physics for undergraduates, the courses that his graduate student Jack Steinberger felt so lucky to work on as a teaching assistant. Fermi taught quantum mechanics, thermodynamics, and nuclear physics. He taught virtually every discipline required for an undergraduate physics major and also taught seminars on special topics and high-level research seminars for advanced undergraduates and graduate students. The only period in which he chose not to teach classes, for reasons that are not entirely clear, was from the fall of 1947 through 1948, picking up again in 1949.

One particular course he taught became something of a legend. In the period between 1952 and 1953, he lectured on nuclear physics. His teaching assistants, Jay Orear, Arthur Rosenfeld, and Robert Schluter, prepared mimeographed notes of the lectures for distribution to anyone interested. Word spread that these notes were available, first through the Chicago community and then, incredibly, to other campuses. Requests came in from far and wide until the overburdened secretaries at the physics department office could no longer keep up with the demand. Anticipating a major problem, the three graduate students approached the editors at the University of Chicago Press to see whether they would be willing to lend a hand. The editors were delighted to do so and the mimeographed sheets were compiled into the textbook *Nuclear Physics,* which sold well for the next three decades. Known among physicists as "Orear-Rosenfeld-Schluter," its authors never claimed that Fermi wrote it and for his part Fermi never reviewed the manuscript. It was successful simply because the authors took notes that conveyed brilliantly Fermi's methodical clarity. The book quickly became a classic.

The students who attended his lectures were fortunate and they knew it. Harold Agnew later recalled a moment in the spring of 1954 when he was passing the physics department building. Suddenly, he heard a huge commotion emanating from the building, loud enough to attract the attention of a nearby campus security officer. The two of them dashed into the building to see what was going on. Other faculty members were emerging from their offices, equally concerned. The din was coming from one of the large lecture halls. Upon opening the door to the room, they found some hundred undergraduates on their feet, applauding and cheering Fermi, who had just delivered his final classroom lecture of the term. There had never before been such a disturbance on the normally sedate campus, Agnew noted.

Lucky indeed were the undergraduates who studied with Fermi, but luckier still were those who received their PhDs from him.

One of the most interesting graduate student stories relates to two young Chinese nationals who, in November 1945, escaped war-torn China and made their way to the United States determined to study either with Fermi or Wigner. One was Chien Ning Yang. (He later adopted the American name "Frank.") The other was Tsung-Dao (T. D.) Lee. They arrived at Pupin Hall at Columbia looking for Fermi but were met with blank stares. They then traveled to Princeton, where they were told that Wigner was not taking on students until the following year. They heard, however, that Fermi was about to start up a new institute in Chicago. They traveled there and before long were accepted into the program. Lee was admitted as a graduate student despite never having received an undergraduate degree.

The two worked closely with Fermi over the next several years.

Lee went on to do a thesis directly under Fermi, studying the inner processes of white dwarf stars. He still recalls his days with Fermi with fondness. Early on during his thesis research, he and Fermi were discussing white dwarfs and Fermi asked Lee if he knew what the temperature was inside the sun. Lee gave him the answer and Fermi asked how he derived it. Lee admitted that he had looked it up, offering as an excuse that the calculations were quite tedious and in any case could not be done on a slide rule because the units involved extended from well below the range to well above the range on available slide rules. In a burst of creative enthusiasm, Fermi suggested that together they build a slide rule large enough to be capable of doing the calculations. It turned out to be six feet in length, and unwieldy, but it did the job.

In the end, Yang's thesis supervisor was Edward Teller, but given Yang's talent it was inevitable that he would spend time with Fermi. Yang collaborated with Fermi on an important paper analyzing whether the pion was a fundamental particle in the same sense as the muon or whether it was composed of other, smaller particles. The paper suggested that it might be a combination of a nucleon and an antinucleon—for example, a neutron and an antiproton. It was a fruitful collaboration and pointed the way to further research. We now understand the pion to be a combination of a quark and an antiquark. Yang has written widely of his experiences during this formative period in his life and summarizes Fermi this way: "He had both feet on the ground all the time. He had great strength, but never threw his weight around. He did not play to the gallery. He did not practice one-upmanship. He exemplified, I always believe, the perfect Confucian gentleman."

None of Fermi's other Chicago graduate students had quite the struggle these two had in getting to Chicago, but each had particular stories of life as a graduate student under Fermi.

Geoffrey Chew, Fermi's young colleague at Los Alamos, enrolled at Chicago after the war and began his PhD work with Teller. Given Fermi's ongoing research at Argonne, Chew assumed that Fermi would only take on experimental students. However, when in mid-1947 Teller informed Chew that Fermi had decided to take on two theory students while waiting for the cyclotron to be built, Chew jumped at the chance and shifted from Teller to Fermi. That same day Chew ran into Marvin "Murph" Goldberger, a fellow theory student, and Chew informed Goldberger that Fermi was looking for one more theory student. The words had barely left Chew's mouth when Goldberger turned and dashed off to find Fermi. He, too, was accepted. Chew and Goldberger worked together on a theoretical study of how neutrons behave inside the deuterium nucleus.

Fermi decided for both Chew and Goldberger when their work was ready for publication, a requirement for the PhD at Chicago, and arbitrarily split the joint work into two separate pieces, effectively separating the conjoined twins without harming either one. Chew later recalled this generosity with great affection. Chew is particularly vivid in his recollection of Fermi's enjoyment in explaining things, especially to people who did not "get it" at first. He also has described Fermi as "the last man who knew everything," not simply because of his ease with both theory and experiment but also because of his mastery of every aspect of the physics of his day, from astrophysics to geophysics, from particle physics to condensed matter physics.

Jack Steinberger, whose PhD thesis resulted in an important new understanding of beta decay, gives Fermi enormous credit for taking him under his wing and helping him develop into a strong experimentalist. "I am completely indebted to Fermi," he said in 2014, "for my understanding of what a physicist should be like as a model. Even more impressive was his kindness to me." According to Steinberger, Fermi emphasized that "in order to understand something one needs to study it thoroughly; that one should not require reassurance but rather have confidence in one's own ability; and that one should be open to all students, that it should not matter whether a student is bright or average or below average, they should all be treated with respect."

In the later crop of students, Arthur Rosenfeld was notable for being at the top of the cohort in the basic exam. In fact, Fermi later told him that Rosenfeld had, unbeknown to him, discovered an error in the solution set to the exam. Rosenfeld "screwed up his courage" and asked Fermi if Fermi would be his thesis adviser. The memory of Fermi smiling at him and saying, "So, I've been expecting you," remained a high point for Rosenfeld, even more than sixty years later. When Fermi decided to lecture on nuclear physics, Rosenfeld was one of his teaching assistants. Rosenfeld recalls that solving the problem sets Fermi gave to his students so that Rosenfeld and his fellow teaching assistants could grade them was "an education in itself." It was Rosenfeld who alerted Fermi to the Shepley-Blair book and helped draft the press release issued in Fermi's name defending Los Alamos against charges of foot-dragging on the Super. Eventually, when Luis Alvarez called Fermi from Berkeley looking for talent to bring to Berkeley as a post-doc, Fermi offered up Rosenfeld as "the second best student I have ever had."* Alvarez pressed Fermi to tell him who the top student was, but that was a secret Fermi was unwilling to share. Rosenfeld went on to a notable career in experimental physics at Berkeley, but during the energy crisis of 1973 he shifted his focus to energy efficiency technology and policy, becoming over time one of the most influential experts in that field. In 2006, Rosenfeld received the Enrico Fermi Award from President George W. Bush, and in 2011 President Barack Obama awarded him the National Medal of Technology and Innovation.

Rosenfeld's fellow student, Jay Orear, was intensely affected by his years with Fermi. He spent a good deal of time talking to Fermi about experimental technique and particularly about the use of probability and statistics in experimental design and evaluation. He took extensive notes of these conversations and published them in 1958, not long after Fermi passed away. He has written with unusual affection about his time with Fermi and the impact Fermi had on his career. Perhaps

* Considering the quality of Fermi's graduate students, being his second best student ever was quite an accomplishment.

most significantly, Orear organized a series of Fermi alumni reunions at Cornell, which a large group of former students and colleagues attended at regular intervals. Unfortunately, Orear passed away in 2014 and the reunions no longer take place.

Another graduate student was Owen Chamberlain, who came to Chicago in 1946 and agreed to pursue an experimental thesis at Fermi's urging. He later recalled that his first meeting with Fermi was at Los Alamos. Segrè, with whom Chamberlain would later share a Nobel Prize for the discovery of the antiproton, introduced the two. To Chamberlain's surprise, Fermi was quietly puttering away in a corner.

> I knew exactly what a Nobel Prize winner looked like and Ernest Lawrence fit the bill exactly: big, with a voice that echoed down the hall. Here was this little man sitting motionless in the corner of the very small room.

Chamberlain's subsequent work with Fermi had a huge impact on the young student. In early 1954, at the end of a short cover letter accompanying some experimental measurement taken by Segrè and Chamberlain, Chamberlain adds, completely unprompted, a lovely tribute to his former professor: "I am very grateful for the time and effort you have invested in me in the past. If I am to be regarded as a decent physicist, it is mostly because of your training."

Jerome Friedman was Fermi's last graduate student. A future Nobel laureate, Friedman's experimental thesis with Fermi was within weeks of being finished when Fermi passed away. This could have spelled catastrophe, because finding a faculty member to sign off on a thesis he or she had not supervised is no small feat. Fortunately, as the impact of Fermi's death became apparent, John Marshall realized that Friedman was in this difficult position and agreed to sign off once the work was completed. Friedman tells of his last interaction with Fermi, the day Fermi returned from the summer school in Italy in September 1954. They saw each other from a distance and Fermi waved hello to Friedman from down the hall; however, Friedman barely recognized Fermi, who was quite thin and haggard. Friedman was unaware of the health difficulties Fermi began to experience during the Varenna summer session and was surprised by Fermi's condition. The small, tightly knit particle physics community, usually fueled by gossip, was reluctant to spread the news of the rapid, catastrophic decline of one of its greatest members.

There were other graduate students as well, some fifteen altogether. The list reads like a who's who of postwar particle physics. Perhaps the only comparable graduate program during this period was Rabi's at Columbia. By general consensus, though, the brightest of all of Fermi's graduate students was Richard Garwin, the young man from Cleveland who pitched up in Fermi's office and helped build the analog computer that calculated solutions for Schrödinger equations. Everyone knew he was bright. Fermi is reputed to have said that Garwin was "the only true genius" he ever met. Apocryphal or not, the usually reticent Fermi sent a letter to the employment director at Los Alamos in the spring of 1949 recommending Garwin for summer work. Fermi was lavish in his praise of Garwin, describing him as "one of the most gifted students with whom I have ever come in contact," which was high praise indeed coming from Fermi, who rarely if ever offered praise, even to his closest colleagues. Among the whole group of Fermi's students, Garwin was the only one offered a position at the institute after graduation, another sign of the esteem in which Fermi held him. He did not stay in the field, instead pursuing an extraordinarily creative and productive career at IBM's Watson Labs as well as a central role in advising presidents and government agencies on science and defense policy issues. This ensured that Garwin's influence—and by association Fermi's—would be felt in a much broader setting than that of his fellow students, with the possible exception of Arthur Rosenfeld. Garwin has won numerous highly prestigious awards, including the Enrico Fermi Award in 1996 and the Presidential Medal of Freedom in 2016.

Other graduate students were heavily influenced by Fermi, even though he was not their formal adviser. Yang was one of them; Guarang Yodh and future Nobel laureate James Cronin were two others.

Yodh, a native of India, was a student of Anderson but got to know Fermi during his graduate years. His own experimental work focused on pion scattering. One day, Fermi approached Yodh while the graduate student was adjusting a particle counter and said, impishly, "That is not the way it is done in the Mahabharata." Fermi had just been reading an

English translation of the classic Indian epic in his reading group and had fun teasing the young Indian student. Yodh recalls with warmth and affection the many parties he attended at the Fermis' home, and the clarity and rigor of his teaching.

Cronin, a strapping native Chicagoan, would have liked to study directly with Fermi. Unfortunately, Orear, Rosenfeld, and Schluter had filled up the available slots, so Sam Allison became Cronin's formal supervisor. Cronin nevertheless spent quite a lot of time with Fermi getting advice and guidance. There was also a social aspect to these relationships, an aspect that could sometimes lead to difficult situations. Cronin recalled with humor a moment when the institute staff were involved in a friendly game of baseball. No game, however informal, was friendly for Fermi, who played this game, as he did all games, with a determination to win. Cronin was on the mound, facing Fermi. It suddenly occurred to Cronin that if he threw a wild pitch and hit Fermi, he might go down in history as the man who killed the father of the nuclear age. He was torn, though, because Fermi demanded to be taken seriously in any athletic competition. Cronin found a way to pitch the ball just far enough away to walk Fermi without Fermi realizing what Cronin did. It was an unlikely dilemma for a graduate student.

Cronin was also one of those students invited by Fermi to lunches in Hutchinson Commons, where Fermi would playfully pose problems that kept students thinking on their feet.

Cronin considered Fermi the most influential person in his professional development. In fact, when it came time to organize a centennial celebration of Fermi's birth at the University of Chicago, it was Cronin who took the lead in setting it up, inviting speakers, and editing a fine volume of papers to commemorate the occasion.

HARRIETT ZUCKERMAN, A SOCIOLOGIST WHO STUDIED THE INFLUENCE of Nobel Prize–winning scientists among future Nobel laureates, concludes that in this respect Fermi stands alone, at least in the United States. When one looks at a "family tree" of those who are linked, either directly or indirectly, to Fermi, that conclusion is only reinforced. Some five of his direct students went on to win Nobel Prizes: Chamberlain, Friedman, Lee, Segrè, and Steinberger. Two others, Cronin and Yang, also won Nobel Prizes and though not formally Fermi students, both credit him with providing career-enhancing inspiration and guidance. Many others went on to prominent and important careers in the field. It is an astonishing record, comparable perhaps only to that of Sommerfeld and Rutherford. Valentine Telegdi judges Fermi's teaching to be his greatest contribution in the Chicago period. It is difficult to argue with him.

CHAPTER TWENTY-FIVE

TRAVELS ABROAD

FERMI TRAVELED OVERSEAS ONLY TWICE IN THE EIGHT YEARS between his return to Chicago in 1946 and his death in 1954. The first trip extended from August through October 1949 and included visits to Basel, Switzerland, and Lake Como in the Italian Alps for a series of conferences and ended with a series of lectures in Rome and Milan. The second trip was to attend two summer schools held during 1954, one in the French Alps and the other at a villa along Lake Como. Once again, given the lack of private diaries and personal letters, one can only speculate why he returned to Europe so infrequently. He was certainly busy at Chicago and spent his summers preoccupied with classified research at Los Alamos. In fact, before he left for Europe in 1949, he spent much of the summer at Los Alamos, involved in the Super and other classified research projects. He may have felt that taking a chunk of time to spend in Europe, in the era prior to easy and convenient air travel, was simply not worth it. A more interesting possibility is related to the classified nature of his research. When he lived and worked in Rome, he had an easy and completely open professional relationship with all his colleagues and could speak to them about his work. Even after the war was over, so much of his work was still classified that he must have struggled to censor his conversations with physicists like Amaldi with whom he was once so close. Amaldi clearly felt the struggle, too, and it may have been partly responsible for his insistence, in the bylaws of CERN, drawn up in the late 1940s, that the new multinational European lab would never engage in classified research.

Whatever the reason, the trips to Europe enabled Fermi to catch up with old friends and discuss unclassified work he was doing in Chicago on the origins of cosmic rays and pion-proton scattering.

AT THE 1949 BASEL CONFERENCE, THE SUBJECT WAS HIGH-ENERGY particle physics. The list of almost two hundred attendees was impressive. Along with Fermi, a host of Italians attended, including groups from Rome, Florence, and Milan. A group of Germans also attended, including Heisenberg and his partner in the German atomic weapons program, Walther Bothe. Other notables included Pauli, Frisch, Placzek, Segrè, Pontecorvo, Alfven, Rabi, Schwinger, Meitner, Kronig, Rosenfeld, Telegdi, and Racah. Fermi was not one of the scheduled presenters in Basel, but he attended presentations on experimental methods, particle theory, and QED. Fermi also swam a mile in the Rhine River, a bracing experience even in the summer. His old friend Bruno Pontecorvo joined him.

Lake Como was Fermi's next stop. He had fond memories of the 1927 Como conference, so this was a particularly emotional way for Fermi to return to his native land. He was especially pleased to be greeted at the station by many of his old friends, some of whom he had not seen in a decade. The topic of the conference was cosmic-ray physics. Fifty-odd papers were presented. Major physicists from around the world were in attendance, and Fermi's Italian colleagues were there in force: Amaldi, Segrè, Pontecorvo, Occhialini, Piccioni, Bernardini, and Wataghin were among the many Italian physicists who came to present work and renew their friendships with Fermi. A whole new generation of physicists came as well, eager to see their legendary compatriot in the flesh. It was a heady time for Enrico and Laura, who were delighted to see old friends. Fermi took on the challenge of playing Pontecorvo in tennis during breaks in the sessions. Segrè records pleasure in watching Fermi try hard, without success, to beat the athletic and handsome Pisan.

The Fermis moved on to Rome, where a lecture series had been planned by the newly revitalized *Accademia dei Lincei*, under the directorship of Fermi's old Roman mentor and promoter, the mathematician

Guido Castelnuovo. Castelnuovo had arranged for Fermi to present six lectures in Rome and another three in Milan. The lectures covered a wide range of topics, at varying levels of sophistication.

The time in Rome also provided an opportunity to see relatives. The Fermis visited the old Capon home in Via dei Villini, where Laura's eleven-year-old nephew, Giorgio Capon, remembers meeting his illustrious uncle for the first and only time. Giorgio's parents expected him to be "seen and not heard" in the presence of Enrico and Laura. Giorgio, who went on to become a respected physicist in his own right, found his famous uncle unpretentious and engaging. Fermi also met with his sister Maria, and the two presumably had some form of reconciliation over Enrico's role in the Manhattan Project. When the lectures were finished at the end of October 1949, the Fermis departed for Chicago. Rome was no longer their home. They were, for better or worse, Americans, and upon their return they settled back into their normal life in Chicago.

The second trip to Europe, in the summer of 1954, covered slightly different territory. They traveled with Giulio, who had just finished a relatively happy year at Oberlin. Laura and Enrico decided to bring him along to see the sights and enjoy the fresh mountain air.

Their first stop was Paris, where they met up with Stan Ulam and his wife, who happened to be spending the summer in Europe as well. The plan was to drive part of the way south with the Ulams in a plain, rented Fiat, but when the president of Fiat heard—how he heard is unclear—what Fermi was planning, he insisted on offering the use of a free vehicle for the trip, a very zippy eight-speed car that Fermi allowed Ulam to drive around the streets of Paris. Heading south, they stopped at a small inn near Avalon, some 150 miles southeast of Paris, where they had a meal Ulam remembered mainly for the ominous conversation he had with Fermi. They spoke at length about the impact of the Oppenheimer hearings, which, they agreed, would make Oppenheimer a martyr. Ulam knew Fermi had no particular respect for Oppenheimer as a physicist but believed Oppenheimer had been treated badly by the AEC and was irritated that Teller's testimony had been particularly damaging. Ulam asked Fermi what he thought the future held for Oppenheimer, Teller, and the physics community more generally, and to his astonishment

Fermi replied, "I don't know, I'll look at it from up there." He pointed skyward. Later that evening they were discussing the future of particle physics, pion research, and the like, and once again Fermi pointed to the sky and said, "I'll know from up there." This struck Ulam as odd. Fermi was only fifty-three years old, and he knew Fermi to be particularly irreligious. It was, Ulam realized later, at least a premonition of impending doom and perhaps a sign that Fermi understood he was quite ill. The next day the two families went their separate ways, the Fermis to the Alps and the Ulams to the Riviera.

Les Houches was a fairly Spartan physics summer school at the base of Mont Blanc in the French Alps. The invitation to lecture there was simply too attractive for Fermi to ignore, combining two of his favorite activities, teaching physics and hiking in the magnificent mountains above the village. When he was not lecturing on a range of advanced topics, he participated in strenuous outdoor activities with younger physicists, including Roy Glauber, a future Nobel Prize winner. Glauber recalls that Fermi was quite active while at Les Houches and was shocked to learn how weak and fatigued Fermi was at Varenna only a few weeks later. He also recalls that the group needed to buy hiking boots for their excursion. Glauber chose the sturdiest, most expensive ones he could find. Fermi, not surprisingly, bought the cheapest ones, of obviously inferior quality. When Glauber gently teased Fermi about this, Fermi replied that the young Glauber would be using his boots a lot longer than Fermi his.

The session at Les Houches ended in mid-July, and the Fermis returned to Italy, to the summer school at the beautiful Villa Monastero in the scenic village of Varenna on Lake Como. Fermi was a star guest at the conference, although other prominent physicists, notably Heisenberg, were there as well. Enrico and Laura stayed at the villa, in the master bedroom of the villa's former owners. Fermi delivered his talks in a lecture hall now called the Sala Fermi and adorned with a bas-relief sculpture of him. Between sessions, attendees could take walks through the genteel gardens sloping down to the lake, take a refreshing dip in the cool lake, or ride by boat across the lake to Bellagio and elsewhere.

Fermi's lectures at Varenna focused on his recent pion experiments—their production, their scattering, and the analysis of these interactions within the nucleus. Several of the younger attendees took careful notes,

FIGURE 25.1. Heisenberg and Fermi enjoy the sunshine at the Villa Monastero, Varenna, in September 1954. *Photo by Juan G. Roederer. Courtesy of AIP Emilio Segrè Visual Archives, gift of Juan G. Roederer.*

supplemented with tape recordings and very short film clips. By this time, Fermi had spent a year or so digesting the results of his Chicago experiments and some of the questions raised by these results proved extremely fruitful in the years to come. This final paper is perhaps the best single summary of those years of high-energy pion experiments.

At some point during the Varenna sessions, however, Fermi's energy began to flag. It was uncharacteristic for him to avoid hikes and strenuous physical activity, but he did so now, noticeably fatigued after exertions that would have been easy for him just a few weeks earlier, climbing in the French Alps. He also began to have trouble swallowing and lost his appetite. The many photographs and film clips that survive from these sessions do not reveal his discomfort, but we can be sure that by the time the conference ended, he knew something was very wrong.

After Varenna, the Fermis met up with the Amaldis at the small resort village of San Cristoforo, near Trento, in the Italian Alps, where the two families rented a villa for several weeks. Persico joined them there. Giulio Fermi and Ugo Amaldi, Edoardo and Ginestra's son, met and got to know each other a bit. There were still walks and hikes, but it was increasingly apparent to everyone that Enrico was not well. There is a photo of Enrico on a tennis court at the villa with the Amaldis and Persico; Enrico worked hard to look well for the photo, but in fact they were only able to play a few minutes before he called it quits.

Enrico may not have been at his best, but he still had the energy and enthusiasm to jump at the chance for informal teaching. Edoardo asked him about his current work and Enrico told him of the important research that was being done with some of the world's first digital computers. Eduardo apparently showed interest, because over several evenings Enrico conducted an intimate seminar—for Edoardo, Giulio, and Ugo—on how to program these new computers in machine language. At the end of each evening, he would give everyone a problem to solve by the next session the following evening. Ugo, who fancied himself quite a strong mathematician, was frustrated to discover that his father took to programming much faster than did he. Ugo does not recall how Giulio fared, but by this point Giulio was a strong mathematician himself, so he probably did well. This informal seminar would be the last class Enrico Fermi would ever teach.

Ugo recalls a dinner during that short vacation when everyone realized just how ill Fermi was. Sometime during the meal, Fermi started to gag. Rising abruptly, he scrambled to the bathroom, where he was sick. To Ugo and everyone else at the table, this was a clear sign that something was seriously wrong with Fermi.

The Fermis returned to the United States through New York and dropped Giulio back at school for his sophomore year at Oberlin. While there, they met Giulio's good friend Robert Fuller. Giulio had made light of his father's importance and rarely if ever spoke about what it was like to be the son of such a great physicist. In fact, he avoided any discussions of his life prior to Oberlin. Fuller knew exactly who Enrico

Fermi was, though, and although they met for only a few minutes, that meeting has stayed with him for more than sixty years. It was not apparent to Fuller that Fermi was ill, though Fuller had never met Fermi when the man was full of the energy and vigor that characterized him throughout most of his life. What Fuller noticed, though, were Fermi's eyes—bright, piercing, darting this way and that, taking in everyone and everything. Fuller had the feeling that Fermi was looking right through him. Even in his increasingly serious condition, Fermi was engaging and engaged, intensely curious about this young man who was to become Giulio's lifelong friend.

CHAPTER TWENTY-SIX

HOME TO DIE

FERMI ARRIVED BACK IN CHICAGO IN SEPTEMBER 1954 IN TIME to celebrate his fifty-third birthday. Two months later he was dead.

Soon after returning, Fermi determined to find out what was wrong with him. He made an appointment with Laura to visit doctors at the university's Billings Hospital on campus. The illness must have been brewing for a long time. Ulam notes that Fermi developed a nervous tic of "swallowing hard," which Ulam noticed during summer work. He later regretted not urging his friend to see a doctor. Fermi's illness was otherwise completely asymptomatic until the Varenna conference and by that time it was far too late to do anything about it.

At Billings, the first doctor who saw Fermi suggested a series of X-rays. He examined them, saw nothing of importance, and told him to come back in a month. The chief of medicine, who was on vacation when Fermi was examined, was concerned when he returned and heard that someone had examined the legendary professor in his absence. Reviewing the X-rays, he decided that Fermi should see the senior surgeon, a well-known thoracic specialist named Lester Dragstedt. Dr. Dragstedt explained to Fermi that the symptoms pointed either to "congestion of the esophagus" or metastatic cancer. Exploratory surgery was necessary. If it turned out to be the former, surgery would be long and complicated, but Fermi would recover. If Dragstedt found metastatic cancer of the esophagus or stomach, he would close Fermi up quickly; nothing could be done to save him.

According to Chandrasekhar, Fermi went into surgery, awoke, and realized that the surgery was shorter than expected. He asked the doctor if there were metastases and the doctor confirmed this, telling him that, in his view, Fermi had some six months to live. With that, says Chandrasekhar, Fermi went back to sleep.

Leona Libby writes that she was in Fermi's hospital room when Laura returned from the postoperative meeting with Dragstedt. Laura did not have to speak. Libby read the bad news on Laura's face.

FERMI RETURNED HOME SHORTLY AFTER THE SURGERY. TOLD HE HAD six months, he began to dictate a book on nuclear physics he had long been planning, but the illness sapped him of strength and after a desultory start he gave up.*

The children were informed. Giulio came back from Oberlin. Nella, who was living in Chicago, helped her mother during the next few weeks as Fermi's condition grew worse. Herb Anderson and Sam Allison knew, as did other close colleagues at the institute. Outside of Chicago, the tragic news was initially restricted to a very small circle. Laura wrote letters to the Amaldis, Persico, and other close Italian friends, keeping them abreast of developments. Others found out slowly through the physics grapevine. At Cornell, Hans Bethe knew, and word got to MIT, where Bruno Rossi had found a position after the war. Rossi wrote in anguish to Allison for advice on how to proceed, because he suspected he was not supposed to know. Segrè found out in an uncharacteristically incoherent phone call from Allison.

The day after surgery, Anderson and Chandrasekhar were among the first to visit. Chandrasekhar has written movingly about that moment:

---

* Wali, *Chandra,* 270. Yang later wrote that Fermi said he had only months left to live, and Leona Libby recounts that when he returned home, he told Laura to rent a hospital bed for him only through November. The doctors may well have told Fermi that he had six months, but clearly Fermi thought he knew better, and he did. Yang, CPF II, 674.

> It was of course very difficult for us to know what to say or how to open a conversation when all of us knew what the surgery had shown. Fermi resolved the gloom by turning to me and saying, "For a man past fifty, nothing essentially new can happen and the loss is not as great as one might think. Now you tell me, will I be an elephant next time?"

Chandrasekhar recalled years later that Fermi showed discomfort only once, snapping at Laura, who was wondering when Segrè would arrive at the airport and who should go pick him up. A weary Fermi, clearly in pain, barked, "Do we have to discuss this here?"

Yang wrote later of the visit he and Gell-Mann paid to Fermi shortly after the operation. According to Yang, Fermi was reading a collection of inspirational stories of men who had overcome overwhelming challenges and misfortunes. He explained that he intended to write a book on nuclear physics. "Gell-Mann and I were so overcome by his simple determination and his devotion to physics that we were afraid for a few moments to look into his face."

Segrè, too, arrived before Fermi left the hospital. He found Fermi sitting quietly, Laura at his side. His old friend was absorbed in measuring the amount of fluids he was consuming by counting off the drops of his intravenous drip while observing a stopwatch. Segrè recalls that Fermi discussed his imminent demise with "Socratic serenity." Fermi asked Segrè to summon Teller to Chicago: "What nobler deed for a dying man than to try and save a soul?" he quipped with a smile, clearly referring to Teller's performance six months previously at the Oppenheimer hearings. The hearings were still very much on Fermi's mind and he wanted to confront Teller, his old friend and intellectual sparring partner, about his reprehensible behavior, which was splitting the scientific community. Fermi wanted to convince Teller to "shut up and disappear from the public eye for a long time." He also told Segrè that a priest, a pastor, and a rabbi had come by his hospital room, offering blessings that Fermi did not reject: "It pleased them and it did not harm me." Fermi also spoke to Segrè of his pleasure with Laura Fermi's newly published memoir of their marriage, *Atoms in the Family*, which he hoped would get a commercial boost with his death. Segrè recounts with some emotion how this visit left him:

> At the end of the afternoon I left. When I got out of the hospital, I felt ill; the emotional upheaval produced in me by the visit was too much for my constitution. I could scarcely stand, and I remember going into the first bar I came across to fortify myself with cognac, something exceedingly rare, perhaps even unique, in my life.

Teller arrived a few days later. Knowing of Fermi's unhappiness with the book by Shepley and Blair, Teller may have expected a scolding. In preparation for his bedside meeting with Fermi, he drafted a paper, "The Work of Many People," which was intended to give credit to the work of Los Alamos and repudiate the Shepley-Blair book. According to Teller's biographers, Fermi read the manuscript and told Teller to publish without delay. Teller seemed to think the promise of quick publication cleared the air, that "what happened had not disturbed our friendship."

Though Teller's biographers do not mention any discussion of Oppenheimer, it seems likely that the subject arose. Teller's attack on Oppenheimer at the hearings in April was a significant element in a controversy that Fermi feared would split the physics world. Teller's testimony alone did not doom Oppenheimer, but it hurt him and enraged Oppenheimer's supporters. Whether Fermi actually gave Teller absolution for the way he treated Oppenheimer will never be known.

Teller's biographers correctly point out that Fermi's death had a double impact on Teller. Teller lost a dear friend, someone with whom he truly enjoyed working during the Manhattan Project and beyond. Shortly after Fermi's death Teller recalled with affection how Fermi played with his five-year-old son, Paul. Laura confirmed in later years that Fermi always considered Teller one of his closest friends. As Teller's biographers note, the loss was compounded by the fact that Fermi was the one person with sufficient influence to bridge the gap between Teller and his detractors and curtail Teller's banishment from the community. Now there was no one to do so and the rift embittered Teller until his death in 2003.

Back home, Laura tended to Enrico, with help from Giulio and Nella. Leona Libby was a frequent visitor, helping out when she could. She reported later that he received visits from priests, more as insurance than out of any belief in the afterlife:

> He spoke of his approaching death as a great experience, but he asked wistfully if I thought there was anything valid in the idea of an afterlife. He was really cross about dying. I came out after each visit and drove home with tears streaming down my face.

The APS was having a November meeting in Chicago, and by that time Fermi's condition had weakened considerably. Many of his old friends came to visit. Wigner described Fermi as "so composed by death's approach he seemed superhuman. Ten days before his death he told me 'I hope it won't take too long.' He reconciled himself perfectly to his fate." Others had the same impression and remarked on his stoicism. Back from his summer in France, Ulam visited Fermi several times, once with fellow Los Alamos colleague Nick Metropolis. On one of these occasions Fermi discussed his impending death with scientific detachment. True to his essential character, Fermi calculated the odds of surviving long enough for doctors to discover a cure for him at 100:1, extrapolating from his belief that doctors would certainly discover a cure within twenty years. Ulam and Fermi turned to the idea that future generations, possessed with the cure for many fatal illnesses, might use genetic material to resurrect the dead and cure them. Fermi objected that such techniques would not be able to re-create the memories of former lives. After the final visit with Metropolis, Ulam was moved to tears and recalls quoting Plato's simple eulogy of Socrates to Metropolis: "That now is the death of one of the wisest men known."

Word of Fermi's illness made its way to official Washington. Soon his many friends were pushing a bill through Congress funding an AEC prize, awarded by the president, for Fermi's contributions to nuclear science. It came with a check for $25,000. Fermi was notified of his prize on November 16, 1954, but was too ill to travel to receive it. (The check arrived after his death.) During a visit from Maria Mayer and Harold Urey's wife Frieda, they discussed the prize. Frieda recalled later that Fermi was more proud of the publication of Laura's *Atoms in the Family* than he was of having received this extraordinary recognition from the

AEC. Mayer later recalled that he endured his illness "with the greatest grace imaginable."

By that time, the pain from the encroaching cancer was rapidly becoming intolerable and Laura, who maintained a stoicism equal to her husband's, began to worry about the prospect of Enrico lingering in agony for many months, as the doctors originally suggested he would. She need not have worried. He lived only another two weeks. Enrico Fermi died at 2:30 a.m. on November 28, 1954, having suffered a heart attack in his sleep.

Two of his friends, Chandrasekhar and Ulam, have written extensively of the comparison between the way Fermi approached death and the way Fermi's great friend John von Neumann approached his own death from cancer a little over two years later, also at the age of fifty-three. Von Neumann found it impossible to accept that his brilliant mathematical mind would be snuffed out at death. Born a nonobservant Jew and having converted to the Catholic faith to please a fiancée who wanted to marry in the church, von Neumann turned to Catholicism as a source of comfort and solace, although it seems to have provided him neither.

Fermi, in contrast, accepted with rare equanimity the fact that he would no longer exist, an acceptance in keeping with his generally realistic, perhaps even pessimistic, view of life. For Fermi, science completely replaced the function of religion and he died much as he had lived, without any obvious need for metaphysical or religious speculation on what happens after death. It was enough for Fermi to know that his life would end, that at that moment his unique, extraordinary mind would flicker out, but that his work would live on.

A private graveside service for the immediate family took place the next day at Oak Woods Cemetery, about a mile south of the campus. Surprisingly, a Lutheran chaplain attached to the University of Chicago Clinics by the name of Granger Westberg delivered a brief but moving prayer, including the following inspirational words:

> We thank Thee for him whom we this day hold in remembrance, who looking upon the face of nature and seeing order in its variety, law in its

constancy, has sought to teach men to live upon earth with reverence toward life.

We thank Thee for his courage as a lonely explorer into the realm of fact and his eagerness to share his insights in the fellowship wich [sic] advances all scientific discovery.

We are grateful for his sense of responsibility for the products of his work, for his sensitivity to the whole meaning of what it was he was doing.

A simple headstone marks the grave.

The university organized a public memorial service for the following Friday, December 3, 1954, at Rockefeller Memorial Chapel on campus. Sam Allison presided and spoke of the institute's debt to Fermi:

Actually, the Institute is his Institute, for he was its outstanding source of intellectual stimulation. It was Enrico who attended every seminar and with incredible brilliance critically assayed every new idea or

FIGURE 26.1. Fermi grave at Oak Woods Cemetery, Chicago. *Photo by Susan Schwartz.*

> discovery. It was Enrico who arrived first in the morning and left last at night, filling each day with his outpouring of mental and physical energy. . . . We may have seen his physical energy before, or his basic balance, simplicity, and sincerity in life before, but who in his lifetime has ever seen such qualities combined in one individual?

To illustrate what he considered the archetypical Fermi moment, Allison recounted the story of how Fermi calculated the effect of high altitude on the accuracy of Arthur Compton's watch. Emilio Segrè spoke next, giving a brief, but compelling account of the years in Rome, particularly the special time in 1934 when the boys of Via Panisperna worked on neutron bombardment. He concluded with some general comments about Fermi's character:

> He had had all the honors a scientist can have, none excluded. He was part of great councils, and for a large number of scientists his word was final. I have not mentioned these facts because for him they were really unimportant. Nothing altered his simplicity, which did not arise from false modesty—indeed he knew quite well how much he was intellectually above other men—but from charity. Nothing altered his unceasing interest in Science and his will to work humbly and indefatigably on the study of nature. If he had foreseen the cruel destiny that was to deprive us of him so unexpectedly early, he could not have husbanded his time to give more than he gave.

Anderson's tribute extolled Fermi's qualities as a teacher in addition to his abilities as a physicist. "His needs were few," he noted. "Chalk, a blackboard, and an eager student or two were enough for a start. Teaching was an essential part of his method." He described the deal Fermi would eagerly make—in exchange for the student correcting his English and teaching him his beloved Americanisms, Fermi would teach physics. He recounted Fermi's approach to problem solving:

> To explore the mysteries of nature with Enrico Fermi was always a great adventure and a thrilling experience. He had a sure way of starting off in the right direction, of setting aside the irrelevancies,

> of seizing all the essentials and proceeding to the core of the matter. The whole process of wresting from nature her secrets was for Fermi an exciting sport which he entered into with supreme confidence and great zest. . . . He was the center of our Institute around whom all revolved and for whom we all tried to do something good enough to win his praise.

It must have been difficult for these three men, who defined themselves professionally and personally almost entirely in terms of their relationship to Fermi, to stand up before the assembled crowd and speak coherently. Many in the audience shared this sense that their careers carried meaning largely because of their relationship with Fermi. In his sudden, unexpected passing, one sees how profound his impact was on those around him. For example, not long after his death his colleagues at Argonne put their recollections of him on record in a two-disc edition published as *To Fermi with Love*. One searches in vain for a similar tribute to any of the twentieth century's other great physicists. The passions he inspired had a dark side, to be sure. In reading memoirs and reminiscences of those who worked with him, one sometimes gets the sense of jockeying for position, of a discreet (or sometimes not so discreet) competition for the unofficial title of "closest to Fermi." Anderson, Allison, Segrè, Amaldi, Libby all laid claim to the Fermi legacy. Each was right, in his or her way, to do so.

In each case, however, the relationship never rose above that of teacher-student, master-apprentice. Looking back over his entire career, Fermi had only a handful of peers, mainly the Europeans who created the field of modern physics in the 1920s and 1930s. It would be hard to argue he belongs in the same group as Einstein or Bohr, although he would certainly have made the case, without arrogance, but with a robust sense of his own capabilities. Born and Schrödinger were older, but they were peers, as were Heisenberg, Pauli, Wigner, Bethe, Dirac, Pauli, and perhaps Teller. Von Neumann certainly was, and Ulam, as well. These were the people against whom Fermi measured himself. Oppenheimer was not in this group, nor was Lawrence. One might argue that Rabi was, and perhaps Felix Bloch. Others, like Alvarez, Gell-Mann, and Feynman would make their mark on physics but were of the next generation.

Hans Bethe, a friend since the early days in Rome, surely spoke for many in an emotional condolence letter to Laura written on November 30, 1954, shortly after he heard the news:

> Dear Laura,
>
> when [sic] I saw you on Friday I had no idea the end was so near. I heard it when arriving here Sunday morning.
>
> Rabi has said it right: There is no one like Enrico, and there will not be another for a hundred years. He was so much to all of us. Already in the last few weeks, I thought often: this I would like to tell Enrico, and that I would like to ask his advice on. And then I remembered, that this will no longer be possible.
>
> Enrico has taught me more than anyone else. You have described it very well in your book. He taught me to separate a problem into its parts, to understand its essentials rather than to plow through and I know that Enrico was its father.
>
> No one could leave a greater hole among physicists than Enrico. But no one, also, would leave behind more friends who will carry on in his tradition, or at least try to do so. His was a short life, much too short for all of us, but it was also a very full life. To be one of the great physicists of all time, to have restarted physics in Italy and a great school in this country, and to be loved by everyone he knew—you cannot ask for more.
>
> Remember, Laura, that we are your friends. We cannot replace Enrico to you, but if you ever need anything, we shall be there. We hope that you will remain in the circle of physicists, and that we shall see you as often as before.
>
> As always,
> Yours,
> Hans

Bethe was right. Those who knew and loved Fermi, those who studied under him, and those who worked alongside him did indeed carry on his work, in ways that no one, not even Fermi himself, could have predicted. He left a distinctive legacy, one that continues to unfold as we explore and clarify our understanding of nature and its innermost workings.

CHAPTER TWENTY-SEVEN

FERMI'S LEGACY

THE STORY OF ENRICO FERMI DOES NOT END WITH HIS DEATH. The Institute for Nuclear Science, though remaining an important center for physics research, changed in subtle but important ways, reflecting the loss of its most inspirational scholar. As a family man, a husband and a father, his influence has continued for generations. Fermi's teaching had profound and lasting impact on everyone who came into contact with him and those who studied with him went on to distinguished careers shaped in important ways by their exposure to Fermi. His scientific work set the agenda for postwar physics for decades to come.

FERMI'S DEATH HAD AN IMMEDIATE IMPACT ON THE INSTITUTE. His spirit had infused the place. His easy collegiality, the intensity of his passion for physics, his broad range of interest across the entire field, all defined the character of the institute and the work done there. Now Fermi would no longer walk the halls, asking people what they were up to and trying to help. Now colleagues looking for advice on how to approach a problem would no longer be able to stick their heads in his doorway and ask a quick question. Inevitably, the institution began to change.

Sam Allison stayed, as did Herb Anderson and James Franck, and together they continued the work of the institute. The young Valentine Telegdi stayed on as well, eventually becoming the Enrico Fermi

Professor at the university. However, important staff members drifted away—the Marshalls, Willard Libby, Harold Urey, the Mayers, and the younger ones, like Gell-Mann and Garwin. World-class physics continued to be pursued at the institute, now named for its most famous resident, but it was never the same. Indeed, no institution could have survived such a loss intact.

THOUGH SHE BRAVELY HID IT FROM THOSE WHO VISITED DURING THE final days, Laura was shattered by her husband's unexpected, painful death. His passing in the early morning hours of November 28, 1954, brought her relief that he was no longer suffering, but it was relief laced with sorrow. They had known each other since she was an adolescent, and their marriage, though not ideal, had been a twenty-six-year adventure. She was forty-seven years old, both children were out of the house, and she was still young enough to make a new life for herself. Although

FIGURE 27.1. Laura Fermi the writer. *Courtesy of the University of Chicago Regenstein Library, Special Collections Research Center.*

she never remarried, she certainly did make a new life, in a career she had already tried out—writing.

The book she wrote in Rome in 1936 with Ginestra Amaldi must have whetted her appetite, because in 1953 she began work on a memoir of her marriage to Enrico. In the 1954 memoir *Atoms in the Family,* one immediately recognizes a distinctive voice, comfortable with the English language and slightly arch in her pithy and sometimes hilarious observations about the people she met during her marriage. She could not have known that the timing of the book's release, in the last month of Enrico's life, would be commercially providential. His illness is mentioned nowhere in the book, but his death helped sell it. She knew that Enrico was enormously proud of the book, and its reception encouraged her to write subsequent books on a variety of subjects: the peaceful uses of atomic energy (*Atoms for the World,* 1957, and *The Story of Atomic Energy,* 1961); the rise and fall of Benito Mussolini (*Mussolini,* 1961); Galileo's contribution to world science (*Galileo and the Scientific Revolution,* 1961, with Gilberto Bernardini); and the influx of talented and important immigrants to the United States (*Illustrious Immigrants,* 1968). She even drafted an unpublished novel about the women of Los Alamos and was at work on a study of women in the Italian Renaissance when she passed away. She never claimed to be a scholar, but she was proud of her efforts writing popular history and science with a clear voice and strong insights.

She eventually moved to an apartment by the lake, but she remained active in the Hyde Park community for many years, helping to establish the Cleaner Air Committee of Chicago, which fought the use of coal for heating that had led to dangerous levels of air pollution in the city. She also championed restrictions on the sale of handguns in Chicago through an organization called the Civic Disarmament Committee. Leona Libby recalls Laura's involvement in an anti–nuclear power campaign in California, in support of her old friend Frieda Urey. Laura kept up with the circle of friends she made with Enrico and also took part in commemorations and tributes, far and wide. As a frail sixty-six-year-old, she attended the dedication of Fermilab in suburban Chicago in May 1974. It must have moved her that the largest physics lab in the country was named after her late husband.

She traveled often to Italy to see her family and old friends, particularly the Amaldis. But she never returned permanently to her native land. She missed her homeland terribly and continued to find the "immense plains" of the United States too empty for her liking. She wondered, in 1954, if she would ever be "Americanized." Yet Chicago was her new home and there she remained. There must have been something too compelling about the United States to leave it and return home. Perhaps it was the openness of the culture, the dynamism of the society, or even the fact that she had become herself a bit of a celebrity in her new land.

At the age of seventy she succumbed to pulmonary congestion and died the day after Christmas in 1977. Buried like her husband in Chicago's Oak Woods Cemetery, her gravestone bears the single word *Writer* inscribed below her name and the years of her life. She may not have chosen that epitaph herself—Nella coordinated everything relating to the burial—but it was almost certainly the way she wanted to be remembered, the role of which she was most proud.

There is, however, one aspect of her burial that is somewhat unconventional. The issue dates back to when Enrico died in 1954. At that time the family chose a plot between two existing graves, with no room for Laura. It would have been easy enough to buy a double plot, as most married couples do. When Laura passed away, the family chose to inter her in a plot some three hundred yards from Enrico's. The urge to jump to a conclusion that it reflects something dark about their marriage is strong, but perhaps should be resisted. More likely neither Laura nor Enrico thought very much at all about where they were to be buried, nor cared. Or perhaps having lived in Enrico's shadow during his life, in death she and her surviving family members wanted to emphasize Laura's independent life.

Enrico's sister Maria outlived him, but not by much. She spent most of her life in the house built by her father in Via Monginevro, having been widowed early by the death of her husband, Renato Sacchetti, who died of influenza. She had three children—Gabriella, Giorgio, and Ida—and was in regular correspondence with Laura after Enrico's death. She died in a plane crash on June 26, 1959, on her way to a conference on contemporary Italian literature to which Laura had invited her. She is buried in the cemetery in the village of Olgiate Olona, just northwest of Milan, where the aircraft went down—apparently according to wishes she had previously expressed to be buried where she died.

When her father died, Nella was already in her early twenties, a recent graduate of the University of Chicago. She went on to marry and raise two children, Alice and Paul. She received a Master of Fine Arts degree and taught art for many years at the Lab School, which she had attended as a child. Intellectually curious, she also went on to earn a PhD in educational psychology with a thesis entitled "Baby Bust and Baby Boom: A Study of Family Size in a Group of University of Chicago Family Wives, 1900–1934." In later years she pursued a certification in financial planning and made a new career in this field.

In an interview conducted for a CBS documentary in the early 1990s, she talked about her life with her famous parents and how it took her until the age of forty-five to come to terms with her father's fame. Nella regretted not reverting to the family name when she divorced her husband, Milton Weiner, in 1965. For her, Los Alamos was a great adventure, but she confessed to an irrational sense of guilt for the bombings of Hiroshima and Nagasaki. She also understood the complexities of the moral dilemma faced by her father and his colleagues as they worked to design and build the terrible weapon.

Nella's daughter Alice (who later changed her name to Olivia Fermi) remained close to her grandmother Laura. In her last years, Nella would urge Olivia to "Put your grandmother Laura first, ahead of Enrico." The words are somewhat opaque, but we can guess at their meaning: Enrico was a star that the world would remember and it was important to foster Laura's legacy. Nella felt Laura was important in her own right and was unjustly overshadowed by her husband.

Nella contributed to a commemorative event organized at Cornell by Jay Orear in 1991, regaling the audience with anecdotes of her childhood with her father. She died in 1995, a victim of lung cancer. By this time, she had made peace with her father's fame and role in the history of science and was comfortable speaking publicly about him and about their relationship.

Giulio had a harder time of it. He never felt comfortable living in his father's shadow and did what he could to distance himself from Enrico. In his adolescence he changed his name to Judd and used an American pronunciation of his last name, "FIR-mee" rather than "FAIR-mee," rarely speaking of his famous father. He chose not to pursue physics, although he had the innate ability to do so. Instead, he pursued a rarified program in pure mathematics, leaving Oberlin early for a mathematics PhD at Princeton. His Oberlin friend Robert Fuller joined him there. After Princeton Judd did a post-doc at Berkeley, where he met and married Sarah Duncan Pietsch, an artist and a literary scholar. They moved to Washington, DC, where Judd held a position at the Institute for Defense Analysis, a distinguished defense policy think tank, for a decade. He eventually became bored in Washington and took a position working in the lab of Nobel Prize winner Max Perutz at Cambridge University, where he developed mathematical models of complex proteins. The severe depression of his adolescence never returned, but he was a quiet and shy man, who actively withdrew from the limelight, happy to contribute to the work of others, but with little interest in generating his own projects. A heavy smoker, he died of a heart attack in 1995 at the age of sixty. His health problems may have been aggravated by the stress he felt when the British government insisted that researchers at government-funded science labs present research projects of their own or else find new work.

Whereas Nella resembled her father, Judd had a greater resemblance to his mother, although his eyes were Enrico's—hazel gray and, for those who knew Enrico, very familiar. Richard Garwin recalls lecturing at Cambridge in the 1980s and noticing someone in the audience who reminded him of Fermi. The eyes, he thought, were unmistakable. It was Judd, who afterward introduced himself as Enrico's son, much to Garwin's astonishment and pleasure.

Of the following generation, Nella's daughter, Olivia, and Judd's daughter, Rachel, have been most publicly involved with their grandparents' legacy. A psychotherapist living in Vancouver, Olivia has embraced the family history. She has two blogs—one about the Fermi family and another about nuclear policy issues. Rachel pursued a career in photography and fine arts but also embraced her grandfather's legacy. In 1995, she produced a comprehensive photographic history of the Manhattan Project, *Picturing the Bomb,* with coauthor Esther Samra. She lives in the north of Scotland with her artist husband and their family.

For both Olivia and Rachel, Laura looms large in childhood memories. Neither had been born when Enrico died, so they know him only through family anecdotes passed down by their grandmother and their parents. Yet the legacy of the Manhattan Project fascinates and disturbs them. For this generation of Fermis, Fermi-Dirac statistics, beta decay, and pion-nucleon scattering all take second place to their grandfather's work on the atomic bomb.

AFTER HIS DEATH, FERMI'S CELEBRITY PROMPTED THE USE OF HIS name to commemorate him in any number of ways. The Chicago institute now bears his name, as does the huge national laboratory built in the early 1970s fifty miles west of the city. A space telescope designed to focus on gamma-ray sources in deep space is called the Fermi Telescope. Nuclear reactors in the United States and in Italy are named after him. Countless towns in Italy have streets and plazas named after him. Train stations in Rome and Turin bear his name.

Among all these tributes he would probably be most proud of the prize bearing his name that the US AEC awards annually, first granted to Fermi on his deathbed. The US Department of Energy describes it in the following terms:

> The Fermi Award is a Presidential award and is one of the oldest and most prestigious science and technology honors bestowed by the U.S. Government. The Enrico Fermi Award is given to encourage excellence in research in energy science and technology benefiting mankind; to recognize scientists, engineers, and science policymakers who have given unstintingly over their careers to advance energy science and technology; and to inspire people of all ages through the examples of Enrico Fermi, and the Fermi Award laureates who followed in his footsteps, to explore new scientific and technological horizons.

Recipients of this honor include an impressive array of men and women in science: in chronological order, von Neumann, Lawrence, Wigner, Seaborg, Bethe, and Teller were the first recipients, beginning in 1956. The award of the prize to Oppenheimer in 1963 was widely seen as an act of contrition by the US government for the way it treated him in 1954. The award of the prize to the trio of scientists who discovered fission—Hahn, Meitner, and Strassman—in 1966 was a belated recognition of Meitner's crucial contribution to the discovery. More recently, Wheeler, Zinn, Bradbury, Agnew, Peierls, Anderson, Alvarez, Weisskopf, Garwin, and Rosenfeld have all received the prize named after their friend and mentor, as have many other luminaries. Many consider it the highest honor they have received.

Several tributes by friends and colleagues are particularly noteworthy. In addition to the warm 1955 recorded tribute, *To Fermi with Love,* the mid-1960s saw the production of a full-length documentary, *The World of Enrico Fermi*. One of the most eminent science historians, Harvard's Gerald Holton, brought the project together with the help of the Canadian National Broadcasting Company. Dozens of colleagues and friends, as well as Laura, discuss Fermi's life.

One of the most lasting and beautiful of the tributes to Fermi was the development and publication, in two volumes, by the University of Chicago Press in collaboration with the *Accademia dei Lincei* in Rome, of *The Collected Papers*. The organizing team of Amaldi, Anderson, Persico, Rasetti, Segrè, Cyril Smith, and Wattenberg, with Laura's active participation, reset every journal paper and article in beautiful typeface and provided valuable introductions to many of the major papers in the volumes. In a predigital world, culling all the papers, deciding which papers were sufficiently important for inclusion, and ensuring that nothing of importance was missing was an enormous undertaking. There are some 270 papers in all, as well as a brief biography written by Segrè and several useful appendices, including a list of his many honors and a chronology of his life. It should not be considered a complete set of his work, however, because some of his important papers and lectures were only subsequently declassified, but it was as comprehensive as possible at the time.

And then there are his scientific legacies, including discoveries related to the weak interaction, the strong interaction, Fermi-Dirac statistics, and computational physics.

THE WEAK INTERACTION HAS BEEN A RICH SOURCE OF DISCOVERIES. Neutrino physics is an enormous field in itself, and the research into the weak interaction had dominated much of particle physics, resulting in more than a dozen Nobel Prizes, including the recent discovery of the Higgs boson. Most interesting, perhaps, is the deep connection between the weak interaction and the electromagnetic interaction first posited by Sheldon Glashow, Mohammad Abdus-Salaam, and Steven Weinberg, a first step in the pursuit of one of physics' holy grails, the unification of all forces. For this work they shared the 1979 Nobel Prize.

Since Fermi's passing, the exploration of the strong force holding the atomic nucleus together has had an equally distinguished history. Fermi's pion scattering experiments led the way to further discoveries about the force that holds the nucleus together. Ultimately, this has resulted in the "quark" theory of matter, first proposed by Fermi's Chicago colleague Murray Gell-Mann. Quarks are the fundamental building blocks of neutrons, protons, pions, and many other subatomic particles. Quarks are bound to each other by bosons that physicists call "gluons." The interaction between quarks and gluons defines the strong interaction that holds these particles, and the nucleus itself, together. Fermi's early pion work was the first step in this direction, and the quark theory, together with our understanding of the electro-weak interaction, constitutes a synthesis called the "Standard Model" of particle physics. The Standard Model explains a lot of the observable world, but it leaves many questions unanswered, questions with which theorists and experimentalists still grapple.

What might Fermi have accomplished if he had lived longer? It is difficult to say. He understood the complexity of the results from his pion scattering experiments in 1951 and 1952, but what he made of it is a different matter altogether. Subsequent advances in organizing and understanding the elementary particle "zoo" relied on group theory. Never a fan of group theory, he learned only as much as he needed to understand the quantum theory work of von Neumann and Weil. It is hard to imagine him willingly diving into group theory in the way required to have come up with the quark theory. Yet doing so may have been more important to him than, say, the QED renormalization completed by others so successfully in the late 1940s. If he decided a problem was sufficiently important, he would invest the time required to solve it, as he did with beta decay.

He was, at heart, a conservative physicist and would have felt uncomfortable with some of the early revelations regarding the weak interaction, particularly the completely unpredicted and revolutionary discovery in 1956, by his former students Lee and Yang, that it did not obey the rules of mirror-image symmetry. In looking to solve specific problems, Fermi rarely if ever chose revolutionary approaches. Yet he certainly would have been fascinated by the theoretical work and the experimental discoveries exploring the weak interaction and would have been an active participant in neutrino physics.

A MORE STRAIGHTFORWARD LEGACY IS THAT OF THE FERMI-DIRAC statistics.

Unlike the beta decay paper (not precisely correct) and the pion experiments (suggestive, but only a stepping stone to our understanding of the strong force), Fermi-Dirac statistics are as valid today as they were when first developed in late 1925 and early 1926. Any analysis of the energy distribution within a system of particles that obey the exclusion principle—gas, liquid, solid, or plasma—involves Fermi-Dirac statistics. In the words of physicist Henry Frisch, if we did not have Fermi-Dirac statistics, physics would be "in the stone age." The use of Fermi-Dirac statistics is so universal and pervasive across different fields of physics that it is virtually meaningless to give examples. There will always be a debate about which of Fermi's contributions to physics is his greatest, but those who favor Fermi-Dirac statistics point to the fact that it is still used the way Fermi presented it to the world in 1926.

WHEN FERMI FIRST BEGAN USING COMPUTERS, HE WAS INTERESTED in simulating physical processes. The computers were primitive and only the simplest of problems could be represented in a program. Analysis of data generated by detectors was all done by hand and eye.

Fermi would not recognize the world of computational physics today.

Computer simulations have become an essential part of any hypothesis testing in particle physics and of predicting the outcome of any given experiment. Computers also sift through petabytes of data generated by complex electronic detectors, looking for key signatures that indicate the presence or absence of a specific interaction. Computational physics has become a field of its own; most physicists are familiar with its techniques and some physicists specialize in it as a subfield. The discovery of the Higgs boson would have been impossible without advanced computational techniques. They are also central to many other physics specialties, for example astrophysics, where computational advances have enabled the modeling of complex processes involved in phenomena such as supernovae or the Big Bang.

Perhaps more interestingly, the Monte Carlo simulation techniques Fermi helped to pioneer, even prior to the invention of the electronic computer, are used wherever systematic simulations can shed light on solutions to complex problems, ranging from engineering and genetics to defense policy and nuclear strategy to finance and economics, even to law and social policy. They have been used to explore whether Joe DiMaggio's fifty-six-game hitting streak should be considered an intrinsically rare event. Reflecting the essence of Fermi's unique way of thinking about problems, the Monte Carlo technique is perhaps the single most important area where Fermi's direct influence can be felt in the world outside physics.

In Italy, Fermi's legacy was championed by Edoardo Amaldi. He directed the physics program in Rome and supported Bernardini's plans to build a cyclotron in Frascatti, a Roman suburb. He fostered interest in the field by continuing to edit and update Fermi's textbook for high schools. Edoardo also played a central role in pan-European physics, including the creation of CERN and the European Space Agency. Dismayed by the gulf that secrecy created between him and his old teacher and friend, he wrote bylaws prohibiting CERN from engaging in classified research. Fermi could not have wished for a better keeper of the flame.

The museum of physics at the University of Rome's department of physics keeps the memory of Fermi alive among young people there, and Via Panisperna, under the auspices of the Italian Physics Society, is being converted to another museum, planned for completion in 2018. It will enable a new generation of Italians to wander the halls where Fermi launched the Rome School, where he conducted classes and seminars, and where he first bombarded uranium with slow neutrons.

The summer school in Varenna, named after Fermi, continues. The Italian Physics Society presents an annual prize, also named after him, to major figures in Italian physics and, more recently, to non-Italians as well.

The centenary of Fermi's birth gave rise to major celebrations throughout Italy, resulting in some of the best commemorative volumes devoted to his life and work. Italian historians continue to illuminate aspects of his life and work, often providing a useful corrective to received wisdom.

After Fermi's death, Amaldi shipped Fermi's notebooks and other archives to the Domus Galilaeana in Pisa, convinced they belonged alongside Galileo's archives. In so doing, he demonstrated the esteem with which Fermi's fellow countrymen viewed him. Sixty-odd years later, it is a decision that few would second guess.

THE TWO LEGACIES STEMMING FROM FERMI'S MANHATTAN PROJECT work—the atomic bomb and the nuclear reactor—are perhaps more difficult to evaluate clearly today.

The legacy of the use of nuclear weapons is, not surprisingly, greatest in Japan, two of whose cities were obliterated by these weapons. More than two hundred thousand people perished in these attacks, and Japan has been in the forefront of the anti-nuclear movement. In 2016, President Obama became the first US leader to visit these cities, paying tribute to those who died. In response, Japanese Prime Minister Abe became the first Japanese leader to visit Pearl Harbor. Memories linger, even if enmities do not.

Over the past seventy years, at least eight other countries have learned the secret behind building nuclear weapons. Fermi understood that if it proved possible to make such a weapon, other countries would eventually do so. He had a pessimistic view of human nature and assumed that people would eventually use whatever weapons were available to prosecute warfare. Fortunately, he has been wrong, at least until now. Aside from the initial use against Japan, no nation has used nuclear weapons against another. But as Fermi would surely have agreed, there is no law of physics preventing this from ever happening again.

Aboveground nuclear testing was frequent in the 1950s and early 1960s, leading to an ever-higher level of ambient radiation around the world. As a 2009 report of the Centers for Disease Control and Prevention made clear, even the first nuclear test at Trinity site had unintended, catastrophic fallout effects on local populations, livestock, and farming. Later on, trace levels of radioactive isotopes like strontium 90 found in food and milk throughout the country led to a wave of public health concern. Though aboveground tests have ceased, we live in an environment contaminated by the radioactive residue of these tests, residue that will last for centuries.

The prospect of all-out thermonuclear conflict has receded, owing largely to the end of the Cold War, but the threats of nuclear proliferation and nuclear terrorism loom larger today than in years past. North Korea increasingly rattles a nuclear saber in its fraught dealings with its neighbor to the south, as well as with the United States. Terrorism remains a serious threat. Some nuclear powers are host to active insurgent movements, and nuclear security is almost certainly not as tight as one would wish. The capture of even one nuclear weapon by an insurgent group bent on terror would be catastrophic. Policy analysts barely considered such possibilities when Fermi was alive. Today they dominate national security thinking.

Is this state of affairs truly a legacy of Enrico Fermi? If Fermi had perished somewhere in the Atlantic during the voyage on the *Franconia* over New Year's 1938–1939, the Manhattan Project would have eventually moved forward, perhaps more slowly and certainly in different ways. Facing the situation they did, US political leaders decided to go ahead with the Manhattan Project, a decision neither surprising nor obviously evil. The decisions surrounding the use of the bomb in the wake of Germany's surrender were made at the highest level, with only cursory attention paid to the views of scientists. President Truman knew that he alone bore responsibility for the decision to use these weapons against enemy cities.

This is not to let Fermi or his colleagues "off the hook" in any sense. For better or worse, they changed the future for us all. The Manhattan Project scientists have, however, assumed a greater burden of guilt than they deserve. If history is to judge Fermi and his colleagues for their wartime work, it should be with a more nuanced perspective that appreciates the situation they faced and their motivations for participating.

The other great legacy of that wartime project is the nuclear reactor. Fermi may not have invented the atomic bomb, but he and Szilard most certainly did invent the nuclear reactor. Some 450 electric power reactors operate worldwide, with 60 more under construction. About one hundred are in the United States. However, only five US reactors have come on line since 1990. The decline in the use of nuclear power in the United States and elsewhere is mainly the result of high-profile nuclear accidents around the world, including those at Three Mile Island in 1979, Chernobyl in 1986, and the 2011 accident at Fukushima. These three notorious accidents, spread out over a period of three decades, have done as much as anything to kill the prospects of nuclear energy in the United States. Yet the technology advances. Engineers at Argonne Lab, Fermi's old stomping ground, have designed modern fast breeder reactors that produce fuel as they consume it. An initial load of fuel could, in principle, last one thousand years and produces no greenhouse gases at all. South Korea, where the nuclear allergy is less severe, is in the process of building such reactors in consultation with Argonne engineers. The future of nuclear energy in the United States is uncertain at best. Whether future generations of Americans will reconsider the option of pursuing safe nuclear energy remains to be seen.

Reactors have purposes other than providing nuclear energy. They provide the main way modern medical radioisotopes are created. Corbino was prescient. He understood the commercial value of the slow-neutron discovery for medical purposes. He never anticipated, however, that his greatest student would invent a virtual radioisotope factory. By irradiating specific elements inside research reactors, medical physicists create substances that can be used to trace the presence of cancer as well as to treat cancer once it is found. Countless lives have been saved through these techniques. More than a dozen research centers around the world produce these isotopes in reactors that can trace their lineage directly to those dusty piles of graphite and uranium that rose up from the floor in the basement of Schermerhorn Hall at Columbia and the squash court at Stagg Field at Chicago, under the watchful eye of Enrico Fermi.

FERMI'S PATH TO GREATNESS DIFFERED FROM THAT OF EINSTEIN, Bohr, Planck, or any of the others who created modern physics. It started from a profound confidence that he could solve any problem thrown his way and that nature would ultimately disclose her most precious secrets to his probing mind. That confidence was based on an incredibly solid foundation of knowledge laid down early in his life through intense and disciplined effort, under the guidance of mentors and professors who understood and cultivated his greatness. He understood that there were no shortcuts to deep understanding and was willing to make a radical commitment to gain that understanding.

The sturdy foundation of his knowledge informed his taste in the types of problems he studied. He had an unerring instinct for the "next big thing," and the decisions he made about his own research agenda set the agenda for the field at large. This instinct led him to focus on the way in which the exclusion principle could be integrated into statistical mechanics. It then led him to focus on solving the beta decay crisis using quantum field theory. Next it led him to neutron experiments that would open up the field of nuclear physics. Finally, he understood that an accelerator that could produce beams of high-energy pions could be used to explore the nature of the strong force. That superb instinct, married to a foundation in the basics second to none, produced a wealth of lasting achievements.

He also believed that anyone could learn what he knew. He believed this quite literally and lived his life devoted to that belief. In the process of digesting physics in his own way so that those around him could grasp it, he developed a technique of stripping problems to their bare essentials and leading his students through step-by-step solutions, ignoring complexities that would obscure the essence of the problems. This conviction ensured that the way he thought about physics influenced future generations of physicists.

A constant theme throughout his career was the central role of probability and chance in his analytic framework. He became a deep student of probability early on, driven perhaps by the loss of his brother, a low-probability event with profound personal consequences. In the world of quantum theory all physicists must understand probability, but Fermi placed it front and center in his research, returning to it time and again—in the Fermi-Dirac statistics, in the Thomas-Fermi model of the atom, in his pen-and-paper analysis of neutron diffusion, and in his pioneering use of Monte Carlo methods to simulate physics problems. The Fermi Paradox—the conclusion that if life existed elsewhere in the universe, they should have visited us by now—is a classic example of how he could break down almost any problem into a series of probabilistic assumptions.

He sacrificed much for physics. He was willing to compromise his political beliefs in exchange for the freedom to pursue physics without interference. He was willing to put his family life second to his career, with unhappy consequences. He may even have sacrificed his life, if we believe that his exposure to radiation had any relationship to his ultimate demise.

Confidence born of innate ability and a strong foundation, a firm belief in his ability to solve any problem, an instinct for important research, an unshakeable faith in his ability to make others understand, a fascination with the role of probability and chance at the core of how the world works, the willingness to make enormous sacrifices for science—all these made Fermi who he was and contributed to his ability to make a lasting impact. Like all of us, however, scientists are prisoners of the era into which they are born. To have had the impact of Einstein, it helped to be born during a period when some of the deepest problems of physics had come to the fore. If Einstein had been born a century earlier, he may have achieved much, but certainly nothing of the magnitude of general relativity. It was Fermi's great good fortune to have been born during a period in which the quantum revolution was unfolding.

And yet that great good fortune had a darker side. As Fermi realized when he was young, one implication of the twentieth-century revolutions in physics was the enormous energy locked inside matter, energy that could blow to smithereens the first physicist to unleash it. He could hardly have realized that it would be his fate to be that first physicist. If every great gift has a price to be paid for it, this was certainly a major one: the field he loved, and pursued with such passion for his entire adult life, uncovered a secret of nature that gave man the ability to destroy the world. Another price was one with which other driven professionals are familiar: the neglect of family relationships in pursuit of compelling career objectives. In many ways Fermi paid for his great gift, but this was an inevitable cost of that gift and the time in which he lived.

GEOFFREY CHEW AND UGO AMALDI HAVE BOTH DESCRIBED Fermi AS "the last man who knew everything." Obviously, he did not know everything. His knowledge of science beyond physics was superficial, and his knowledge of history, literature, art, music, and much else besides was limited, to say the least. He was not a universal genius.

He did, however, know everything about physics. In his day that was rare enough. Chandrasekhar marveled that Fermi, with no prior background in astrophysics, could jump into the field relatively late in life and make significant contributions. He should not have been surprised.

Fermi loved all aspects of physics and he lived at a time—perhaps the last time—when it was possible for someone with the proper background and innate ability to master all of physics. Fermi did so, not only across all subdisciplines of the field—astrophysics, nuclear physics, particle physics, condensed matter physics, even geophysics—but across theory and experiment. In this, he was truly unique. He saw physics as an integrated whole, comprehensible through a handful of powerful analytical tools he worked hard to master. Today physicists rarely talk across subdisciplines, and when they do they have increasing difficulty understanding each other. Even in Fermi's day, theorists and experimentalists had trouble seeing the world from each other's perspective. Today the problem is compounded by the magnitude and complexity of experiments and the increasingly sophisticated mathematics involved in cutting-edge theory. To be a world-class researcher in any subdiscipline today requires enormous commitment. A particle physics experiment might take a decade or more to conduct and involve many thousands of physicists, none of whom would have the time to explore other areas of physics, however motivated they might be. This problem is true for every subdiscipline of the field.

Fermi was certainly the last man who knew everything about physics, the study of matter, energy, time, and their relationships—the way the physical world works. He knew everything about how the physical world worked across subdisciplines and across theory and experiment as far as physicists were able to know these things during his lifetime. Our knowledge has evolved since he died, shaped by theory and experiment in ways that would have delighted Fermi had he lived. Even so, for one person to master all the physics of his day was a unique achievement. We may never see another like him.

ACKNOWLEDGMENTS

This project has been one of the great adventures of my life and my wife, Susan, was by my side the whole way, as always. We traveled far and wide in search of Enrico Fermi and met an enormous number of people who have been incredibly helpful and generous. This is my opportunity to thank everyone. Please forgive me if I have omitted anyone inadvertently.

My first thanks go to a group of people who worked with Fermi when they were young and who generously shared their memories of him. These include Geoffrey Chew, James Cronin, Jerome Friedman, Richard Garwin, Arthur Rosenfeld, Jack Steinberger, and Courtney Wright. Geoffrey Chew was the first to suggest the title of this book. Tsung-Dao Lee and Chen Ning Yang, colleagues of my father, were supportive and encouraging; many years ago, Lee posed the question that led to my father's muon neutrino experiment. The support of these scientists for this project has been a special privilege and has brought my subject vividly to life. Sadly, James Cronin and Arthur Rosenfeld have since passed away.

My archival research began at the University of Chicago's Regenstein Library. I am grateful to Julia Gardner, Head of Reader Services, Special Collections Research Center; Eileen Ielmini, Assistant University Archivist; and their respective staffs who were all extremely helpful. Christine Colburn was also helpful in the photographic archives. Outside of Chicago, Roger Blomquist, Patricia Canaday, and David Hooper at Argonne National Lab were generous with their time. Lance Friedmann and Sari Gluckin were amazing hosts during our frequent visits to Chicago.

Staff at the three major Manhattan Project sites were extremely helpful. I would like to thank Alan Carr, Rebecca Collinsworth, Barbara Lemmick, Heather McClenehan, and Glen McDuff in Los Alamos, Barbara Penland and Steven Stow in Oak Ridge, and Russel Fabre and the B-Reactor team in Hanford.

At Columbia University, I am grateful to William Zajc, who sponsored me for a Visiting Scholar position, giving me crucial access to the university's digital portal for scholarly resources for two years. Bill is a former colleague of my father who currently holds the I. I. Rabi Chair of Physics that my father held in the 1990s. Bill was a crucial sounding board and invaluable resource throughout this project, including last-minute comments that saved me from making several mistakes. I am grateful for the many hours spent with him. I am also grateful to Christopher Laico and Tom McCutchon, who provided special access to the Columbia University archives.

In College Park, Maryland, I'd like to thank Greg Good, the director of the American Institute of Physics Center for the History of Physics, as well as the director of the AIP's Neils Bohr Library, Melanie Mueller, and her colleagues Amanda Nelson and Audrey Lengel. Also in College Park, National Archives staff Rebecca Calcagno, Tab Lewis, and Laurel Macondray, were helpful.

We spent one month in Rome during the fall of 2015 where I was a Visiting Scholar at the American Academy. Founded in 1894 and set in a magnificent villa atop Gianicolo Hill overlooking all of Rome, the Academy has been home to countless scholars, artists, writers, and musicians. Our apartment in the greenhouse at the Villa Aurelia was indeed a luxurious base camp. Thanks to the entire staff, including President Mark Robbins, Executive Director Kim Bowes, and community members Gianpaolo Battaglia, Christine Begley, Paola Gaetani, Denise Gavio, Lindsay Harris, Sebastian Hierl, Peter Miller, Laura Offeddu, Cristina Puglisi, and to everyone else who made our stay so pleasant and productive. We note with sadness the passing of Administrative Director Pina Pasquantonio, who was particularly helpful during our stay. We owe gastronomic thanks to Chris Behr and his team at the Rome

Sustainable Food Project, founded by Alice Waters, with whom we enjoyed two memorable meals. I am also indebted to William Higgins, who first introduced me to the Academy, and to Eli Gotlieb, a former Rome Prize winner, who helped me with the application process.

In Rome, many people gave their time and expertise to the project. Luisa Cifarelli, president of the Italian Physics Society, hosted us at the 2015 Enrico Fermi Award dinner. She moved mountains to get us a tour of the Via Panisperna site, which was in the midst of major construction in preparation for its reopening as a museum in 2018. Giovanni Battimelli opened the doors to the vast Amaldi archives at La Sapienza and gave us unrestricted access during our stay in Rome. He continued to answer questions long after we returned to the United States. Francesco Guerra and Nadia Robotti have also been invaluable resources. They spent a full day with us in Rome, discussing our project and presenting their own highly informed, and sometimes iconoclastic, perspectives on Fermi and his colleagues. They also continued to provide valuable insights during the period following our visit. We will never forget the spectacular Sardinian dinner to which they treated us at a restaurant near La Sapienza. Paola Cagiano and Alessandro Romanello served as helpful and generous guides through the archives of the Accademia dei Lincei and the Reale Accademia d'Italia. Alessandro also joined us for a wonderful lunch near the Villa Corsini, and then drove us back to the American Academy one afternoon in the pouring rain. Also in Rome, Laura Fermi's nephew, Giorgio Capon, and his wife, Teresa, welcomed us into their home, the same home in which Laura grew up, and shared memories of Enrico and Laura. Others who were helpful along the way include Sandro Bettini, Mauro Canali, Adele La Rana, Giovanni Organtini, Marta Pepe, Francesca Salvatore, and Andrea Trentini.

In Pisa, Roberto Vergara Caffarelli was extraordinarily generous with his time and insights and remained helpful throughout the project. Luca Galli and Giovanni Signorelli were brilliant archival and culinary guides; our morning *caffé crema* remains a fond memory. In addition, I would like to thank Maura Beghè, Monica Biondi, Alessandro Corsi, Chiara Letta, Anita D'Orazio, Umberto Parrini, and Maddalena Taglioli.

In Geneva, we were welcomed by the physicists at CERN, who are in the midst of one of history's greatest scientific odysseys. Director General Fabiola Gianotti took time out of a hectic schedule to chat about her own work and how it related to Fermi; she also arranged for Steven Goldfarb and Iva Raynova to give us a tour of the ATLAS experiment. We were joined in our meeting with Fabiola by Jack Steinberger, with whom we also enjoyed having lunch in the CERN cafeteria. Ugo Amaldi and his wife, Clelia, hosted us at their lovely home for a leisurely glass of wine and biscotti while he talked about his father, Edoardo, his mother, Genestra, and his own memories of the Fermi family. He independently suggested the title of this book. Toward the end of our visit with him, we were briefly joined by his daughter Silvia, a spitting image of her beautiful grandmother.

Our time in England was quite productive. Our travels took us to the home of Fausta Segrè Walsby and her husband, Tony, in Bristol; to the home of Judd Fermi's widow, Sarah Fermi, in Cambridge; and to the Senior Common Room at Exeter College, Oxford, where we met with physicist and writer Frank Close. We thoroughly enjoyed all of these meetings and appreciate their ongoing help. Many thanks to Elizabeth Greitzer and John Durrell with whom we stayed in London—we felt right at home and wish we could see them more often.

Three professional physicists read the drafts carefully and helped me through the thickets of explaining complicated concepts in ways that the general public can understand and appreciate: William Zajc of Columbia University, mentioned above; Nicholas Hadley of the University of Maryland, an old high school friend who has played a major role in two significant particle discoveries during his career; and Andrea Gambassi of the International School for Advanced Studies (SISSA) in Trieste, who in addition to making some important scientific corrections was kind enough to explain the intricacies of a Scuola Normale Superiore education. All three are distinguished, busy scientists. They all read my drafts voluntarily and provided essential feedback on numerous aspects of the book. I hope they enjoyed reading it as much as I enjoyed discussing it with them. Any mistakes or errors in the physics are, of course, solely my own.

A number of other individuals read the book in part or in its entirety in various drafts and provided invaluable comments and corrections along the way: Joerg Baldauf, Cynthia and Marvin Blynn, Laurie Bruckmann, Camilla Calamandrei, Michael Cohn, Beth Hadley, Howard Jennings, George Minkoff, Rick Peterson, David Rudofsky, Raymond

Rudofsky, and Melanie Shugart. Numerous other people were helpful and encouraging along the way. These include Rinaldo Baldini, Nelson Beebe, Patrizia Bigotti, Glenda Bingham, Luisa Bonolis, Justin Breaux, William Briscoe, Paula Bruni, Viola Buckenberger, Elisabetta Calusi, Antonio de Candia, Roberto Casabuoni, Tina Cordova, Gene Dannen, Evan Faye Earle, Joe Escamillo, Alexis Fama, Susan Fine, Julia Foster, John Fox, Colleen French, Henry Frisch, Brett and Carmel Fromson, Joe Gonzales, Karl Grandin, Susanne Grulich, Kevin Haggerty, Larry Haler, Megan Halsband, Matthew Hopkins, Laurie Innes, Karl-Heinz Kampert, Brian Keelean, John Khadem, Sanford Kingsley, Jasper and Rita Kirkby, Penny Kome, Louis Linfield, Daniel Linke, Jack and Lynne Lloyd, Kathryn Ma, John Marshall III, Daniel Meyer, Michela Minesso, Arnon Mishkin, Julie O'Neill, Borden Painter, Shanice Palmer, Sara Paretsky, AnnaLee Pauls, Christie Peterson, Giancarlo Righini, Sandra Romiti, Kevin Roark, Harry and Carol Saal, Tristan Scholl, Anthony Shugaar, Leon Sigal, Megan Smith, Alessandra Stanley, Andrew Szanton, Donna Thompson, Rachel Trent, David Torney, Anne Vargas, Patrick Waide, Andrew Weston-Hawkes, Fletcher Whitworth, Stan Wojcicki, Kristina Wolff, Hillary Dorsch Wong, Nathan Woods, Lila Yawn, and Rita Zanatta.

The Fermi family has been extremely supportive and helpful during the course of the project. In addition to Giorgio Capon and Sarah Fermi, mentioned above, I am grateful to Olivia Fermi, Rachel Fermi, and Gabriella Sacchetti. Olivia and Rachel have provided a wealth of materials relating to their family, including documents and photographs as well as precious memories. Robert Fuller, Giulio/Judd's oldest and closest friend, also shared personal memories. Some of these discussions touched on delicate matters of family history; I hope I have reflected their perspectives with appropriate respect.

My agent, Michael Carlisle, believed in this project from the beginning and has been a long-time supporter. He and his wife, Sally, have become dear friends. He understands what an author puts into a book and has been a continuous source of wise counsel and support. I am so lucky to work with him and his team at Inkwell Management, including William Callahan, Michael Mungiello, and Hannah Schwartz.

At Basic Books, my editor T. J. Kelleher shepherded this book from the start, with patience, wisdom, insight, and enthusiasm. His team, including Hélène Barthélemy, Sandra Beris, Betsy DeJesu, Kait Howard, Carrie Napolitano, Christina Palaia, and Kelsey Odorczyk, has been extremely helpful.

My family deserves special thanks. My mother, Marilyn, inadvertently kicked this project off with the fortuitous discovery of a batch of my father's papers in 2013 and was unwavering in her enthusiasm and encouragement, as were my sisters, Diana and Betty. My son, Alex, inherited his grandfather's gift in the lab, recognized by his high school when they awarded him the Jack Steinberger Prize for scientific research. Susan's brother, David Rudofsky, outdid himself in proofreading and fact-checking the galleys. Susan and I are both indebted to him. My father-in-law, Raymond Rudofsky, took great joy in the project every step of the way. He was particularly supportive of the decision that Susan and I made to step away from our business in late 2015 so I could devote myself full-time to the project. I am indebted to him for his staunch support and enthusiasm in ways I cannot enumerate.

My father, Melvin, passed away in 2006. Many were the moments when I wanted to consult with him on some complicated aspect of modern physics, as I did when he was alive. He had Fermi's gift for lucid explanation and enjoyed teaching as much as he did research. He would have loved to see this project. I have been guided throughout by his memory and his inspiration. Dad, I miss you every day.

Finally, this book would simply not have been possible without the support of my research assistant, archival photographer, logistics consultant, travel agent, editorial adviser, relentless proofreader, confidante, therapist, cheerleader, and wife. Her name is Susan, and this book is dedicated to her, with infinite love, affection, and gratitude.

CREDITS

The plural term "quotations" refers to permissions obtained to use several quotes throughout the book; the singular "quotation" refers to only one used in the book. Listed here in order of publication name or author's last name.

Quotations by Samuel Allison, Emilio Segrè, and Herbert Anderson in *Physics Today* 8, 1, 9 (1955) reproduced with the permission of the American Institute of Physics.

Quotation from Luis Alvarez, *Adventures of a Physicist* (New York: Basic Books, 1987), reproduced with permission of Walter Alvarez.

Quotations from American Institute of Physics (AIP), "Oral History Interviews: Franco Rasetti and Enrico Persico," interviewed by Thomas S. Kuhn, April 8, 1963; "Oral History Interviews: George Uhlenbeck—Session III," April 5, 1962; "Oral History Interviews: Herbert Lawrence Anderson—Session II," interviewed by Lillian Hoddeson and Alison Kerr, January 13, 1981; and "Oral History Interviews: I. I. Rabi—Session II," interviewed by Stephen White, February 21, 1980, reproduced with permission of the American Institute of Physics.

Quotation from Bernice Brode, *Los Alamos Tales* (Los Alamos, NM: Los Alamos Historical Society, 1997) reproduced with permission of the Los Alamos Historical Society.

Quotations from the *Santa Fe New Mexican* and private letters by Hans Bethe reproduced with permission of the estate of Hans Bethe.

Quotations from Subrahmanyan Chandrasekhar, "The Pursuit of Science," *Minerva* 22, nos. 3/4 (September 1984): 410–420 reproduced with permission of Springer Science and Business Media.

Quotation from Max Dresden, letter in "Heisenberg, Goudsmit, and the German 'A-Bomb,'" *Physics Today* 44, no. 5 (May 1991) reproduced with permission of *Physics Today*.

Quotations from *Enrico Fermi: The Collected Papers* (Chicago: University of Chicago Press, 1962 and 1965) reproduced with permission of the University of Chicago Press.

Quotations from Laura Fermi, *Atoms in the Family* (Chicago: University of Chicago Press, 1954) reproduced with permission of the University of Chicago Press.

Quotation from Leona Marshall Libby, *The Uranium People* (New York: Crane, Russak/Charles Scribner's Sons, 1979) reproduced with permission of John Marshall III.

Quotations from "Memorial Symposium Held in Honor of Enrico Fermi at the Washington Meeting of the American Physical Society, April 29, 1955," *Reviews of Modern Physics* 25, no. 3 (July 1955) reproduced with permission of *Reviews of Modern Physics.*

Quotations from Emilio Segrè, *Enrico Fermi, Physicist* (Chicago: University of Chicago Press, 1970) reproduced with permission of the Estate of Emilio Segrè.

Quotation from Spencer R. Weart and Gertrude Weiss Szilard, eds., *Leo Szilard, His Version of the Facts* (Cambridge, MA: MIT Press, 1978) reproduced with permission of Estate of Leo Szilard.

Quotation from Emilio Segrè, *A Mind Always in Motion* (Berkeley: University of California Press, 1994) reproduced with permission of the Estate of Emilio Segrè.

Quotations from Eugene Wigner, *Symmetries and Reflections* (Woodbridge, CT: Ox Bow Press, 1979) reproduced with permission of the Ox Bow Press.

NOTES

PREFACE

ix **When he was a senior at Columbia:** A former colleague of my father objected to this story, recalling that Jack did not have a formal position on the Columbia faculty at the time. All I can say is that this is the way my father remembered it.

x **Telegdi's paper on Fermi:** Telegdi, "Enrico Fermi" in Shils, *Remembering the University of Chicago,* 110–129.

xi **The biographer is left triangulating:** Since 2013, two full biographies have been published: *The Pope of Physics,* by Emilio Segrè's nephew Gino and his wife, Bettina Hoerlin (Holt, 2016); and Giuseppe Bruzzaniti's *Enrico Fermi: Obedient Genius* (Springer, 2016). The latter is a translation of a book originally published in Italian in 2007.

INTRODUCTION

xv **"puerile" prankster consumed by jealousy:** Magueijo, *Brilliant Darkness,* 53–56, 86–87.

xv **the greatest scientist in Western history:** Orear, *Enrico Fermi,* 64.

xvi **during an interview for the CBC documentary:** TWOEF, Interview with I. I. Rabi, 1.

xvii **Snow put it succinctly:** Snow, *Physicists,* 79.

xix **In terms of influence as a teacher:** Telegdi, "Enrico Fermi," 125. As Valentine Telegdi points out, two other giants of twentieth-century physics, Ernest Rutherford and Arnold Sommerfeld, had records equal to Fermi's as teachers, but Rutherford trained only experimentalists, and

Sommerfeld, only theorists. Fermi's students number as both world-class experimenters and theorists.

xx **Calculations of probabilities:** Mysteriously, he considered a 10 percent probability of an occurrence to be so low that it was not worth taking seriously.

xx **The latter was sometimes misinterpreted:** Peierls, "Nuclear Physics in Rome," in Stuewer, *Nuclear Physics in Retrospect,* 59.

xxi **MAJORANA: There are scientists who "happen":** Cited in Chandrasekhar, "The Pursuit of Science," 410.

xxii **quantum theorist Wolfgang Pauli once quipped:** Wolfgang Pauli to George Uhlenbeck. See Weiner, *Exploring the History of Nuclear Physics,* 188.

xxiii **Nevertheless, it is one worth trying:** Geoffrey Chew, interview with author, May 5, 2014. Ugo Amaldi, interview with author, June 7, 2016.

CHAPTER ONE: PRODIGY

4 **The Ministry of Railroads was a prestigious place:** Schram, *Railways and the Formation of the Italian State.*

5 **He clearly had ability and ambition:** Segrè, *Enrico Fermi,* 4. Laura Fermi, *Atoms,* 14.

5 **Do Not Disturb signs hang:** The Fermis occupied only one of the apartments; however, one senses that the current owners of both apartments get unwanted visits.

6 **Crying was forbidden:** Laura Fermi, *Atoms,* 15.

7 **Guglielmo Marconi, the inventor of radio:** Raboy, *Marconi.* Marconi's claim to having invented radio was disputed at the time, and continues to be. Like all great revolutionary technologies, radio was the result of many people working both independently and collaboratively, but Marconi was a central figure, and his fame would have reached the Fermi household.

7 **When very young he had a bit of a temper:** Libby, *Uranium People,* 15. In her interview for *The World of Enrico Fermi,* Laura Fermi observed that, although Enrico generally learned to control his temper, bad drivers would bring out the worst in him. TWOEF, Laura Fermi interview, 22.

7 **Enrico began to impress teachers:** Vergara Caffarelli, "Enrico Fermi al Liceo," in Bassani, *Fermi, Maestro e Diddata,* 8–15.

9 **It was home for Enrico:** The address was Via Principe Umberto 133, which no longer exists—the street names have changed, and the numbering as well. There is some debate as to whether the building still exists, and if so, where it is located. There is some evidence that it can be found nearby at the current Via F. Turati 48, but some disagree and assert that the building can no longer be found.

9 **Many photographs of the Fermi children:** Vergara Caffarelli, *Enrico Fermi,* 9–15.

9 **"On the appointed morning":** Laura Fermi, *Atoms,* 15–16.
9 **"The blow could not have been heavier":** Ibid., 16.
10 **He did, however, resolve:** Ibid.
11 **"We formed the habit of taking long walks":** Segrè, *Enrico Fermi,* 2.
11 **he couldn't stop talking about it:** Laura Fermi, *Atoms,* 19.
11 **The two boys conducted a variety of experiments:** Ibid.
11 **Persico recalled extended conversations:** American Institute of Physics (AIP), "Oral History Interviews: Franco Rasetti and Enrico Persico," interviewed by Thomas S. Kuhn, April 8, 1963, https://www.aip.org/history-programs/niels-bohr-library/oral-histories/4995.
12 **Fermi particularly enjoyed the traditional Italian:** Segrè, *Enrico Fermi,* 5.
13 **Seven years younger than Alberto Fermi:** On Amidei's education, see Vergara Caffarelli, "Enrico Fermi al Liceo," 8. Data on his employment record are based on his employment file at Ferrovie dello Stato, graciously supplied by Sandra Romiti.
13 **"Is it true that there is a branch of geometry":** Segrè, *Enrico Fermi,* 9ff.
13 **He lent Enrico the book:** The book was considered at the time the definitive treatment of this most abstract of geometric disciplines.
14 **"I became convinced that Enrico":** Segrè, *Enrico Fermi,* 9.
14 **Olga recounts a chance encounter:** The story is recounted by Vergara Caffarelli in "Enrico Fermi al Liceo," 9. Many years later David Lilienthal, a major figure in the development and regulation of atomic energy in the United States, made the same observation. Lilienthal, *The Journals,* 2:128. Giuseppe Bruzzanti picks up the comparison with Galileo throughout his recent biography, *Enrico Fermi: Obedient Genius.*
15 **Amidei believed that Fermi was ready:** Many years later Segrè would stumble upon a two-volume set of Poisson's textbook at a rare bookstore and buy it for Fermi as a forty-fifth birthday gift. EFREG, Box 59.
15 **"I had already ascertained":** Segrè, *Enrico Fermi,* 10. Segrè recounts that a proof offered by Fermi several decades later tracked an obscure section of Poisson. Ibid., 12.
16 ***Treatise on Physics* by Russian physicist Orest Chwolson:** Segrè, *Enrico Fermi,* 12. Jay Orear, many years later, attributed Fermi's phenomenal grasp of the fundamentals of all of physics to his intensive study of Chwolson during this period. Orear, *Enrico Fermi,* 63–64. See also Bernardini, "Enrico Fermi e il trattato," 15ff. Orear and Bernardini suggest that the version he read was in French.
16 **British physicist Joseph John (J. J.) Thomson's:** To this author's knowledge, it was only published in English, meaning that the two young men plowed through some fairly advanced physics in a foreign language.
16 **Fermi also recommended the book:** AIP, "Oral History Interviews: Franco Rasetti and Enrico Persico."
17 **"I studied mathematics with passion":** Segrè, *Enrico Fermi,* 10.
17 ***Annalen der Physik:*** The former, edited by Max Planck, was the journal in which Einstein published his five groundbreaking papers of 1905.

17 When he arrived at university in the fall: AIP, "Oral History Interviews: Franco Rasetti and Enrico Persico."

17 And it did have a direct effect: *Elenco Degli Allievi Dal 1813–1998.* Edizione Provvisoria (Pisa: Scuola Normale Superiore, October 1999), 26–27.

18 it was the most prestigious institution: There is no clear equivalent in the United States. With its miniscule size and extreme selectivity, the Scuola Normale was unique, and remains so to this day.

19 The entrance exam paper Fermi submitted: He also wrote exam papers in algebra and geometry, both of which survive alongside the physics exam in the archives of the Scuola Normale Superiore.

CHAPTER TWO: PISA

22 The mathematicians of the Scuola Normale: Rasetti, in AIP, "Oral History Interviews: Franco Rasetti and Enrico Persico."

22 much of his time in independent study: Segrè, *Enrico Fermi,* 193.

22 Fermi earned perfect grades: SNS Archives, Enrico Fermi Folder.

22 the only courses that involved some work: Pre-med students throughout the world will rejoice to learn that organic chemistry was one of only two science courses at the university in which Fermi did not receive honors. He did, however, receive a perfect grade.

23 His name was Franco Rasetti: Del Gamba, *Il raggazzo di via Panisperna,* is the only full-length biography available on Franco Rasetti.

24 They were saved only: Laura Fermi, *Atoms,* 24.

24 The Anti-Neighbor Society eventually grew: It is unclear when the Anti-Neighbor Society first admitted women to its ranks; Fermi mentions female members in a letter to Persico. Segrè, *Enrico Fermi,* 198.

24 "barring one or two exceptions ugly enough": Segrè, *Enrico Fermi,* 202.

25 They scored a perfect grade: Rasetti in AIP, "Oral History Interviews: Franco Rasetti and Enrico Persico."

26 He dutifully recorded the entire sum: Later he began to compile a reference system that he called his "mechanical memory," a series of carefully organized folders that contained reference material he found useful in his daily work. These are stored in nineteen boxes with folders on subjects as varied as Pion Experiments and Quantum Electrodynamics, Liquids, Solids, and Algebra and Calculus.

26 The 1919 notebook is 102 pages: EFREG, 49:1. Included is a summary of analytical dynamics, the electron theory of matter (a summary, presumably, of his studies on Richardson's "Electron Theory of Matter"), Planck's work on blackbody radiation, Boltzmann's entropy theorem, and gas discharges.

26 With the pen knife: The author is indebted to Laura Offeddu of the American Academy in Rome, whose grandfather, Filippo Ferrari, studied in Pisa with Fermi and related the penknife incident to her.

27 **Rasetti and Nello Carrara were admitted:** Rasetti, in CPF I, 55–56.

27 **lay in the direction of X-ray research:** There had been enormous interest in understanding the structure of crystals by aiming X-rays into them and exposing film to show how the X-rays bounced off of the crystals' internal structure. It was this research program that Fermi and his graduate student colleagues decided to explore.

27 **Rasetti noted that Fermi's disinterest:** Rasetti, in CPF I, 56.

27 **By the time he was ready to present:** CPF I, 1ff.

28 **The third presented an important theorem:** These "Fermi coordinates" have become more or less universally accepted as part of the conceptual apparatus of general relativity theory.

28 **The fourth was a highly successful effort:** This was a solution to the so-called 4/3 problem. Classical mechanics calculates the mass of a rigid spherical charged body as 4 / 3 E / c ; Einstein calculates it as E / c . The difference, Fermi realized, is that rigid spheres do not retain their sphericity as they move in space-time but flatten out perpendicular to the direction they move in; when this is taken into account, the 4/3 factor falls away.

28 **"If we could liberate the energy":** CPF I, 33–34. Author's translation.

29 **Einstein had consulted Levi-Civita:** Years later, when asked what he liked about Italy, Einstein is said to have replied: "Two things: spaghetti and Levi-Civita."

29 **So impressed was Levi-Civita:** Bonolis, "Enrico Fermi's Scientific Work," 320.

29 **So fretful was he:** Segrè, *Enrico Fermi,* Appendix I, 200–201.

29 **nor was the thesis published:** The paper was published in part in 1926, but the full thesis itself was lost in the Scuola archives and discovered after Fermi's death, in 1959. CPF I, 227ff.

29 **Years later his wife would claim:** Laura Fermi, *Atoms,* 26.

CHAPTER THREE: GERMANY AND HOLLAND

31 **In the midst of this turmoil:** A full treatment on the rise of fascism in Italy can be found in Bosworth, *Mussolini's Italy;* Hibbert, *Il Duce;* Smith, *Mussolini*; and Lyttleton, *The Seizure of Power.* Laura Fermi also published a more impressionistic, less scholarly, but quite readable book, Laura Fermi, *Mussolini.*

32 **Laura Fermi enjoyed a course:** Laura Fermi, *Atoms,* 40.

33 **With each new role he further cemented:** Segrè, *Enrico Fermi,* 30, is particularly good on this point.

33 **"Do you think he may go":** Laura Fermi, *Atoms,* 29–30. As with all verbatim conversations recollected years later and reported through the eyes and ears of those who were not there, we might doubt the precise words recorded by Laura Fermi, but the sentiment rings true to what we know about both Fermi and Corbino. They were anti-fascist but also instinctively conservative individuals who were even more worried by

the violence that would result if the government resisted Mussolini at this point. Laura Fermi notes that Enrico considered emigrating at that time, but adds "the fact that sixteen years almost to the day he left Italy for the United States does not make a prophet of him." Ibid., 31.

34 **Born was a slightly shy, somewhat formal:** Born, *My Life & My Views,* is a short, well-written memoir.

35 **Though his letters home to his father:** Laura Fermi, *Atoms,* 31–32. For letters during this period, see the sheaf held at the University of Pisa's physics department library, donated by Maria's daughter Gabriella Sacchetti after Maria's death. Letters dated February 28, 1923; March 21, 1923; and June 9, 1923.

35 **Segrè puts some of the blame:** Segrè, *Enrico Fermi,* 32–33.

35 **"a stored-up, never forgotten bitterness":** Libby, *Uranium People,* 15–16, 36.

38 **Her name was Laura Capon:** Much of what follows is based on Laura Fermi, *Atoms,* 38ff.

40 **"You should have seen Laura":** Libby, *Uranium People,* 28.

40 **"He shook hands and gave me":** Laura Fermi, *Atoms,* 1.

40 **"There was an easy self-reliance in him":** Ibid.

40 **"the first afternoon I spent with Enrico":** Ibid.

41 **Segrè notes that Fermi in later:** Segrè, *Enrico Fermi,* 33.

42 **"He was younger than I was":** American Institute of Physics, "Oral History Interviews: George Uhlenbeck—Session III," April 5, 1962, https://www.aip.org/history-programs/niels-bohr-library/oral-histories/4922-3.

42 **Much of Fermi's success in Leiden:** Letter from Fermi to Maria Fermi, Leiden, October 27, 1924. In Fermi folder, Department of Physics library, University of Pisa.

42 **Ehrenfest, a close friend of Einstein:** Kumar, *Quantum,* 169.

42 **Fermi kept a group photo of Einstein:** Private communication with Roberto Vergara Caffarelli, whose book, *Enrico Fermi, Immagini e domumenti* (2001), includes the photo on page 53.

42 **Einstein clearly impressed the young man:** Letter from Fermi to Maria Fermi, Leiden, October 27, 1924. In Fermi folder, Department of Physics library, University of Pisa.

43 **Fermi expressed annoyance:** Telegdi in Orear, *Enrico Fermi,* 98, and in Shils, *Remembering the University of Chicago,* 126. Also, Geoffrey Chew, interview with author, May 6, 2014.

43 **Fermi got to work quickly:** CPF I, 134–137.

CHAPTER FOUR: QUANTUM BREAKTHROUGHS

44 **The breakthroughs in quantum theory:** What follows over the next two chapters is an extremely simplified summary of a highly complex history. Readers with an interest in finding out more can go to many sources, including Van Der Waerden, *Sources of Quantum Mechanics;*

Kumar, *Quantum;* and Fernandez and Ripka, *Unravelling the Mystery of the Atomic Nucleus.*

45 Physicists began to call the heavy central: "Heavy" in a relative sense only; most of an atom's mass is in the nucleus, but a nucleus is extremely small—for a hydrogen atom, on the order of one-millionth of a billionth of a meter across (a distance known as a fermi).

46 although in most important details: Bohr thought electrons orbited in circular orbits, but we know now that they don't really "orbit" at all—they seem to be everywhere at once, and the shape of the "orbit" can be quite exotic, certainly not circular. Hence the standard term "orbital."

46 One such phenomenon was called: For this discovery, Zeeman shared the 1902 Nobel Prize in Physics with Hendrik Antoon Lorentz.

46 That he continued to keep abreast: CPF I, 134. Published in Italian in 1925.

47 With the great theorist Arnold Sommerfeld: Sommerfeld's productive career included many important discoveries, most notably the constant that describes the relative strength of the electromagnetic force, what physicists call the "fine structure constant." Over more than three decades at the University of Munich he sponsored the PhD work of a number of future Nobel Prize winners, including not only Pauli but also Werner Heisenberg and Peter Debye, and influenced the early studies of a number of others, including Haldan Hartline, Linus Pauling, and Isidor Isaac (I. I.) Rabi. Nominated for the Nobel Prize himself some eighty-four times—more than any other physicist—he sadly never won it.

47 As a theorist Pauli was perhaps: See Rudolf Peierls's colorful and readable brief biography in Peierls, *Atomic Histories,* 3–17.

47 Pauli also had a legendary mean streak: Miller, *137: Jung, Pauli.* See also Gross, "On the Calculation of the Fine-Structure Constant," 9.

48 Pauli was born Catholic: Heisenberg, *Physics and Beyond,* 86–87.

48 He had a deep mystical streak: Meier, *Atom and Archetype.*

48 He was also a man of obsessions: Kumar, *Quantum,* 164; Pauli, "Remarks on the History of the Exclusion Principle," 214.

49 angular momentum in opposite directions: Quantum spin should not be confused with classical spin. Quantum spin is the intrinsic angular momentum of the particle; the particle should not be thought of as spinning like a top, although a spinning top is perhaps the most obvious example of angular momentum.

49 It was picked up again: See Kumar, *Quantum,* 169–175, and S. A. Goudsmit, "The Discovery of the Electron Spin," http://www.lorentz.leidenuniv.nl/history/spin/goudsmit.html.

50 "the strangest man": Farmelo, *The Strangest Man.*

50 Paul Adrien Maurice Dirac was the baby: Kumar, *Quantum,* 198–199; Farmelo, *The Strangest Man,* 86–87, 101. Dirac saw the connection between Heisenberg's ideas and Poisson brackets used in classical mechanics.

50 Dirac was, in some sense: Heisenberg, *Physics and Beyond,* 86–87.
51 His PhD thesis was impressive: Kumar, *Quantum,* 197.
51 Almost immediately, Dirac was propelled: Though he never contributed fundamentally in experimental fields, he pursued an engineering degree prior to his shift to physics and always showed a strong interest in the experimental side of the field.
51 His was the first PhD degree: Kumar, *Quantum,* 230.

CHAPTER FIVE: OF GECKOS AND MEN

53 Perched on a low hill: The elders of Florentine physics, including Antonio Garbasso, the head of the institute who was also, surprisingly, the mayor of Florence, chose the site because of its connection to Galileo, who lived the last years of his life in the Villa "il Gioiello" (the jewel), a ten-minute walk uphill from the Istituto di Fisica. The villa still exists and now belongs to the University of Florence. My thanks to Andrea Gambassi for this information.
53 Now he was delighted to take on: Casalbuoni, Frosali, and Pelosi, *Enrico Fermi a Firenze,* 29ff.
53 The notes for his mechanics lectures: Often students who had learned a bit of calligraphy prepared class lecture notes and sold them for some extra pocket money. The lovely notes on Fermi's lectures were preserved by two of his students, Bonanno Bonanni and Paolo Pasca, and have been published recently in a fascinating volume on Fermi's years in Florence. See ibid., 73ff.
53 The two pranksters noted: Laura Fermi, *Atoms,* 37–38.
54 Together they hatched a plan: AIP, "Oral History Interviews: Franco Rasetti and Enrico Persico." Rasetti claimed that the target of the prank was the wife of the janitor, who served food at the cafeteria, but it would have been difficult to frighten her without also frightening all the other young women serving there.
54 A paper he wrote in January 1924: CPF I, 124.
54 He was struggling with the application: Belloni, "On Fermi's Route to the Fermi-Dirac Statistics"; Parisi, "Fermi's Statistics," 67–74; Sebastiani and Cordella, "Fermi toward Quantum Statistics," 71–96.
54 "The quantum states of an atom": Bethe, *The Santa Fe New Mexican,* January 6, 1955, 2.
54 "On the Quantization of a Perfect": Belloni, "On Fermi's Route to the Fermi-Dirac Statistics," discusses the meteorological aspects of the story.
55 It took him the rest of the year: CPF I, 178–195.
55 The paper had an immediate impact: It may well not have been Fermi's primary purpose in doing so, but the implications for understanding degeneracy were clear to theorists around the world and provided the first, immediate applications. Belloni, "On Fermi's Route to the Fermi-Dirac Statistics."

55 **Word spread as far as England:** Farmelo, *The Strangest Man*, 101.

55 **Dirac developed an interest in the problem:** Ibid., 105. Also Kragh, *Dirac*.

56 **"In your interesting paper":** Quoted in Kragh, *Dirac*, 36.

56 **"When I looked through Fermi's paper":** Ibid.

56 **As is the case with many important developments:** Schuking, "Jordan, Pauli, Politics, Brecht, and a Variable Gravitational Constant," 26.

57 **Fermi's candidacy fell victim:** Glick, *The Comparative Reception of Relativity*, discusses the challenges the acceptance of relativity theory faced throughout Europe.

58 **Levi-Civita, of course, was one:** Segrè, *Enrico Fermi*, 42.

58 **Pontremoli, to Milan:** Pontremoli went on to establish the Institute of Physics at the University of Milan. He died tragically young, at the age of thirty-two, during the ill-fated *Italia* dirigible flight to the North Pole led by General Umberto Nobile. Pontremoli was on board to help measure the magnetic field of the earth as well as to detect cosmic rays. On its return trip to base camp, the dirigible crashed onto the ice. Some of the unlucky crew survived, but some, like Pontremoli, were never found and were presumed dead.

58 **"That Fermi had been intensely interested":** Segrè, *Enrico Fermi*, 39–40.

58 **as Amaldi once suggested:** See TWOEF, Amaldi interview.

59 **"he moves with complete assurance":** Quoted in Segrè, *Enrico Fermi*, 45. Emphasis added.

CHAPTER SIX: FAMILY LIFE

63 **The years Fermi spent in Rome:** Holton, "Striking Gold in Science," covers this period and points out the fact, disconcerting though it may be, that all these achievements took place under, and with the support of, a repugnant political system.

64 **On Saturdays, Fermi returned:** Laura Fermi, *Atoms*, 42ff., is the only account of what must have been quite remarkable gatherings. Given the overlapping professional interests of these men, it would seem likely that many important developments in algebraic geometry and other fields were first hashed out leisurely at Saturday discussions in Castelnuovo's parlor.

64 **young woman named Ginestra Giovene:** Laura Fermi, *Atoms*, 45. She would also become Laura Fermi's closest friend.

64 **Fermi would also play "director":** Laura Fermi, *Atoms*, 41, and interview with Ugo Amaldi, June 8, 2016. Fermi at one point suggested that Edoardo Amaldi play Greta Garbo, a suggestion Amaldi accepted "with good grace."

66 **"We could see nothing of interest":** Laura Fermi, *Atoms*, 66.

66 **Alexandria in the *Enciclopedia Italiana:*** Laura Fermi, *Atoms*, 67. The *Enciclopedia Italiana* was a favored project of Mussolini, who tapped

industrialist and cultural patron Giovanni Treccani to edit it. It brought together outstanding scholars throughout Italy in every field; it included important nonfascists on its authors list. When Fermi received an offer to become a professor in Zurich in 1928, Mussolini persuaded him to stay by arranging a consulting job for him on the *Enciclopedia* to supplement his income.

66 **"You mean to say":** Laura Fermi, *Atoms,* 10–11.

68 **"He wanted a tall, strong girl":** Ibid., 52.

68 **Its egg-yolk yellow color:** Ibid., 52–53.

71 **"Patiently I learned the mathematical instruments":** Ibid., 59–60.

74 **Published in 1936 and aimed at:** EFREG, 57:2.

74 **The project worked and by 1941:** Goudsmit, "The Michigan Symposium in Theoretical Physics," 178–182.

74 **The two Dutchmen invited Fermi:** CPF I, 401–445.

76 **he mispronounced a few words in such a wonderful:** After the war, a young Fausta Segrè, Emilio's daughter, corrected Fermi's pronunciation of the word *battleship* as "bottle-sheep." Fausta Segrè Walsby, interview with author, June 2, 2016.

76 **"There seemed to be a total incomprehension":** Laura Fermi, *Atoms,* 80–81.

76 **She loved it there:** Ibid., 94ff.

77 **"There is no democracy in physics":** Greenberg, *Politics of Pure Science,* 43.

77 **emphasized by Italian scholar Giovanni Battimelli:** Battimelli, "Funds and Failures," 169–184.

CHAPTER SEVEN: THE ROME SCHOOL

80 **This "Rome School" of physics:** Holton gives the best overview of the school and its development, in three classic studies: "Striking Gold in Science"; *The Scientific Imagination,* 155–198; and *Victory and Vexation,* 48–64. See also Segrè, *Enrico Fermi,* 46–100; also, *passim,* Bernardini et al., *Proceedings of the International Conference "Enrico Fermi and the Universe of Physics"*; Bernardini and Bonolis, *Enrico Fermi;* and Segrè, "The Rome School," in Stuewer, *Nuclear Physics in Retrospect*, 35–62.

80 **Its hallmarks were a balance of theory:** Holton, "Striking Gold in Science."

82 **Fermi invited Segrè to attend the 1927:** Laura Fermi, *Atoms,* 44.

83 **He also had a fiery temper:** Laura Fermi writes (Laura Fermi, *Atoms,* 46) that during one seminar session in Fermi's office, Segrè got himself so worked up because the other attendees wouldn't let him speak that he pounded Fermi's desk and left a hole in it. This author has examined the desk and has seen no such hole, so either there was another desk or perhaps Laura Fermi was exaggerating; regardless, it is clear he had a temper.

84 **Majorana was by all accounts the brightest:** Magueijo, *Brilliant Darkness,* gives a thorough, if somewhat idiosyncratic, perspective on Majorana's life and work.
85 **"Fermi's seminar was always improvised":** Segrè, *Enrico Fermi,* 51–52.
85 **Ginestra wrote that students learned:** Laura Fermi, *Atoms,* 46.
86 **"that moved slowly but knew no obstacles":** Segrè, *Enrico Fermi,* 56.
86 **"When the four came together":** Laura Fermi, *Atoms,* 44.
86 **"certain playfulness, a naïve love":** Ibid., 45.
86 **When Ginestra first encountered the private seminar:** Ibid., 46.
87 **Segrè tells the story of a member:** Segrè, *Enrico Fermi,* 52.
88 **One problem that Corbino and Fermi faced:** Segrè is at pains to emphasize that this did not seem to be an explicit objective for Fermi, but there is little doubt, given the sequence of events, that Corbino had this as a conscious objective and shared these thoughts with Fermi. It is inconceivable that Fermi did not share Corbino's objective.
89 **most important presentation was that of Bohr:** Bohr believed that every phenomenon of physics could be viewed in two complementary ways: the classic example is light behaving as both a particle and a wave.
89 **"May I introduce the applications":** Pauli to Rasetti, October 30, 1956 (Amaldi Archives). It might have been more apt to say "May I introduce the exclusion principle to its most important application."
90 **Rasetti, reflecting on the importance:** American Institute of Physics, "Oral History Interviews: Franco Rasetti and Enrico Persico," interviewed by Thomas S. Kuhn, April 8, 1963, https://www.aip.org/history-programs/niels-bohr-library/oral-histories/4995.
91 **Lo Surdo had made:** Leone, Paoletti, and Robotti, "A Simultaneous Discovery."
91 **Corbino and Lo Surdo were personal rivals:** The author thanks Francesco Guerra and Nadia Robotti for this and other points regarding Lo Surdo. Private communication with author, February 27, 2016.
91 **Lo Surdo may not have won:** Laura Fermi, *Atoms,* 69ff. She refers to Lo Surdo by a pseudonym, "Mr. North."
92 **The story does not hold up:** Guerra and Robotti, private communication with author, February 27, 2016.
92 **He also wanted, somewhat naïvely:** In these early years he was far more tolerant of dissent than he was in later years. However, many distinguished figures of Italian culture wanted nothing to do with him.
92 **"INCIPIT VITA NOVA":** EFDG, 2.
93 **Fermi was by far the youngest:** https://en.wikipedia.org/wiki/Royal_Academy_of_Italy.

CHAPTER EIGHT: BETA RAYS

96 **gave rise to density functional theory:** Density functional theory is a computational technique that uses electron density to describe the

electron structure of many-bodied systems. It is a key technique in modern condensed matter physics, chemistry, and materials science. See "Density Functional Theory," *Wikipedia,* last modified January 9, 2017, accessed January 30, 2017, https://en.wikipedia.org/wiki/Density_functional_theory.

96 **"The Calculation of Atomic Fields":** For his part, Thomas failed to credit either Fermi or Dirac.

96 **Some Italian historians:** They point out that Thomas's work was limited to an electron in its ground state, whereas Fermi's approach was more general. Francesco Guerra and Nadia Robotti, private communication with author, February 3, 2017.

97 **The eponymous Maxwell's equations:** Forbes and Mahon, *Faraday, Maxwell, and the Electromagnetic Field,* for a comprehensive discussion of Faraday and Maxwell and their achievements.

99 **He was a mere twenty-four:** A few years later Dirac pulled together all his thoughts on the subject in a book called *The Principles of Quantum Mechanics* (Oxford, 1930, revised four times). It stands even today as one of the most important treatments of the subject.

99 **Fermi understood the importance:** Segrè, *Enrico Fermi,* 55.

99 **The private seminars at Via Panisperna:** Ibid. See also CPF I, 305.

100 **This description of how an atom:** CPF I, 401–402.

100 **"a masterpiece, instructive and refreshing":** Wilczek, "Fermi and the Elucidation of Matter," 42.

100 **"Many of you, like myself":** Bethe, "Memorial Symposium," 253.

100 **"He disliked complicated theories":** Wigner, *Symmetries and Reflections,* 254.

102 **Beta radiation seemed to violate:** It seemed to violate other conservation laws as well, most notably quantum spin conservation. See Guerra, Leone, and Robotti, "When Energy Conservation Seems to Fail."

103 **Some physicists surrendered:** Bohr, "Atomic Stability and Conservation Laws," 119–130.

103 **He called this imaginary particle:** Amaldi later claimed credit for suggesting the name to Fermi, who adopted it at once, and the rest of the world followed suit. Segrè suggests that the name "neutrino" was first used by Fermi at the Rome conference of October 1931; however, Chadwick had yet to discover the neutron, so there would have been no need to differentiate Pauli's particle from Chadwick's.

103 **"better not to think about":** Cited in Pauli's letter, "Open Letter to the Group of Radioactive People at the Gauverein Meeting in Tübingen," Physics Institute, Zurich, December 4, 1930, http://microboone-docdb.fnal.gov/cgi-bin/RetrieveFile?docid=953;filename=pauli%20letter1930.pdf.

104 **"A Tentative Theory of Beta Rays":** In recognition of the importance of this paper, there is an enormous amount of literature on its contribution to modern physics. For a quick sample, see Konopinski, "Fermi's

Theory of Beta-Decay"; Cabibbo, "Fermi's Tentativo and Weak Interactions," 305–316; Cabibbo, "Weak Interactions," 138–150.

104 **When the neutron is changed:** Fermi's original paper did not identify the neutrino emerging from the conversion of a neutron to a proton as an antineutrino; that part of the theory came a bit later.

104 **Fermi sent the paper to the Italian journal:** CPF I, 551–590.

104 **This story is so central:** "Fermi's Interaction," *Wikipedia,* last updated July 19, 2017, https://en.wikipedia.org/wiki/Fermi%27s_interaction#History_of_initial_rejection_and_later_publication.

105 **In fact, no such public statement:** This author has reviewed the digital online archives and has found no such statement; the archivist of *Nature* confirmed in private communication that none was ever printed.

105 **It is unfortunately impossible:** This was confirmed both by the current archivist of *Nature* and by Laura Garwin, the daughter of Fermi's graduate student and colleague Richard Garwin, who worked on the staff of *Nature* for several years and who searched unsuccessfully for the correspondence regarding the beta decay paper. Richard Garwin, interview with author, May 22, 2014.

105 **A more logical British publication:** It might be objected that because Fermi was not a fellow of the Royal Society he could not submit the paper to its *Proceedings,* but the *Proceedings* regularly accepted papers submitted by fellows on behalf of nonfellows.

105 **A white lie:** This interpretation has been suggested by Francesco Guerra and Nadia Robotti, private communication, September 21, 2015.

105 **In the 1970s, Fermi's future:** Yang, "Reminiscences of Enrico Fermi," 243.

CHAPTER NINE: GOLDFISH

107 **Around 1930, however, Fermi sensed:** For a good accounting, see Guerra and Robotti, "Enrico Fermi's Discovery of Neutron-Induced Artificial Radioactivity: The Influence of His Beta Decay Theory."

108 **To Fermi, the nucleus presented:** Holton, *Scientific Imagination,* 163ff.

108 **"Italy will regain with honor":** Ibid., 164–165.

108 **Unfortunately, Corbino was never able:** See Battimelli "Funds and Failures," 169ff.

109 **The massive 575-page book:** Rutherford, Chadwick, and Ellis, *Radiations.*

109 **a young Pisan named Bruno Pontecorvo:** Guerra and Robotti, "Bruno Pontecorvo in Italy."

110 **In another carefully considered step:** This process is discussed in Segrè, *Enrico Fermi,* 58, and in Holton, *Scientific Imagination,* 167.

112 **The problems associated with beta decay:** Segrè, *Enrico Fermi,* 70, suggests that Fermi began calling Pauli's phantom particle a "neutrino" in these discussions. However, there was no need at the time of the Rome conference to find a way to distinguish Pauli's particle from Chadwick's particle; Chadwick's discovery lay several months in the future. Most historians date the use of the term "neutrino" to sometime in 1932.

112 **It was a productive meeting:** Segrè, *Enrico Fermi,* 68.

113 **Robert Millikan at Caltech:** Millikan won the 1923 Nobel Prize for measuring, in an exquisitely precise experiment, the electric charge of a single electron.

113 **he received the startling news:** This is exactly the kind of short report that *Nature* published during this period, in contrast to Fermi's lengthy beta decay paper.

114 **Wick was an insightful theorist:** Guerra and Robotti, "Enrico Fermi's Discovery of Neutron-Induced Artificial Radioactivity: The Influence of His Theory of Beta Decay."

114 **and had located a small sample of radium:** Radium releases gamma rays as part of its radioactive signature. A thorough review of Fermi's work with neutron sources can be found in Guerra and Robotti, "Enrico Fermi's Discovery of Neutron-Induced Artificial Radioactivity: Neutrons and Neutron Sources."

114 **In an act of extraordinary generosity:** For his role in providing radon, he received an appropriate nickname: *La Divina Provvidenza* (Divine Providence).

116 **One example is the story:** Laura Fermi, *Atoms;* Guerra and Robotti, "Enrico Fermi's Discovery of Neutron-Induced Artificial Radioactivity: The Influence of His Beta Decay Theory," 398–399.

116 **Another story, told by Segrè and Rasetti:** See Guerra and Robotti, *Enrico Fermi e Il Quaderno Ritrovato,* 128. It is possible, of course, that Fermi did not record the first few experiments, which were unsuccessful, but this seems unlikely, given his thoroughness.

118 **"Your results are of great interest":** Rutherford to Fermi, April 23, 1934. EFDG, Box IV L7.

118 **"What did you think I was president":** CPF I, 641.

118 **Only a German physicist, Ida Noddack:** Noddack, "Uber das Element 93."

118 **Her suggestion was ignored:** Fermi had done calculations that seemed to show the impossibility of nuclear fission; he was working with data that proved to be wrong. See "Emilio Segrè's Interview," Voices of the Manhattan Project, June 29, 1983, http://manhattanprojectvoices.org/oral-histories/emilio-segr%C3%A8s-interview. Brought to my attention by William Zajc.

119 **They had split the uranium atom:** He would not learn about his mistake until January 1939.

120 **By Saturday, October 20, 1934:** Segrè recalls it as October 22, 1934, but the notebooks at Domus Galilaeana clearly indicate that Fermi began his work on October 20, 1934.

120 **"I will tell you how I came to make":** Chandrasekhar, "The Pursuit of Science," 415.

121 **His notebooks suggest:** Holton, *Victory and Vexation in Science,* 58–59; Orear, *Enrico Fermi,* 32.

121 **Fermi conducted the paraffin experiment:** That so many people happened to be at Via Panisperna on that particular day is one of the complicating factors for an historian—they all had subsequent recollections of these events, some details of which do not match up.

123 **The report is dated October 22, 1934:** CPF I, 751–752.

123 **Yet, in the end, it was Fermi:** See Holton, *Victory and Vexation in Science,* 58–59, for a discussion of the way Fermi's explanation illuminates the way scientific breakthroughs are made.

124 **This method was later christened:** Segrè, cited by Metropolis, "The Beginning of the Monte Carlo Method," 128.

125 **Yet they made the astonishing discovery:** Chapter 3 of Miriam Mafai's book *Il lungo freddo: storia di Bruno Pontecorvo, lo scienziato che scelse l'URSS* reports how three of the Panisperna boys looked back on this error some twenty years later. Segrè: "God, in his inscrutable will, made us blind in front of the phenomenon of the fission"; Amaldi: "We made an historical mistake"; and Pontecorvo: "We were just unlucky." Thanks to Andrea Gambassi of SISSA-Trieste for calling my attention to this, and for the translation.

CHAPTER TEN: PHYSICS AS SOMA

127 **These papers became a foundation:** CPF I, 837–1016.

128 **In December 1934, skirmishes:** The Italian general in charge of the operation, Pietro Badoglio, wrote a self-aggrandizing account of the invasion, *The War in Abyssinia.* For a more balanced view, see Barker, *Rape of Ethiopia.*

129 **In 1936, George Pegram:** His lectures can be found in Enrico Fermi, *Thermodynamics.*

129 **Bloch and Fermi drove back across:** Laura Fermi, *Atoms,* 114.

130 **"Physics as soma":** Segrè, *Enrico Fermi,* 90.

131 **The building that would house:** Painter, *Mussolini's Rome,* 64–65.

131 **The brilliant Ettore Majorana:** The strange mystery of Ettore Majorana's disappearance continues to fascinate historians of science. See Magueijo's *A Brilliant Darkness.* A controversial novel by an Italian author and politician, Leonardo Sciascia, is loosely based on the case. Occasionally, the Italian authorities announce that they are reopening the investigation, but to date no definitive resolution has occurred.

132 **Fermi wrote a eulogy:** CPF I, 1020.

133 **Street lamps were torn down:** Some of these lamps can still be seen in the vicinity of the Ostiense station.

133 **"It satisfied my ambitions":** Laura Fermi, *Atoms,* 113.

134 **The beautiful Margherita Sarfatti:** Cannistraro and Sullivan, *Il Duce's Other Woman.*

134 Many, like Laura Fermi's family: Laura Fermi, *Atoms,* 106ff.

135 Fermi's file with the fascist political police: Archivio Nazionale, Ministero dell Interno, Polizia Politica, Enrico Fermi folder.

135 He hoped that the Italian authorities: It was not long after he arrived that Italian authorities began to suspect he would never return. EFREG, 16:4.

135 One could apply to the regime: Thanks to Mauro Canali for this information and for access to the Archivio Nazionale.

136 Jews, including Laura's father: Giorgio Capon, interview with author, September 16, 2015. Fermi's sister Maria offered to help the admiral hide out, but he rejected the offer, believing to the end that the Germans would not touch an admiral of the Italian Navy, even though he was Jewish.

136 He also was never able to obtain: Battimelli, "Funds and Failures," 169ff.

CHAPTER ELEVEN: THE NOBEL PRIZE

138 The concern of the Swedish Academy: Laura Fermi unedited interview transcript, TWOEF, 6.

138 Fermi received thirty-six nominations: "Nomination Database," Nobelprize.org, http://www.nobelprize.org/nomination/archive/show_people.php?id=2955. Thanks to Karl Grandin for guidance on this website. Fermi received two more nominations postwar, presumably for work that was not recognized by the Nobel Committee in 1938.

139 The prize came with a twenty-three-karat: "Prize Amount and Market Value of Invested Capital Converted into 2015 Year's Monetary Value," Nobelprize.org, updated December 2015, http://www.nobelprize.org/nobel_prizes/about/amounts/prize_amounts_16.pdf.

140 "I was determined to be of good cheer": Laura Fermi, *Atoms,* 121–122.

140 Fermi had been awarded the Nobel Prize: "The Nobel Prize in Physics 1938," Nobelprize.org, http://www.nobelprize.org/nobel_prizes/physics/laureates/1938/.

141 the solution was clear: Vergara Caffarelli, *Enrico Fermi,* 67–71.

141 Laura never mentioned: In her interview for "The World of Enrico Fermi," there is a moment when she talks about her passport not having been stamped with a *J* for Jew, but she does not offer any more details. Even in the 1960s she clearly was not proud of what she had to do to leave Italy. Laura Fermi unedited interview transcript, TWOEF, 29.

141 The final preparations were fraught: Laura Fermi, *Atoms,* 128–129.

143 The ceremony, on Saturday, December 10: The award ceremony can be seen on YouTube: "Enrico Fermi arriva a Stoccolma con la sua famiglia per ritirare il Premio Nobel per la Fisica," YouTube video, 1:29, December 21, 1938, posted by CinecittaLuce, June 15, 2012, https://www.youtube.com/watch?v=rno28bDjsd8.

143 **Perhaps, as Laura Fermi suggests:** Laura Fermi, *Atoms,* 132. Leona Libby, on the other hand, suggests that Fermi almost landed on Mrs. Buck's lap. *Uranium People,* 8.

144 **Fermi chose, understandably, to focus:** CPF I, 1037–1043.

CHAPTER TWELVE: THE NEW WORLD

149 **The transatlantic journey:** Laura Fermi, *Atoms,* 139.

149 **Neither he nor Laura:** "Ocean Travelers," *New York Times,* January 2, 1939, 27.

149 **Fermi was not particularly interested:** He did, however, occasionally raise his voice in song—as when, according to Hans Bethe's wife, Rose, he led a group of physicists in a rousing chorus of "My Darling Clementine." See "Rose Bethe's Interview," Voices of the Manhattan Project, June 11, 2014, http://manhattanprojectvoices.org/oral-histories/rose-bethes-interview.

150 **teach the younger Giulio English herself:** In time the children would teach their parents American idioms, hastening Enrico and Laura's Americanization. Laura Fermi, *Atoms,* 151.

150 **King's Crown on West 116th Street:** "420 West 116th Street," *Wikicu,* updated February 2, 2014, http://www.wikicu.com/420_West_116th_Street.

153 **small scientific team, closeted away:** Point made by Sparberg, "A Study of the Discovery of Fission."

153 **Frisch informed Bohr:** Frisch, *What Little I Remember,* 116.

153 **Frisch extracted a promise:** In 1945, the Royal Academy of Science of Sweden retroactively awarded Hahn the 1944 Nobel Prize for this work. No one doubts Hahn deserved the recognition, but the Nobel Committee's decision to ignore the contributions of Strassmann and Meitner became the subject of an extended and sometimes rancorous debate within the physics community. In 1966, perhaps in an effort to correct the Nobel Committee's rather obvious lapse, President Johnson presented the three of them with the prestigious Enrico Fermi Award, given by the Atomic Energy Commission for exceptional achievement in scientific areas associated with the development of nuclear energy. See Crawford, Sime, and Walker, "A Nobel Tale of Postwar Injustice."

CHAPTER THIRTEEN: SPLITTING THE ATOM

156 **John Wheeler, a twenty-seven-year-old:** Later in his career Wheeler was the dissertation adviser for a young and brilliant Richard Feynman, encouraging Feynman to solve one of the great puzzles of QED. Later still, he coined the phrase "black hole" to describe what happens when a star collapses and becomes so dense that nothing, not even light, can escape its gravitational pull. Throughout his career he was a man with an extraordinary gift for metaphor.

156 **The conference was scheduled:** Squire, Brickwedde, Teller, and Tuve, "The Fifth Annual Washington Conference on Theoretical Physics," 180–181.

156 **The Fermis persuaded Bohr:** Wheeler, *Geons, Black Holes, and Quantum Foam,* 14ff.

157 **I. I. Rabi, visiting from Columbia:** Ibid. This is at variance with Segrè, *Enrico Fermi,* 106, and other sources, all of which have both Rabi and Lamb attending. Also, Bohr is often given credit as the Princeton lecturer who first informed US physicists. The exact sequence of events is not clear from the historical record, but this author relies on Wheeler, who was at the center of the events at Princeton.

157 **Lamb reported with presumed understatement:** Wheeler, *Geons, Black Holes, and Quantum Foam,* 17.

157 **"probably a scientist not discovering fission":** Allison, "Enrico Fermi, 1901–1954," 129.

157 **Such blame would be unfair:** Laura Fermi, *Atoms,* 157.

158 **classic case of cognitive dissonance:** Festinger, "Cognitive Dissonance." See also Pearson, "On the Belated Discovery of Fission," who finds it astonishing that no one took Ida Noddack's proposals more seriously at the time.

158 **"Let me tell you about fission!":** American Institute of Physics, "Oral History Interviews: Herbert Lawrence Anderson—Session II," interviewed by Lillian Hoddeson and Alison Kerr, January 13, 1981, https://www.aip.org/history-programs/niels-bohr-library/oral-histories/24508-2. Anderson says that he couldn't really understand Bohr's explanation, but Fermi made the phenomenon quite clear. Because Fermi obviously knew about fission, it is clear that the Anderson-Bohr meeting took place after Fermi had been informed by Lamb.

158 **Anderson had found a mentor:** Laura Fermi, *Atoms,* 150.

159 **Fermi was happy to throw himself:** CPF II, 1.

159 **used the train ride back:** Close, *Neutrino,* 53–55, gives an accessible account of the work.

160 **They had seen fission:** Squire, Brickwedde, Teller, and Tuve, "The Fifth Annual Washington Conference on Theoretical Physics," 180–181. It remains one of the great puzzles of the story that the news of fission had not traveled beyond Princeton and Columbia by the time the Washington Conference began.

160 **he was front-and-center:** See GW Astrophysics Group, "Washington Conferences on Theoretical Physics," George Washington University and the Carnegie Institute of Washington, http://home.gwu.edu/~kargaltsev/HEA/washington-conferences.html.

161 **On January 16, 1939, one week:** Weart and Szilard, *Leo Szilard,* 53ff.

161 **"Since it was a private meeting":** Ibid., 54.

162 **"Ten percent is not a remote possibility":** Ibid.

163 **"the line was drawn":** Ibid.

164 **paper he published with Anderson:** CPF II, 5.

165 **Pegram opted for sending papers:** Pegram authorized *Physical Review* to publish both papers in April in response to Joliot-Curie's definitive decision to continue publishing his own results.

CHAPTER FOURTEEN: FERMI MEETS THE NAVY

166 **"fit of house-cleaning enthusiasm":** Laura Fermi, *Atoms,* 162.
166 **"This morning I had a telephone":** Cronin, *Fermi Remembered,* 54–55.
167 **Laura was mystified by the letter:** Laura Fermi, *Atoms,* 163.
168 **"There's a wop outside":** Rhodes, *Making of the Atomic Bomb,* 295.
169 **"the excess in the number":** Schuyler minutes quoted in Strauss, *Men and Decisions,* 237.
169 **"these experiments show more neutrons":** Ibid.
169 **"In the small samples used":** Ibid.
170 **Fermi made it clear:** Ibid.
170 **Years later, Szilard dismissively suggested:** Weart and Szilard, *Leo Szilard,* 56.
170 **"Who is this man Fermi?":** Ibid.
170 **Unfortunately for the Navy scientist:** Philip Abelson wrote a fine account of Gunn's career for the National Academy of Sciences: Philip H. Abelson, "Ross Gunn: 1897–1966," in *Biographical Memoirs* (Washington, DC: National Academies Press, 1998), http://www.nasonline.org/publications/biographical-memoirs/memoir-pdfs/gunn-ross.pdf.
171 **Rabi was one of them:** American Institute of Physics, "Oral History Interviews: I. I. Rabi—Session II," interviewed by Stephen White, February 21, 1980, https://www.aip.org/history-programs/niels-bohr-library/oral-histories/24205–2.
171 **Rabi would tell his biographer:** Rigden, *Rabi,* 83. Rabi later found out that Fermi was one of the people who nominated him for his 1944 Nobel Prize in Physics. Einstein was another.
173 **"It remains an open question":** CPF II, 13.
173 **Anderson later noted several important points:** CPF II, 11.
173 **Fermi estimated that this phenomenon:** CPF II, 13.
174 **"This was the first, and also the last":** CPF II, 11.
175 **"There was actually not much to do":** Dresden, "Heisenberg, Goudsmit, and the German 'A-Bomb,'" 93–94.
176 **Some three decades later Heisenberg:** Heisenberg, *Physics and Beyond,* 169ff.
176 **"That's a great pity":** Ibid., 171.

CHAPTER FIFTEEN: PILES OF GRAPHITE

178 **Hydrogen has two heavy isotopes:** The German uranium project would, to its detriment, rely on heavy water, made with deuterium, for its moderator.
179 **Fermi and Szilard came to the idea:** Cronin, *Fermi Remembered,* 56ff.

179 **The image of a carload:** There are many accounts of this fascinating tale of Hungarians in search of Einstein. Perhaps the best is Rhodes, *Making of the Atomic Bomb,* 303–307. This author retraced their drive during the summer of 2016 and, like them, had trouble finding Einstein's house.

180 **"Some recent work by E. Fermi":** Letter from Albert Einstein to Franklin D. Roosevelt, August 2, 1939, http://www.dannen.com/ae-fdr.html.

180 **but Roosevelt was already engaged:** Kaiser, *No End Save Victory,* recounts the efforts of Roosevelt prior to Pearl Harbor to ready the United States for war.

180 **sometimes irritating cheerleader:** Szilard once berated members of the team for planning to take a weekend off, reminding them that the Germans were hard at work on their own nuclear weapons. When they canceled their weekend plans, Szilard then announced that he was leaving town for the weekend. Lanouette, *Genius in the Shadows,* 221.

181 **"High Energies and Small Distances":** See Enrico Fermi, "High Energies and Small Distances in Modern Physics," Charles M. and Martha Hitchcock Lectures, University of California, Berkeley, http://gradlectures.berkeley.edu/lecture/high-energies/. See also CPF II, 29.

181 **"he could calculate almost anything":** CPF II, 31.

181 **Colleagues sometimes noted his tendency:** Segrè, "Nuclear Physics in Rome," 59.

181 **The game was ideal for someone:** Laura Fermi, *Atoms,* 147–148.

182 **"the first time when I started climbing":** CPF II, 1000.

183 **"a huge amount" of graphite:** Ibid.

183 **A second experimental project:** Fermi, "The Development of the First Chain-Reacting Pile," 22.

184 **The actual pile they built:** CFP II, 129ff. Each brick was 4 inches by 4 inches by 12 inches—approximately 15 pounds of graphite.

185 **Undaunted, Fermi and his colleagues:** CPF II, 112.

CHAPTER SIXTEEN: THE MOVE TO CHICAGO

188 **In January 1942, Compton brought:** Hewlett and Anderson, *History of the United States Atomic Energy Commission, Volume I,* 54–55.

188 **Fermi was unable to attend:** Laura Fermi, *Atoms,* 168.

188 **Compton reports that Fermi immediately:** Compton, *Atomic Quest,* 82.

189 **"He is supposed to have left Italy":** "Freedom of Information Act—Federal Bureau of Investigation (FBI)—1," Leo Szilard Papers, MSS 32:2, Special Collections and Archives, UC San Diego Library, http://library.ucsd.edu/dc/object/bb6964105q.

189 **"as long as Fascist Party retains control":** FBI File 062-HQ-59521, released under Freedom of Information Act request 1358450–000, dated April 18, 2017. The file is of interest in several respects. An October 19, 1940, report states that the Fermis' Leonia neighbors were canvassed for their views. The agent erroneously reports that the Fermis were naturalized citizens. It also identified a "Mrs. H. C. Eurey," no doubt Harold Urey's wife, who was eager to let the agent know that the Fermis were dedicated to the process of Americanization, with English dictionaries "all over the house."

189 **Enrico and Laura began to feel:** Laura Fermi, *Atoms,* 172.

190 **Later on in Chicago:** At Los Alamos the opening of letters continued, leading to a traumatic moment for Nella, when a letter she had written to a friend in Chicago was inadvertently seen by Fermi's driver, John Baudino—it had been opened by censors, and she was quite upset. Laura Fermi, "The Fermis' Path to Los Alamos," in Badash, Hirschfelder, and Broida, *Reminiscences,* 96.

190 **If the Nazis conquered:** Laura Fermi, *Atoms,* 171.

190 **There was just enough time:** CPF II, 137–151.

191 **He suffered severe burns:** Laura Fermi, *Atoms,* 187.

192 **The illness, called berylliosis:** Ibid., 188.

192 **"The problem interested me":** Teller, *Legacy of Hiroshima,* 37.

192 **The idea that the sun is powered:** Close, *Neutrino,* 53–55.

193 **He brought two young Columbia graduate:** Wattenberg, "The Fermi School," 88.

193 **Wattenberg recalls playing chess:** Ibid.

194 **University of Chicago was an extraordinary:** Shils, *Remembering the University of Chicago.*

194 **Anderson and Libby soon discovered:** Libby, *Uranium People,* 2.

194 **Harold Agnew, then a student:** Agnew, "Scientific World Pays Homage to Fermi," 8.

195 **"Fermi would like to show superendurance":** CPF II, 328.

196 **"If we brought the bomb to them":** Weart and Szilard, *Leo Szilard,* 147.

197 **abandoned Chicago football stadium:** Stagg Field was effectively abandoned in 1939 when the university's president, Robert Maynard Hutchins, disbanded the school's famous and popular football program.

197 **Fermi told his wife:** Laura Fermi, *Atoms,* 191.

197 **If natural uranium is exposed:** See, for example, https://fas.org/blogs/fas/2013/09/where-does-the-plutonium-come-from/ (accessed May 5, 2017). Also, http://nuclearweaponarchive.org/Library/Plutonium (accessed May 5, 2017). The process is the result of beta decay stimulated by the overcrowding of neutrons when they are absorbed by uranium nuclei.

198 **Laura arrived with the children:** Laura Fermi interview, TWOEF, 27. Stein founded the Chicago investment advisory firm of Stein Roe & Farnham and after the war moved to Winnetka, on Chicago's North Shore. The Fermis and the Steins came to be socially friendly after the

war. See also Monica Copeland, "Sydney Stein Jr.," *Chicago Tribune,* October 4, 1991, http://articles.chicagotribune.com/1991–10–04/news/9103300669_1_mr-stein-stein-roe-farnham-susan-stein.

198 **With an Italian family occupying:** Laura Fermi, *Atoms,* 174.

198 **The Fermis soon began to entertain:** Laura Fermi, *Atoms,* 174. Rhodes, *Making of the Atomic Bomb,* 428.

199 **a flattened, roughly spherical, ellipsoid shape:** It also was the optimal configuration given the amount of material available for the project. CPF II.

199 **The more he thought about it:** See Allison, "Initiation of the Chain Reaction."

201 **Beginning in September 1942:** CPF II, 216–230.

201 **"run quick-like behind a hill":** Ibid., 217. Also significant in these lectures is an analytical precursor to the Monte Carlo method, illustrated in a "probability tree" of the life of a hundred neutrons in a reactor. See ibid., 225.

202 **Owing to the new authority:** So great was Groves's authority that he was able to commandeer the entire national stockpile of silver for use in electrical wiring when the stockpile of copper was depleted. The silver was returned, almost to the ounce, at the end of the war.

202 **Fermi had a sense of humor:** Wilson and Serber, *Standing By and Making Do,* 71. Jane Wilson, a Los Alamos wife, assures the reader the story is apocryphal, but it has more than a whiff of authenticity.

204 **Compton thought about it:** Conant and Groves only found out about the plan on November 14, 1942. Conant turned white at the news; Groves immediately grabbed the phone to confirm that the pile could not be built at Argonne. But work had already begun under Stagg Field. Hewlett and Anderson, *History of the United States Atomic Energy Commission, Volume I,* 108. This meeting on November 14, 1942, was also the moment when Conant's S-1 committee learned of the tendency of plutonium to fission spontaneously, a fact that Berkeley and Chicago scientists learned earlier in the year.

CHAPTER SEVENTEEN: "WE'RE COOKIN'!"

205 **"Back-of-the-Yards" kids:** Wattenberg, "December 2, 1942," 22ff., esp. 26.

208 **At 9:45 a.m., Fermi instructed:** What follows is based on Allardice and Trapnell, "The First Pile"; Wattenberg, "December 2, 1942."

208 **"Every time the intensity leveled off":** Wattenberg, "December 2, 1942," 31.

209 **The instrumentation had recorded:** Allardice and Trapnell disagree with Wattenberg on what rod actually fell into the pile. Wattenberg recalls it was the zip rod, whereas Allardice and Trapnell report that it was another of the safety rods controlled electronically. They agree, however, that Fermi chose this moment to take a lunch break.

209 **At about 3:25 p.m. Fermi ordered:** Segrè, *Enrico Fermi,* 129, records the time as 2:20 p.m., but the preponderance of accounts report 3:25 p.m., including both the official account of Allardice and Trapnell, and Rhodes.

210 **"We're cookin'!":** Holl, *Argonne National Laboratory,* 19.

211 **Twenty-eight minutes into criticality:** Allardice and Trapnell, "The First Pile"; Segrè, *Enrico Fermi;* and Laura Fermi, *Atoms,* 129, 197. Rhodes, *Making of the Atomic Bomb,* 440, reports only 4.5 minutes, but that may be at peak power, just shy of one watt.

211 **The room was quiet:** Libby, *Uranium People,* 122.

212 **Szilard recounts that he shook:** Weart and Szilard, *Leo Szilard,* 146.

212 **Compton took Greenewalt with him:** Compton, *Atomic Quest,* 144.

212 **Compton rang up Conant:** Ibid.

212 **Laura Fermi had been planning:** Laura Fermi, *Atoms,* 176ff.

213 **"Are you making fun of me?":** Laura Fermi, *Atoms,* 179–180. Libby, *Uranium People,* 129.

212 **Laura grilled Enrico about:** Laura Fermi, *Atoms,* 180.

213 **"The first self-sustained atomic":** Compton, *Atomic Quest,* 149.

213 **"Do we then exaggerate":** Wigner, *Recollections of Eugene P. Wigner,* 241.

214 **"almost to the exact brick":** Allardice and Trapnell, "The First Pile," 44.

214 **He may have had some underlying:** Laura Fermi suggests that this was indeed on his mind. Laura Fermi, *Atoms,* 197.

214 **He played with his slide rule:** These operations with the slide rule may have also, in some way, steadied his nerves at a particularly emotional moment for him—never one to willingly show his emotions, he sought the comfort of familiar routines. Thanks to William Zajc for this suggestion.

215 **Bothe and Heisenberg decided to use:** Powers, *Heisenberg's War,* 197.

215 **long and exhaustive 1936 paper:** CPF I, 892ff.

CHAPTER EIGHTEEN: XENON-135

217 **In February 1943, after the labor:** MED, Book IV, 3.9.

217 **the pile, originally dubbed CP-1:** MED, Book IV, 3.12ff.

217 **Fermi reconfigured it so it:** Libby, *Uranium People,* 148.

219 **When Leona suggested they might:** Ibid., 163.

219 **Fortunately for everyone involved:** Several years later, Fermi graduate student Richard Garwin reported to his wife, Lois, that he might have to help deliver Libby's second child, because she did not let her advanced pregnancy deter her from lifting heavy canisters of gas in the lab. Like her first child, her second was delivered in the hospital. Garwin, interview with author, May 22, 2015.

219 **Nella was twelve years old:** Raw footage of CBS interview with Nella Fermi for the fiftieth anniversary of the pile, 1992; Olivia Fermi, private communication with author.

220 Also, Giulio was upset: Sarah Fermi, conversation with author, June 1, 2016, Cambridge, England.

220 Fermi told him not to bother: Segrè recounts a conversation between Allison and Fermi on the subject of this textbook. Allison, who liked to tease Fermi as much as Fermi liked to tease Allison, commented that Fermi's PhD thesis on X-ray diffraction could not have been particularly good, because it was not referenced in Compton/Allison. Fermi, quick to rise to the bait, replied that this only indicated the inadequacy of the famous textbook. Segrè, *Enrico Fermi,* 245n42.

220 "As a student of Compton's": Alvarez, *Alvarez,* 117.

221 In time, the two of them: Arthur Rosenfeld, interview with author, May 7, 2014.

221 That awe was not confined: Allison, "A Tribute to Enrico Fermi," 9.

222 A large cube of graphite: Rhodes reports 1,248, but MED, Book IV, part 2, 4.2, reports 1,260.

222 Into these channels rods of uranium metal: MED, Book IV, part 2, 4.2–4.3.

224 The story of how Oppenheimer: Rhodes, *Making of the Atomic Bomb,* 121, 451ff. In retrospect, the traditional narrative is only part of the story. Activists like Marian Naranjo have worked hard to bring to light the relationship of indigenous peoples to the mesa and the surrounding areas, a relationship that was sadly not taken into account in the acquisition of the land and its subsequent use. See Dennis J. Carroll, "Santa Clara Activist Works to Find Balance among Disparate Cultures," *Santa Fe New Mexican,* January 10, 2015, http://www.santafenewmexican.com/life/features/santa-clara-activist-works-to-find-balance-among-disparate-cultures/article_202f5cf1-7ec9-5fb4-b5c7-7aec1fbf574c.html.

224 "I announce to you with greatest joy": Ulam, *Adventures of a Mathematician,* 162.

224 When the entire group was assembled: Serber, *Los Alamos Primer.*

225 "I believe your people actually *want*": Davis, *Lawrence and Oppenheimer,* 182.

225 A letter from Oppenheimer to Fermi: Malcolm W. Browne, "U.S. Weighed Use of Radioactive Poison in '43, Oppenheimer Letter Shows," *New York Times,* April 19, 1985, http://www.nytimes.com/1985/04/19/us/us-weighed-use-of-radioactive-poison-in-43-oppenheimer-letter-shows.html.

225 Oppenheimer was the product: Bird and Sherwin, *American Prometheus;* Monk, *Robert Oppenheimer.*

226 "Emilio, I am getting rusty": Segrè, *Enrico Fermi,* 134.

227 "Not a philosopher": Davis, *Lawrence and Oppenheimer,* 266. Davis's book on Lawrence and Oppenheimer was controversial from the moment it was published, with some people, notably Jane Wilson, Alice Smith, and Frank Oppenheimer, doubting the veracity of large sections of the book—even direct quotations. That said, the quotation attributed to Oppenheimer regarding Fermi has the ring of authenticity and in this author's judgment accurately reflects what Oppenheimer thought of his Manhattan Project colleague. See Wilson, "Lawrence and Oppenheimer," 31–32; Smith, "Dramatis Personae," 445–447; and Oppenheimer, "In Defense of the Titular Heroes," 77–80. I am grateful to William Zajc for the references.

227 **Leona Libby recalls a dinner party:** Libby, *Uranium People,* 109.

228 **The project was vast:** MED, Book IV, vols. 4–6.

228 **The first reactor to go live:** The absence of A, C, and E reactors is an odd artifact of the planning and construction of the Hanford facility.

229 **Wigner momentarily forgot:** Wigner, *Recollections of Eugene P. Wigner,* 237.

230 **According to Wheeler, the plan:** Wheeler, *Geons, Black Holes, and Quantum Foam*, 55. "John Marshall's Interview," Voices of the Manhattan Project, 1986, http://manhattanprojectvoices.org/oral-histories/john-marshalls-interview.

232 **Xenon-135 has a half-life:** See "Shielding of Neutron Radiation," Nuclear Power, http://www.nuclear-power.net/nuclear-power/reactor-physics/atomic-nuclear-physics/fundamental-particles/neutron/shielding-neutron-radiation/. Xenon poisoning would rear its ugly head in another context some thirty-two years later at a Soviet reactor at Chernobyl. The reactor began to die down, and instead of checking for xenon poisoning, the operators pulled the control rods entirely out of the core, creating a major meltdown and the world's worst nuclear catastrophe.

CHAPTER NINETEEN: ON A MESA

234 **Laura and the children were:** Laura Fermi, *Atoms,* 200–236. Her account of the time she spent at Los Alamos during the war is one of the highlights of her book, *Atoms in the Family,* and must reading for anyone interested in how the families of the scientists lived during that intense period. She describes, for example, one of her first weekends at Los Alamos, before Enrico arrived, being driven on an expedition to Frijoles Canyon along with the Peierlses to see the old Native American ruins there; the driver was a young man named Klaus Fuchs, who was later to confess to having been a Soviet spy throughout the Manhattan Project.

234 **housing along "Bathtub Row":** Laura Fermi shows no awareness that her husband might have pulled rank like this, and in her writings was careful to insist that the assignment of living quarters was an arbitrary process controlled by the impersonal rules of the Army. It is, however, difficult to believe that Enrico could not have pressed for better accommodations if he had chosen to do so. Laura Fermi, *Atoms,* 230. See also Laura Fermi, "The Fermis' Path to Los Alamos," 93.

235 Nella recalls these days: Raw footage of CBS interview with Nella Fermi for the fiftieth anniversary of the pile, 1992; Olivia Fermi, private communication with author.

235 Laura eventually found work: Laura Fermi, *Atoms,* 227–228, 232–233.

236 prefissioning Segrè's microsamples: MED, Book VIII, Vol. 2, sec. 6.23, VI-8. One solution, obviously, might have been putting the plutonium through the same isotope separation process that was being pioneered for uranium at Oak Ridge. However, plutonium was far rarer, and the mass differential between Pu-239 and Pu-240 was one-third the mass differential between the uranium isotopes. And at this point in the project, no one was certain that the untested Oak Ridge projects would produce meaningful quantities of enriched uranium on time.

238 The Los Alamos reorganization: MED, Book VIII, vol. 2, sec. 9.

238 Research into implosion: The young physicist Seth Neddermeyer, who championed the implosion concept under Parsons, was sidelined, a result of personality conflicts with Kistiakowsky.

238 determined how quickly "prompt" neutrons: MED, Book VIII, vol. 2, sec. 6.57. Prompt neutrons are prompt indeed; they occur almost simultaneously with the fission, which occurs so fast that it has never been precisely measured but certainly in less than a billionth of a billionth of a second. See MED, Book VIII, vol. 2, sec. 6.70ff. See Aurel Bulgac, Piotr Magierski, Kenneth J. Roche, and Ionel Stetcu, "Induced Fission of Pu within a Real-Time Microscopic Framework," March 25, 2016, https://arxiv.org/pdf/1511.00738.pdf, for recent work on this subject. Thanks to William A. Zajc for bringing this to the author's attention.

239 Segrè describes Joan: Segrè, *Enrico Fermi,* 141.

239 The story of the British project: See Farmello, *Churchill's Bomb,* for a recent history of this fascinating aspect of World War II.

240 German refugees Rudolf Peierls: Close, *Half Life,* 82ff.

240 Peierls admired Fermi greatly: Segrè, "Nuclear Physics in Rome," 59.

241 "We had a meeting with him": Feynman, *"Surely You're Joking,"* 132. A more detailed description of this encounter is at American Institute of Physics, "Oral History Interviews: Richard Feynman—Session IV," interviewed by Charles Weiner, June 28, 1966, https://www.aip.org/history-programs/niels-bohr-library/oral-histories/5020–4.

241 "Only if that child is Fermi": Ibid.

242 He emigrated in 1933: After the war he dabbled in economics, creating—with his colleague Oscar Morgenstern—the entire field of game theory.

242 He joined the war effort early: CPF II, 437.

243 "We're plowing": Wright, "Fermi in Action," 182.

243 Fermi enjoyed using them himself: Fermi to Pegram, October 18, 1944, in Pegram Papers, Box 1, Fermi folder.

243 **Fermi also used the newest wave:** "Computing and the Manhattan Project," Atomic Heritage Foundation, http://www.atomicheritage.org/history/computing-and-manhattan-project.

244 **Segrè recounts a moment:** Segrè, *Enrico Fermi,* 140.

244 **He also began to give lectures:** See EFREG, 45:6, for notes on the neutron physics course he gave during 1945.

244 **Geoffrey Chew, a twenty-year-old:** Chew, interview with author, May 6, 2014.

245 **"I see, so it's a battle of wits!":** Segrè, *Enrico Fermi,* 140.

246 **Chew tells the story:** Chew, interview with author, May 6, 2014. This type of parlor game must have been quite popular among the Fermi circle; after the war, Fermi student Jay Orear recalls playing a similar game. Orear, "My First Meetings with Fermi," in Cronin, ed., *Fermi Remembered,* 202.

247 **"He said in his mild and reasonable voice":** Brode, *Tales of Los Alamos,* 79.

247 **"tickling the dragon" experiments:** MED, Book VIII, vol. 2, sec 15.7ff. Even reading the somewhat dry description of the experiment is enough to make the heart race faster.

247 **The configuration of the high-explosive:** Alex Wellerstein, "What Did Bohr Do at Los Alamos?" Restricted Data, the Nuclear Secrecy Project, May 11, 2015, http://blog.nuclearsecrecy.com/2015/05/11/bohr-at-los-alamos/. See also Hoddeson, Henriksen, Meade, and Westfall, *Critical Assembly,* 317.

248 **"I think Fermi began to be very worried":** Ibid.

248 **Niels Bohr and his son Aage:** Pais, *Niels Bohr's Times,* 495ff., tells the harrowing adventures of the Bohrs as they escaped Denmark in the late summer of 1943, their stopovers in Sweden and London, their arrival in New York, and their time at Los Alamos.

249 **Fermi, confronted with this elegant:** Fermi apparently preferred an initiator code named "grape nuts," which remains classified.

249 **a lot of money would have been spent:** This was an observation he would make explicitly, to Groves's irritation, just prior to the Trinity test in mid-July 1945. Libby, *Uranium People,* 225.

250 **The Germans had come nowhere near:** Bernstein, *Hitler's Uranium Club.*

250 **Their decision initially to pursue:** Powers, *Heisenberg's War;* Baggot, *First War of Physics.*

250 **So did the president himself:** It is difficult to imagine a vice president today being kept in the dark about a multi-billion-dollar defense project; however, FDR was famously secretive throughout his presidency and may not have felt the need to bring Truman into the picture, because he had never been particularly close to the former senator from Missouri and had only met with him on a few occasions prior to Roosevelt's death on April 12, 1945.

251 **Compton later recalled the meeting:** Compton, *Atomic Quest,* 219ff. See also "Notes of the Interim Committee Meeting, Thursday, 31 May 1945," http://nsarchive.gwu.edu/NSAEBB/NSAEBB162/12.pdf.

251 **Fermi limited his participation:** "Notes of the Interim Committee Meeting," 4.

252 **What Stimson thought of Fermi:** Stimson and Bundy, *On Active Service,* makes no mention of the meeting at all, and Fermi neither discussed nor wrote about the meeting.

252 **Stimson may have been concerned:** As Leon Sigal suggests, the decisions about using the bomb were really the purview of the Targeting Committee, and the Interim Committee's main task was organizing nuclear research postwar. If Stimson could also use it to undermine the resistance of scientists to using the bomb against Japan, that would be an added benefit. Sigal, "Bureaucratic Politics & Tactical Use of Committees."

252 **He drafted a letter, eventually signed:** James Franck, Donald J. Hughes, J. J. Nickson, Eugene Rabinowitch, Glenn T. Seaborg, J. C. Stearns, and Leo Szilard, *Report of the Committee on Political and Social Problems Manhattan Project "Metallurgical Laboratory" University of Chicago, June 11, 1945 (The Franck Report),* US National Archives, Record Group 77, Records of the Chief of Engineers, Manhattan Engineer District, Harrison-Bundy File, folder 76, http://www.dannen.com/decision/franck.html.

253 **Just 350 words in length:** Atomic Heritage Foundation, "Science Panel's Report to the Interim Committee, June 16, 1945, Top Secret: Recommendations on the Immediate Use of Nuclear Weapons," http://www.atomicheritage.org/key-documents/interim-committee-report-0.

253 **How Oppenheimer crafted this "consensus":** Compton, *Atomic Quest,* 239–241.

253 **Oppenheimer's secretary, Anne Wilson Marks:** Wyden, *Day One,* 170–171. All subsequent accounts rely entirely on Compton; however, it is difficult to discount the story told by Marks, because she had no particular axe to grind and her memory is consistent with subsequent Fermi behavior.

254 **morning of Sunday, June 17, 1945:** The report to Stimson and the Interim Committee is dated June 16, 1945; either Oppenheimer postdated it or prepared it in advance of Fermi's agreement, or Wyden's book is off by a day.

254 **foreign-born national who only recently:** They became naturalized US citizens on July 11, 1944. Laura Fermi, *Atoms,* 175.

254 **Groves stamped it "secret":** "A Petition to the President of the United States," US National Archives, Record Group 77, Records of the Chief of Engineers, Manhattan Engineer District, Harrison-Bundy File, folder 76, http://www.dannen.com/decision/45–07–17.html. Leon Sigal, *Fighting to a Finish,* 204.

CHAPTER TWENTY: AN UNHOLY TRINITY

256 **Fermi arrived by car a few days:** Allison, "Scientific World Pays Homage to Fermi," 8.

256 **Fermi got himself into trouble:** Victor Weisskopf reports that one of his colleagues took Fermi's joke so seriously that he had a nervous breakdown shortly before the explosion. Weisskopf, *Joy of Insight,* 149.

257 **He chose a viewing site:** There is some confusion about exactly how far he was from the test. He reported that he was ten miles from the blast. "Trinity Test, July 16, 1945, Eyewitness Accounts—Enrico Fermi," US National Archives, Record Group 227, OSRD-S1 Committee, Box 82 folder 6, "Trinity," http://www.dannen.com/decision/fermi.html. On the other hand, L. D. P. King later reported that they were ten thousand yards from the blast. L. D. P. King interview transcript for TWOEF. Weisskopf agrees with Fermi. *Joy of Insight,* 151.

259 **Fermi announced to those within earshot:** Weisskopf recalls that Fermi announced a more accurate 20-KT yield, but Fermi's own recollection was that he estimated it at 10 KT. "Trinity Test, July 16, 1945, Eyewitness Accounts—Enrico Fermi," http://www.dannen.com/decision/fermi.html; Weisskopf, *Joy of Insight,* 152.

259 **Within an hour of the detonation:** Rhodes, *Making of the Atomic Bomb,* 677.

259 **"It was like being at the bottom":** "Trinity Test—1945," Atomic Heritage Foundation, http://www.atomicheritage.org/history/trinity-test-1945.

259 **The light from the detonation:** McMillan, *Atom and Eve,* 86.

260 **In Santa Fe, Dorothy McKibbin:** Steeper, *Gatekeeper to Los Alamos,* 106.

260 **Allison later described the trip home:** Allison, "Scientific World Pays Homage to Fermi," 8.

260 **radiation exposure of the indigenous groups:** "Key Findings of CDC's LAHDRA Project: Public Exposures from the Trinity Test," CDC's LAHDRA, http://lahdra.org/pubs/reports/Posters/LAHDRA%20Trinity%20Test%20Poster-%20reduced%20size,pdf.pdf.

260 **contaminating rainwater cisterns:** The explosion distributed some ten pounds of unfissioned plutonium in a wide area surrounding the blast.

261 **Her efforts and those of others:** Los Alamos Historical Document Retrieval and Assessment Project, http://lahdra.org/pubs/pubs.htm. Victor Weisskopf is one of the few participants who indicates that there was a concern about the dangers to local inhabitants, mentioning the town of Carrizozo sixty miles away. It is not clear whether he was aware of the population living much closer to the blast. Weisskopf, *Joy of Insight,* 150.

262 **Some eighty thousand people died:** The calculation of casualties at Hiroshima and Nagasaki remains imprecise and controversial; the numbers here are not definitive.

262 **avoiding a prolonged and bloody battle:** Historians continue to debate the necessity of dropping the bombs to end the war, and some of those who agree that the bomb on Hiroshima might have been necessary argue that the second one against Nagasaki was dropped simply to evaluate its effects on a second city with very different geography from the first. They also debate the degree to which Soviet entry into the war against Japan influenced US decision making on the bomb itself. See, for example, Sigal, *Fighting to a Finish.*

262 **"Our stuff was dropped on Japan!":** Laura Fermi, *Atoms,* 237.

262 **Laura herself was more circumspect:** Ibid., 244–245.

263 **"People of good judgment abstain":** Ibid., 245. This author has been unable to find the original letter in archival files.

263 **Fermi wanted to finish his comprehensive:** CPF II, 440–541.

263 **then-current state of the Super project:** FOIA 09–00015-H, declassified at the request of Professor Alex Wellerstein. See also D. R. Inglis, "Super Lecture No. 1: Ideal Ignition Temperature," http://work.atomlandonmars.com/kf/1945%20-%20Fermi%20Super%20Lectures%20(LANL%20FOIA).pdf.

264 **"In concluding this series of lectures":** Ibid. The document is not Fermi's but rather notes taken by someone who attended the lectures; however, the final paragraph seems quite consistent with Fermi's own sense of humor.

264 **Amaldi was drafted and served:** Ugo Amaldi, conversation with author, June 7, 2016.

264 **When Amaldi returned from the front lines:** Francesco Guerra and Nadia Robotti, conversation with author, September 21, 2015, Rome.

264 **Marcello Conversi and Ettore Pancini, conducted:** Conversi, Pancini, and Piccioni, "On the Decay Process of Positive and Negative Mesons" (1945 and 1947).

265 **Indeed, in one of his 1938 letters to Pegram:** Fermi to Pegram, Brussels, October 22, 1938. Fermi Folder, Papers of George Pegram, Rare Book Collection, Columbia University Libraries.

266 **He was the "go-to" physicist:** When word spread at Los Alamos that Fermi would be arriving in the late summer of 1944, three physicists approached Laura to insist that they see Enrico as soon as he arrived. From the moment he stepped onto the mesa, he was always in demand. Laura Fermi, "The Fermis' Path to Los Alamos," 93.

CHAPTER TWENTY-ONE: RETURN TO CHICAGO

271 **She started a book group:** LFREG, 8:4.

271 **"using the old Italian method":** Segrè, *Enrico Fermi,* 176.

272 **Fermi confided to Segrè:** Segrè, *Enrico Fermi,* 175–176.

272 **Amazingly, Laura even persuaded Enrico:** EFREG 2:7, 1951 pocket diary for May 23.

272 "Never make something more accurate": Nella Fermi in Orear, *Enrico Fermi: The Master Scientist,* 131.

272 It was "Silly Putty": Ibid., 132–133.

272 Giulio had a more troubled relationship: Sarah Fermi, interview with author, June 1, 2016. Robert Fuller, interview with author, June 25, 2016.

273 Fuller notes that Giulio: Robert Fuller, interview with author, June 25, 2016.

273 He posed impishly: Richard Garwin, "Fermi's Mistake?" The photo session was even used for the image of the US postage stamp issued in his honor in 2001. The erroneous formula is just visible at the top left corner of the stamp.

274 replaced Arthur Compton as dean: Compton accepted an invitation to become chancellor at Washington University in St. Louis.

275 December 1945 letter to Bartky: Cronin, *Fermi Remembered,* 111–113.

276 Their work had not yet reached: Monaldi, "Mesons in 1946," covers the situation just prior to the discovery of the pion.

276 "How thick does the dirt": Interview with James Cronin, October 20, 2014.

276 down the hall to where Maria Mayer: Dash, *A Life of One's Own,* 316–317.

277 Mayer later vividly recalled: Ibid., 317.

278 Mayer wanted to include him: He knew that people assumed, incorrectly, that her husband, Joseph, was the main contributor to their coauthored classic textbook, *Statistical Mechanics,* published when the Mayers were at Columbia.

278 It was a generosity that characterized: At the 1963 Nobel ceremony, Jensen said to Mayer, "I have convinced Heisenberg and Bohr. You have convinced Fermi. What do we care about the others? You see, when you have convinced Fermi you have really accomplished something." Quoted in Zuckerman, *Scientific Elites,* 184.

279 Gell-Mann recalls with some frustration: Gell-Mann, "No Shortage of Memories," 151.

279 Fermi was a consultant to the project: *Invention of Cyclotron* (Chicago: Particle University of Chicago, April 16, 2008), http://hep.uchicago.edu/cdf/frisch/p363/InventionOfCyclotron_shiraishi.pdf. See also INSREG, Box 1, for complete documentation of the project. Richard Garwin tells of the regular morning meetings cyclotron engineers held with Fermi while dealing with the electronic challenges of creating the synchrotron. Garwin, "Working with Fermi," 144.

279 Though none of his postwar work: Sociologist of science Harriet Zuckerman points out, however, that Fermi is one of a handful of Nobel laureates whose research after the Nobel Prize was "roughly comparable" to the work he did before winning the prize. In coming to this judgment, she no doubt includes the wartime work done for the Manhattan Project and the postwar work as well. Zuckerman, *Scientific Elites,* 1219–1220.

280 **"This is why it is very important":** Wattenberg, "Fermi as My Chauffeur," 174. It is tempting to see intimations of mortality in this statement, but the more likely explanation is that Fermi simply wanted to put his passenger at ease.

281 **"I'll hold the mirror up":** Ibid., 178.

281 **These experiments, which continued:** CPF II, 568–614.

282 **The Norwegian astrophysicist Hans Alfven:** Allison, "Enrico Fermi, 1901–1954," 133. In Sam Allison's biography of Fermi for the National Academy of Sciences, he claims Fermi took the idea of electromagnetic acceleration from the Alfven lecture. Chandrasekhar later disputed this and wrote to Allison that Fermi had the idea first. Only later did Alfven present his ideas at Chicago. Chandrasekhar letter to Allison, SCREG, 15:6.

282 **Born in India, "Chandra":** A speech in India by Arnold Sommerfeld inspired the work.

283 **Chandrasekhar arrived at the University of Chicago:** Chandrasekhar was personally a bit more formal than Fermi and professionally more concerned about the elegance of his solutions than Fermi was. And yet the two of them became good friends. See Wali, *Chandra,* 19–20.

283 **The two of them hit it off well:** SCREG, 15:6.

283 **the discussions between Fermi and Chandrasekhar:** CPF II, 923ff.

283 **The two men published several papers:** CPF II, 927ff., 931ff., 970ff.

283 **but Chandrasekhar likened him:** CPF II, 923.

284 **Steinberger, a German Jew:** Steinberger, *Learning about Particles,* 1–29, for the story of how Steinberger came to the United States and ended up working with Fermi.

285 **top of Mount Evans in Colorado:** The higher the detectors are in the atmosphere, the more cosmic rays are detected, because the atmosphere serves as a filter for cosmic ray bombardment.

285 **"probably neutrinos," accompany:** Steinberger, *Learning about Particles,* 22.

285 **Neither Steinberger nor Fermi realized:** Ibid., 23.

285 **"new ideas are not always":** Ibid.

285 **three University of Chicago graduate students:** Ibid., 23. Papers making this same observation were published by John Wheeler and Jayme Tiomno, as well as Giampetro Puppi. Steinberger also points out that Bruno Pontecorvo made suggestive observations in this vein as early as 1947, but Fermi seemed either unaware of them or ignored them.

285 **Steinberger recalls that Fermi was:** Jack Steinberger, interview with author, May 5, 2014.

286 **Fermi received an invitation:** His pocket diary for 1947 has June 2 marked for the Ram's Head Inn. EFREG, 2:6.

286 **his friends would often notice:** Libby, *Uranium People,* 21. Long after his torn retina healed, Fermi continued to test his eyesight. His friend Stanislaw Ulam recalled an outdoor dinner in the south of France during the summer of 1954 at which he watched as Fermi moved his head back and forth, studying how a particular star disappeared and then reappeared behind a telephone wire. Ulam, *Adventures of a Mathematician,* 234–235.

287 His presentation was so long: Gleick, *Genius,* 255ff. See also Segrè, *Enrico Fermi,* 174. For a more exhaustive scientific account, see Schweber, *QED and the Men Who Made It.*

287 Feynman's approach was anything: Fermi would in due course appreciate the value of Feynman diagrams. For example, he used them in his January 1954 talk at Columbia, "What Can We Learn from High Energy Accelerators?" EFREG, 40:10.

287 Fermi and Franck jointly nominated: EFREG, 19:1. Fermi and Franck also cited Princeton physicist Freeman Dyson, who brilliantly demonstrated why all three approaches, so different on the surface, solved the same problem.

288 Peierls's observation that Fermi: Segrè, "Nuclear Physics in Rome," 59–60.

288 "never make something more accurate": See Nella Fermi in Orear, *Enrico Fermi: The Master Scientist,* 131.

289 "I must confess my confidence": Hewlett and Anderson, *A History of the United States Atomic Energy Commission,* 432.

289 nuclear program under civilian control: The Atomic Energy Act of 1946 still governs the development of atomic energy and weapons development in this country.

290 help of graduate student Richard Garwin: Garwin, "Working with Fermi," 145.

290 affectionately dubbed "Fermiac": "Fermi Invention Rediscovered at LASL"; Nicholas Metropolis, "The Beginning of the Monte Carlo Method."

291 inconsistent with its expected ergodicity: CPF II, 978. Ulam, *Adventures of a Mathematician,* 226. An ergodic system is one in which at any given time the likelihood of it being in any of all possible states is equal. If a system is more likely to be in a particular state than in any other possible state, it is not ergodic. The paper is considered an early contribution to chaos theory.

291 "Taylor instability": Fermi's papers refer to Taylor alone, but the phenomenon is better known as Rayleigh Taylor instability, because it was first studied by Lord Rayleigh as well as G. I. Taylor. See Libby, *Uranium People,* 210ff., for a good discussion of this subject and its relevance to weapons testing.

292 short volume called *Elementary Particles:* Enrico Fermi, *Elementary Particles.*

292 not to be an exact predictor: Ibid., 79.

292 He also prepared a paper: CPF II, 825.

293 type of particle a "resonance": Brown, Dresden, and Hoddeson, *Pions to Quarks,* 10ff. Pickering, *Constructing Quarks,* 48ff.

293 "Through my illness [berylliosis]": CPF II, 923.

294 **Anderson did have a jealous streak:** Amaldi worked with Fermi from 1927 through 1938, some eleven years. Anderson met Fermi early on in 1939 and worked with Fermi until 1954, almost fifteen years.

294 **Freeman Dyson, the young theorist:** Dyson, "A Meeting with Fermi," 297.

295 **"There are two ways of doing":** Ibid.

295 **"Looking back after fifty years":** Ibid.

CHAPTER TWENTY-TWO: IN THE PUBLIC EYE

297 **the first General Advisory Commission (GAC):** Hewlett and Anderson, *History of the United States Atomic Energy Commission,* 648.

297 **His FBI file makes for interesting:** FBI Case File 116-HQ-1255 at National Archives, College Park, MD.

299 **"To have spent the day":** Lilienthal, *Journals of David E. Lilienthal,* 2:128.

299 **He was outspoken in his support:** Segrè, *Enrico Fermi,* 164.

299 **Teller and Ulam cracked the puzzle:** There are several excellent histories of the development of the hydrogen bomb. See Rhodes, *Dark Sun;* Ford, *Building the H Bomb.* See also US Atomic Energy Agency, *In the Matter of J. Robert Oppenheimer,* passim.

300 **Under significant political pressure:** AEC chairman Lewis Strauss, who took over from David Lilienthal, expressed the straightforward position that the United States must develop a hydrogen bomb before the Russians did. He was not alone.

300 **"the development of these weapons":** Rigden, *Rabi,* 205.

300 **"by example some limitations":** Ibid.

300 **Rabi and Fermi offered a far more:** Ibid., 206–207. The written record of these deliberations is clear, but Lilienthal, who chaired the meeting, records a slightly different recollection. His record: "Fermi, his careful enunciation, dark eyes, thinks one must explore it and do it and that doesn't foreclose the question: should it be made use of?" Lilienthal, *Journals of David E. Lilienthal,* 2:581. But it is clear that in the end, Fermi signed on to Rabi's statement.

301 **"The fact that no limits exist":** Rigden, *Rabi.*

301 **he responded with a volcanic outburst:** Libby, *Uranium People,* 15. This is one of the very few recorded incidents of Fermi losing his temper with colleagues.

302 **Fermi and Ulam decided on a good-natured:** Ulam, *Adventures of a Mathematician,* 218; Wheeler, *Geons, Black Holes, and Quantum Foam,* 209; Ford, *Building the H Bomb,* 104–105.

302 **someone would find a new fatal flaw:** It was this flawed concept of the hydrogen bomb that Teller sold to President Truman when the president overrode the GAC in January 1950. Ford, *Building the H Bomb,* 105.

302 **flash of insight by Ulam:** Called the Teller-Ulam invention, it was far more complicated than this simple description suggests. The concept was for many years highly classified, and aspects of it remain classified to this day. Ulam and Teller rarely spoke to each other after this most unusual collaboration. There was a long-standing dispute, never definitively resolved, over who deserved the lion's share of the credit.

303 **the Ivy Mike test:** The device detonated was the size of a very large house and was designed by Garwin, under the direction of Teller and Ulam. It released some ten megatons of energy, a thousand times greater than the bombs that fell on Japan. Large parts of the atoll of Eniwetok were vaporized in the test. Eniwetok hosted many subsequent tests. The cleanup of the islands in the late 1970s exposed soldiers to dangerous levels of radioactivity. See Dave Philipps, "Troops Who Cleaned Up Radioactive Islands Can't Get Medical Care," *New York Times,* January 28, 2017, https://www.nytimes.com/2017/01/28/us/troops-radio active-islands-medical-care.html.

303 **Fermi and Garwin worked extensively:** Richard Garwin, interview with author, May 22, 2014.

303 **who invented the hydrogen bomb:** The classic account, which unfortunately misses the key role of Garwin in the design of the Ivy Mike hydrogen bomb test in November 1952, is Rhodes, *Dark Sun*. A more recent account, which covers the technical issues in some detail, is Ford, *Building the H Bomb*. An idiosyncratic but colorful account of some of the Fermi-Garwin work is contained in Mayer, "An Indecisive Meeting," Los Alamos Historical Archives M203–62–1-96. See also Ulam, *Adventures of a Mathematician,* 209ff., and Wheeler, *Geons, Black Holes, and Quantum Foam,* 104ff.

303 **When Fermi's four-year term:** Oppenheimer, "Scientific World Pays Homage to Fermi," 8.

304 **Fermi clearly wanted nothing to do:** See EFREG, 9:16. In 2012 Manhattan Project historian Alex Wellerstein discovered the unredacted transcripts of the Oppenheimer hearings of 1954, in which Leslie Groves admits that the sensitivity of the material the Rosenbergs shared with the Soviet Union was not particularly damaging to the United States. Nevertheless, he was in favor of their execution. Wellerstein, "Oppenheimer, Unredacted: Part II."

305 **"dared to hope that you":** EFREG, 14:22. Darrow to Fermi, May 8, 1951.

306 **the APS, under Fermi's leadership:** The files on this issue in the Fermi Archives in Chicago reflect the systematic way in which Fermi canvassed colleagues and duly recorded the views of each person on the executive committee. See EFREG, 16:18, for example.

306 **very few African American members:** There were not many African American members of the APS, but James Van Allen, the University of Iowa astrophysicist who discovered the radiation belts surrounding the earth that now bear his name, had an African American graduate student, Robert Ellis. He faced down the US Navy to bring Ellis onto a segregated Navy vessel that year and may well have wanted to bring Ellis to the APS meeting at Duke. My thanks to Greg Good at the American Institute of Physics for bringing this to my attention.

307 the executive committee prevailed: EFREG, 15:3.

307 an anti-communist "loyalty oath": "The Loyalty Oath Controversy, University of California, 1949–1951: Resolution Adopted by the Regents of the University of California, April 21, 1950," University of California History, http://www.lib.berkeley.edu/uchistory/archives_exhibits/loyaltyoath/regent_resolution.html.

307 few comments we have from Fermi: EFREG, 9:19.

307 APS came to Condon's defense: Wang, "Edward Condon and the Cold War." See also Alice K. Smith, *Peril and a Hope,* passim.

308 case against Oppenheimer: The classic account of the Oppenheimer case can be found in Bird and Sherwin, *American Prometheus,* 487ff. See also Polenberg, *In the Matter of J. Robert Oppenheimer.*

309 "Statements of this kind are bound": A Teller-inspired 1954 book by journalists James Shepley and Clay Blair, *The Hydrogen Bomb* was a thinly veiled attack on Los Alamos and its role in the development of the hydrogen bomb. Fermi's graduate student Arthur Rosenfeld read the book and called Fermi's attention to it. Fermi was reportedly outraged and allowed Rosenfeld to draft a press release under Fermi's name rejecting the book's implications and praising the work of his Los Alamos collaborators. In an interview with the author in May 2014, Rosenfeld remembered the press conference taking place before Fermi went to Varenna, but the book was released in late September–early October 1954, and the document is dated October 4, 1954. Rosenfeld, "Reminiscences of Fermi," 204–205. EFREG, 16:2.

309 A letter in November 1953: Fermi had had interactions with Borden in November 1952, when Borden wrote a letter to Fermi asking how long it might take for the Soviet Union to develop a hydrogen bomb. Fermi stressed that it was virtually impossible to provide anything but a "wild guess," but if pressed he would estimate two to five years. EFREG, 9:17.

310 "What a pity that they took him": Telegdi, "Enrico Fermi, 1901–1954," 126–127.

310 He agreed to serve: EFREG, 17:1.

311 redacted version was released: Wellerstein, "Oppenheimer, Unredacted: Part I" and "Oppenheimer, Unredacted: Part II."

311 "what more do you want, mermaids?": US Atomic Energy Agency, *In the Matter of J. Robert Oppenheimer,* 468.

311 When pressed, Teller said: Ibid., 710.

312 "My opinion at that time": Ibid., 395.

312 "I would not know": Ibid., 397.

313 "I think I might possibly": Ibid., 398.

CHAPTER TWENTY-THREE: A PATENT FIGHT

314 **state of his bank account:** EFREG, 50:5.

315 **He also recorded in detail:** The pocket diaries in EFREG are filled with such notations.

315 **Giulio recalls his father as somewhat stingy:** TWOEF, notes of Giulio Fermi interview.

315 **Fermi approached Macmillan Publishers:** EFREG, 13:2. The correspondence is lengthy and makes for depressing reading.

317 **decision to replace the term *velocity*:** Velocity is a vector—it has magnitude and direction. Speed is a scalar, having magnitude only. The difference to a physicist is major.

317 **The next letter in the file:** Schluter suggests that it was sent in late 1952, but this author could not find the final letter as sent. Schluter, "Three Reminiscences of Enrico Fermi," 206.

317 **"I am sincerely sorry that the plan":** Robertson to Fermi, November 10, 1952. EFREG, 13:2.

318 **contains a seven-digit entry for July 3:** EFREG, 2:2.

318 **Thus began a thirteen-year saga:** The most comprehensive treatment of this subject is Turchetti, "'For Slow Neutrons, Slow Pay.'" D'Agostino and Trabacchi were part of the Italian patent application, but not part of the US application. The US patent itself makes for interesting reading. In it Fermi reviews the results of neutron bombardment on each of the ninety-two elements in the then-current periodic table. When discussing uranium, he speculates he may have created transuranic elements, though by the time the patent was granted he knew he—and the rest of the world—had been mistaken. See also EFREG, 19:2–7.

319 **compensation for its use by the US:** Lanouette, *Genius in the Shadows,* 254. Fermi and Szilard were not the only Manhattan Project scientists who filed patents for war-related intellectual property. The team responsible for creating and studying the first samples of plutonium, led by Seaborg at Berkeley, also filed patents, as did Lawrence, Oppenheimer, and others who were involved in the invention of the "calutron" method of isotope separation.

319 **Fermi and Segrè made several ineffective:** Turchetti, "'For Slow Neutrons, Slow Pay,'" 11.

321 **Irritated, Fermi contemplated resigning:** Turchetti reports Maltese's suggestion that this incident was critical in Fermi's decision not to renew his GAC contract in January 1951. Turchetti, "'For Slow Neutrons, Slow Pay,'" 19n. To be clear, however, the debate that raged over the hydrogen bomb probably served as the main impetus for Fermi's decision not to renew his contract. Also, as previously noted, the Chicago cyclotron was expected to come online and Fermi did not wish to be distracted from an experimental agenda he had waited four years to implement.

321 **Intelligence services quickly discovered:** Reuters, "British Atomic Scientist Believed Gone to Russia," 3. See also Reuters, "Atomic Scientist,

Family Disappear," 34. His next appearance in public, however, came in a Moscow press conference in 1955, where he denied passing atomic secrets to the Soviet Union.

322 **$30,000 was a significant sum:** Ulam recalled that Fermi hoped that the inventors would receive "tens of millions." Ulam, *Adventures of a Mathematician,* 233.

CHAPTER TWENTY-FOUR: BRILLIANT TEACHER, BELOVED MENTOR

323 **One of the very few Fermi lectures:** The recording is in the Niels Bohr Library and Archives at the American Institute of Physics in College Park, Maryland. File AV_7_54_1.mp3.

324 **"On the contrary," Telegdi writes:** Telegdi, "Enrico Fermi, 1901–1954," 123. Telegdi was a brilliant but highly volatile Hungarian physicist. He was the first Enrico Fermi Professor of Physics at Chicago, where he taught for some twenty-five years. He then returned to his alma mater, Eidgenössische Technische Hochschule (ETH) Zurich, where Einstein taught prior to coming to the United States.

324 **The only period in which he chose:** This corresponds with the period of his torn retina.

325 **Harold Agnew later recalled:** Agnew, "Scientific World Pays Homage to Fermi," 8.

325 **One of the most interesting graduate:** Libby, *Uranium People,* 238. Also Yang, "Reminiscences of Fermi," 241ff. Lee explained in later years that the university would accept graduate students who demonstrated a solid knowledge of the "great books" of Western civilization, irrespective of whether they had an undergraduate degree. Fermi and his colleagues apparently persuaded the admissions office that Lee knew the Chinese "classics"—Lao-Tzu, Mencius, Confucius, etc.—and he was admitted. Lee, "Reminiscences of Chicago Days," 198.

326 **"He had both feet on the ground":** Yang, "Reminiscences of Fermi," 242.

327 **would only take on experimental students:** Chew, "Personal Recollections," 188.

327 **decided to take on two theory students:** Perhaps also because he had no classroom teaching obligations during this period.

327 **Chew is particularly vivid:** Chew, interview with author, May 6, 2014.

327 **"I am completely indebted to Fermi":** Jack Steinberger, interview with author, May 5, 2014.

328 **Arthur Rosenfeld was notable:** Arthur Rosenfeld, interview with author, May 6, 2014.

328 **Alvarez pressed Fermi to tell him:** Ibid. See also Rosenfeld, "Reminiscences of Fermi," 203–205.

328 **Rosenfeld went on to a notable:** He passed away January 27, 2017. See Julie Chao, "Art Rosenfeld, California's Godfather of Energy Efficiency,

Dies at 90," Berkeley Lab, January 27, 2017, https://newscenter.lbl.gov/2017/01/27/art-rosenfeld-californias-godfather-energy-efficiency-90/?utm_source=Art+Rosenfeld+Dies+at+90&utm_campaign=Rosenfeld-obit&utm_medium=email.

328 He took extensive notes: Orear, "Notes on Statistics for Physicists," University of California Radiation Laboratory, August 13, 1958, UCRL 8417, https://cds.cern.ch/record/104881/files/SCAN-9709037.pdf.

328 He has written with unusual affection: Orear, *Enrico Fermi.*

329 he would later share a Nobel Prize: The award of the Nobel Prize to Segrè and Chamberlain in 1958 became the subject of some controversy in the 1970s, because another Fermi student, Oreste Piccioni, sued the Berkeley physicists, claiming they had used a crucial idea of his for the experiment and had not added him as a coauthor on the resulting paper. The suit made its way to the US Supreme Court, which decided not to hear the case on the basis of lapsed statute of limitations.

329 "I knew exactly what a Nobel Prize": Chamberlain, "Brief Reminiscence of Enrico Fermi," 187.

329 "I am very grateful": EFREG, 9:19.

329 Fermi waved hello to Friedman: Friedman, interview with author, January 28, 2015.

330 The list reads like a who's who: The full list includes, in chronological order, George Farwell, Geoffrey Chew, Marvin Goldberger, Lincoln Wolfenstein, Jack Steinberger, Owen Chamberlain, Richard Garwin, Tsung-Dao Lee, Uri Haber-Schaim, Jay Orear, John Rayner, Robert Schluter, Arthur Rosenfeld, Horace Taft, and Jerome Friedman. See Telegdi, "Enrico Fermi, 1901–1954," 125.

330 "the only true genius": He almost certainly would have said the same thing about Majorana.

330 usually reticent Fermi sent a letter: Fermi to Doty, April 11, 1949. EFREG, 14:13.

330 Garwin has won numerous: Joel N. Shurkin, *True Genius: The Life and Work of Richard Garwin, the Most Influential Scientist You've Never Heard Of* (2017), is an account of his amazing life and career.

330 One day, Fermi approached Yodh: Yodh, "This Account Is Not According to the Mahabharata!" 251–253. It is perhaps telling that Fermi chose to make lighthearted fun of the book in which Oppenheimer found his famous Trinity test quote, "I am become Death, destroyer of worlds," taken from the Bhagavad Gita. The Bhagavad Gita is one of the two main sections of the Mahabharata.

331 Cronin recalled with humor: Cronin, interview with author, October 20, 2014.

331 Harriett Zuckerman, a sociologist: Zuckerman, *Scientific Elites,* 100. It is, of course, a narrow point of view to judge the success of any particular mentor simply by Nobel laureates mentored. Fermi's non-Nobelist students include a president of Caltech and extremely prominent theorists and experimentalists on both sides of the Atlantic.

331 When one looks at a "family tree": "Physics Tree," Academic Family Tree, http://academictree.org/physics/tree.php?pid=34756.

332 It is an astonishing record: Zuckerman makes this point, as does Telegdi in Cronin, *Fermi Remembered,* 125. Telegdi notes, however, that Sommerfeld trained only theorists and Rutherford, only experimentalists, while Fermi's students were relatively evenly divided, reflecting Fermi's own universal interests.

CHAPTER TWENTY-FIVE: TRAVELS ABROAD

333 A more interesting possibility: I am indebted to Giovanni Battimelli for this suggestion.

334 a host of Italians attended: Attendee list provided by University of Basel.

334 Fermi was not one of the: Summary graciously provided by University of Basel.

334 Fermi also swam a mile: Laura Fermi, *Atoms,* 256.

334 Fifty-odd papers were presented: See *Il Nuovo Cimento* 6, no. 3, Suppl. (January 1949).

335 Giorgio, who went on to become: He still lives in the old family home, with his wife, Teresa.

335 Ulam asked Fermi what he thought: Ulam, *Adventures of a Mathematician,* 234–235.

336 he participated in strenuous outdoor activities: See Glauber, "An Excursion with Enrico Fermi," 44–46. The topics Fermi chose to lecture on were quite varied, reflecting the breadth of his interests and expertise: "Schein showers"—cosmic photon showers; polarization of fast protons; cosmic radiation in spiral galactic arms; pion production in cyclotrons; and stellar structure. See EFREG, 50:4.

336 Glauber chose the sturdiest: Glauber, "An Excursion with Enrico Fermi," 46.

336 Fermi delivered his talks: The villa was restored in 2003 to its original splendor when the De Marchi family lived there, with the artwork and the antique furniture they bequeathed with the villa.

336 attendees could take walks: A delightful two-minute short film on the summer school, showing Fermi and other attendees, can be found at https://www.youtube.com/watch?v=JGs1lM1KvKA—the narration is in Italian, but some snippets of the talks, including Fermi's, are in English. Fermi shows no outward sign of the pain he must have been experiencing.

336 Several of the younger attendees: His former student and Manhattan Project colleague Bernard Feld edited the notes and tape recordings into a coherent paper, published as the final paper in *Collected Papers.* CPF II, 1004ff.

338 **new computers in machine language:** Machine language is the most primitive of all computer programming languages, the only language that early computers understood.

CHAPTER TWENTY-SIX: HOME TO DIE

340 **the first doctor who saw Fermi:** Wali, *Chandra,* 269–270.
341 **Leona Libby writes:** Libby, *Uranium People,* 20.
342 **"It was of course very difficult":** Wali, *Chandra,* 269.
342 **"Do we have to discuss this here?":** Ibid., 270.
342 **Yang wrote later of the visit:** Yang, CPF II, 674
342 **imminent demise with "Socratic serenity":** Segrè, *Mind Always in Motion,* 251–252.
343 **"At the end of the afternoon":** Ibid., 253.
343 **Teller arrived a few days later:** Blumberg and Owens, *Energy and Conflict,* 374–375.
343 **Knowing of Fermi's unhappiness:** EFREG, 16:2.
343 **Teller seemed to think the promise:** Blumberg and Owens, *Energy and Conflict,* 375.
343 **Fermi's death had a double impact:** Ibid., 375.
343 **Shortly after Fermi's death:** Teller, "Scientific World Pays Tribute to Fermi," 8.
344 **"He spoke of his approaching death":** Libby, *Uranium People,* 21.
344 **"so composed by death's approach":** Wigner, *Recollections of Eugene P. Wigner,* 278.
344 **Ulam was moved to tears:** Ulam, *Adventures of a Mathematician,* 237–238.
344 **Fermi was more proud of the publication:** Dash, *Life of One's Own,* 333. Maria Mayer recalls the incident slightly differently, with Enrico saying he was prouder of Laura's book than he was of *anything* he had accomplished.
344 **"with the greatest grace imaginable":** Ibid.
345 **Laura, who maintained a stoicism:** Toward the end, she warned Amaldi and Persico, who wanted to visit, that such a visit would not make sense—because he was under the influence of morphine and spent most of his time sleeping. Amaldi Archives.
345 **his friends, Chandrasekhar and Ulam:** Wali, *Chandra,* 269ff.; Ulam, *Adventures of a Mathematician,* 234ff.
345 **"We thank Thee for him":** EFREG, 7:7.
346 **"Actually, the Institute is his Institute":** Allison, "A Tribute to Enrico Fermi," 9–10.
347 **the archetypical Fermi moment:** Ibid.
347 **"He had had all the honors":** Segrè, "A Tribute to Enrico Fermi," 12.
347 **"To explore the mysteries of nature":** Anderson, "A Tribute to Enrico Fermi," 13.

348 **robust sense of his own capabilities:** See Chandrasekhar, "The Pursuit of Science," cited in Introduction.

348 **condolence letter to Laura:** LFREG, 2:8.

CHAPTER TWENTY-SEVEN: FERMI'S LEGACY

350 **Sam Allison stayed:** Cronin, interview with author, October 20, 2014.

352 **She even drafted an unpublished novel:** Private communication with Olivia Fermi.

352 **She eventually moved to an apartment:** See Olivia Fermi's essay on her grandmother's life: http://fermieffect.com/laura-fermi/laura-fermis-life/.

352 **Leona Libby recalls Laura's involvement:** Libby, *Uranium People,* 190. Libby was highly critical of Laura, noting that Laura supported civilian nuclear power when Enrico was alive. Laura explained that while Enrico was alive, she trusted his competence in these matters; now that he was gone, she had little faith in those responsible for the development of nuclear energy.

353 **She wondered, in 1954:** Laura Fermi, *Atoms,* 152–153.

353 **She is buried in the cemetery:** "Maria Fermi Sacchetti (age 60)," http://www.olgiateolona26giugno1959.org/10_lives/Sac_e.html. See also transcript of interview with Laura Fermi, TWOEF, 17.

353 **raise two children, Alice and Paul:** After her divorce, Alice changed her name to Olivia and adopted her grandparents' family name, Fermi.

353 **interview conducted for a CBS:** I am indebted to Olivia Fermi for providing me with a copy of the unedited interview.

354 **"Put your grandmother Laura":** "Laura Fermi's Life," The Fermi Effect, http://fermieffect.com/laura-fermi/laura-fermis-life/.

355 **His health problems may have:** Sarah Fermi, interview with author, June 1, 2016.

355 **Richard Garwin recalls lecturing:** Garwin, interview with author, May 22, 2014.

355 **She has two blogs:** The Fermi Effect (http://fermieffect.com/) and On the Neutron Trail (http://neutrontrail.com/).

355 ***Picturing the Bomb:*** Fermi and Samra, *Picturing the Bomb.*

356 **"Fermi Award is a Presidential award":** "The Enrico Fermi Award," US Department of Energy, http://science.energy.gov/fermi.

357 ***The Collected Papers:*** Purists may quibble about the editing. Some of the papers were slightly rearranged in order to make more sense within the context of the volume.

357 **Steven Weinberg, a first step:** Weinberg has written extensively and brilliantly for a general audience. See, particularly, his classic *Dreams of a Final Theory*. He recently published a provocative meditation on the flaws in quantum theory: "The Trouble with Quantum Mechanics," *New York Review of Books,* January 19, 2017, http://www.nybooks.com/articles/2017/01/19/trouble-with-quantum-mechanics/.

358 **exploration of the strong force:** See Brown, Dresden, and Hoddeson, *Pions to Quarks,* and Pickering, *Constructing Quarks.*

358 **Subsequent advances:** Group theory suggested that none of these "elementary" particles were actually elementary—that they were composed of truly elementary particles, called quarks.

358 **Never a fan of group theory:** He famously began a seminar on group theory by going through the concepts in alphabetical order. When pressed, he explained that group theory is just a series of definitions, and it made as much sense to go alphabetically through those definitions as it did to develop the field theorem by theorem. Lee, "Reminiscence of Chicago Days," 198–199.

358 **rules of mirror-image symmetry:** Physicists believed that all interactions obeyed left-right symmetry. That is, viewing an image of a particle interaction, one would not be able to tell if it were the actual interaction or a mirror image of that interaction. This turns out not to be the case for weak interactions, where left-handed spin dominates. This asymmetry was first proposed by T. D. Lee and Chen Ning Yang during the summer of 1956 at Brookhaven Lab and confirmed experimentally by Chien-Shiung Wu at Columbia later that year. It won Lee and Yang the 1957 Nobel Prize.

359 **"in the stone age":** Henry Frisch, interview with author, January 9, 2017.

360 **DiMaggio's fifty-six game hitting streak:** Samuel Arbesman and Stephen H. Strogatz, "A Monte Carlo Approach to Joe DiMaggio and Streaks in Baseball," https://arxiv.org/ftp/arxiv/papers/0807/0807.5082.pdf.

360 **The Italian Physics Society presents:** "Enrico Fermi Prize," *Wikipedia,* last updated November 26, 2016, accessed September 12, 2016, https://en.wikipedia.org/wiki/Enrico_Fermi_Prize.

362 **contaminated by the radioactive residue:** This residue has had some unintended consequences for biological studies. One of the by-products of these tests is increased carbon-14 in the air; to the extent that it has since been ingested by all life-forms, this has resulted in complications with carbon-14 dating of modern life-forms.

362 **Some 450 electric power reactors:** These and other figures here are taken from "The Database on Nuclear Power Reactors," International Atomic Energy Agency, https://www.iaea.org/pris/ (accessed September 29, 2016).

363 **Reactors have purposes other than:** "Radioisotopes in Medicine," World Nuclear Association, last updated December 28, 2016, accessed February 2, 2017, http://world-nuclear.org/information-library/non-power-nuclear-applications/radioisotopes-research/radioisotopes-in-medicine.aspx. Amaldi's son, Ugo, has worked extensively in this area and has written of the importance of the slow-neutron work for medical physics. See Ugo Amaldi, "Slow Neutrons at Via Panisperna," 145–168.

364 **The Fermi Paradox:** Robert Gray correctly points out that the "paradox" was not actually about the existence of extraterrestrial life but about the possibility of intergalactic space travel. Fermi's quip—"Where are they?"—indicated that if such travel were possible, aliens would certainly have visited us by now. Of course, if extraterrestrial life does not exist, that would also be an explanation. Gray, "The Fermi Paradox."

365 **"the last man who knew everything":** Geoffrey Chew, interview with author, May 6, 2014; Ugo Amaldi, interview with author, June 7, 2016.

BIBLIOGRAPHY

Anyone who attempts to write a biography of Enrico Fermi immediately discovers that he is stepping into some very big shoes. Laura Fermi knew her husband better than anyone else, and her memoir, *Atoms in the Family,* published the year Fermi died, cannot be improved upon as a portrait of a marriage. It is an engaging, colorful account, but the reader immediately knows that it will be relatively light on the science. For example, the famous paper on beta radiation is discussed in a single, off-hand sentence. In addition, some of the historical incidents—for example, the goldfish pond episode on the weekend of the discovery of slow neutrons—are more difficult to corroborate than they should be.

The other set of shoes that any biographer must fill are those of Emilio Segrè, whose 1970 book, *Enrico Fermi: Physicist,* has remained the classic account for more than four decades. Segrè knew Fermi well during a particularly active phase in Fermi's scientific life, and thus writes with an appropriate level of authority. There are, however, several factors that argue for a more up-to-date treatment of Fermi's life. One is that Segrè's book is a bit intimidating for the general reader who may not know what the Zeeman effect is or may not understand Wigner/Jordan creation and destruction operators. Another is that—as I have tried to argue—Fermi's legacy has grown considerably since Segrè published his work. A third is that new facts of some relevance have emerged since 1970. One clear example is the 1972 revelation of Werner Heisenberg that Fermi tried to persuade him to move to the United States in the summer of 1939, just prior to the war. There are many others.

But still the task of coming up with something new to say, above and beyond what his wife and one of his closest students had to offer, is daunting. Yet the time seems to be right, as several other recent authors have also recognized.

Archival material on Enrico Fermi is scattered among several important libraries. The archives at the University of Chicago's Regenstein Library are truly a thing of beauty. In over sixty boxes one finds an enormous quantity of material, with particular focus on the US portion of Fermi's career. The archive is exquisitely preserved and organized. The other great archive is that held in the Domus Galilaeana in Pisa, Italy, where Edoardo Amaldi deposited everything that existed at the University of Rome directly relating to Fermi in 1955, after Fermi's passing. Other archives of relevance include archives at the physics department of the University of Rome; archives at the *Scuola Normale Superiore* and the University of Pisa, related to his student years; archives at the Villa Corsini, home to all the archives of the *Accademia dei Lincei* and the *Reale Accademia d'Italia;* and the Italian national archives. There is also important material held by the Los Alamos Historical Society, the archives of the American Institute of Physics, the US National Archives, and various other university collections, most notably the Bancroft Library at Berkeley and the Butler Library at Columbia.

In 1986, Richard Rhodes wrote his Pulitzer Prize–winning history of the Manhattan Project, *The Making of the Atomic Bomb,* which remains the classic treatment of an important period in Fermi's career. Richard Hewlett and Oscar Anderson wrote an accessible history of the US atomic program, which despite occasional flaws remains a useful source. The US Army also published an official history, by Vincent Jones, as part of its fine series of histories of the US Army in World War II.

Another extraordinary resource is the declassified official report of the Manhattan Engineering District, declassified in the late 1970s and available at https://www.osti.gov/opennet/manhattan_district.jsp. It is highly technical and is not written with the role of individual participants particularly in mind. In the volumes on the X-10 and the B reactors, Fermi's name is not mentioned, although he clearly had an important role in the design, construction, and operation of both. Another resource, published in 1994, is the journal that Berkeley chemist Glenn T. Seaborg kept from 1939–1946, during which time he played a central role in the Manhattan Project, most notably as the person whose team discovered plutonium and who developed the techniques for separating the metal out of reactor by-products. These journals contain, to my knowledge, the only records of deliberations at the Technical Committee of the Met Lab, where so many important technical decisions were made.

Then, of course, there are the dozens of memoirs written by those who knew Fermi at various points in their lives. Like all memoir material, they are primary in the sense that they are not interpretations of someone else's work, but they must always be regarded with caution. Memories are sometimes unreliable, and almost everyone who came into contact with Fermi had an incentive to portray their relationship with him in the best possible light. Some of the memoirs date from the fifties and sixties; others are collected in volumes celebrating his centenary. The portraits they paint are fairly consistent, which helps the biographer enormously. I found the 2004 volume edited by James Cronin, *Fermi Remembered,* to be particularly helpful, but the reader will find others, listed in the bibliography below.

There is, of course, a vast number of publications in the Italian language. I have listed them here along with the sources I have used in English; for a reader with a passing knowledge of the Italian language and a good Italian-English dictionary, these sources are extremely useful. I have also benefited greatly from the translation app offered by Google.

Fermi himself would have been amused to discover a continued interest in the events of his dramatic, but all-too-brief, life. One imagines him pointing to *The Collected Papers,* insisting that these two volumes constitute his real biography. *The Collected Papers,* in spite of an occasional editorial lapse, is a remarkable tribute prepared by the students and colleagues who knew him best. The commentary on the important papers, written by people like Segrè, Rasetti, Persico, Anderson, and Libby, among others, allows the reader to place the papers in historical and scientific context. The beautiful two-volume set reflects Fermi's enormous productivity well into the 1950s. The first volume, of course, is largely in the Italian language, but the most important papers of that era—the statistics paper and the beta decay paper, for example—can be found on the Internet.

I have written that Fermi left no diaries, but that is not quite accurate. His diaries are his handwritten notebooks, which exist in the archives at the Domus Galilaeana and the Regenstein Library. Every morning he jotted down his thoughts on physics, covering virtually every aspect of physics he could think of, and this fact tells us as much about his personal life as almost anything else. I am particularly excited that the Domus Galilaeana has scanned all the Fermi notebooks in their archives, and will make them available online in due course. This will make an enormous amount of archival material available to scholars worldwide.

What follows is a list of books and articles that I have found helpful in the writing of this book. Some have been more central than others, but all of them have added some particular aspect of the overall portrait I have tried to paint.

ABBREVIATIONS

CPF I/CPF II Edoardo Amaldi, Herbert L. Anderson, Enrico Persico, Franco Rasetti, Emilio Segrè, Cyril S. Smith, Albert Wattenberg, eds. *Enrico Fermi: The Collected Papers.* Chicago and Rome: University of Chicago Press and Accademia Nazionale dei Lincei. Two Volumes: Volume I, 1962. Volume II, 1965.

EFDG Enrico Fermi Archives, Domus Galilaeana, Via Santa Maria 26, 56126 Pisa, Italy.

EFREG The Enrico Fermi Collection. The Special Collections Research Center, Regenstein Library, University of Chicago, 1100 E. 57th Street, Chicago, IL 60637.

INSREG The University of Chicago. Institute for Nuclear Studies. Cyclotron. Records, 1946–1952. The Special Collections Research Center, Regenstein Library, The University of Chicago, 1100 E. 57 Street, Chicago, IL 60637

LFREG The Laura Fermi Papers. The Special Collections Research Center, Regenstein Library, University of Chicago, 1100 E. 57th Street, Chicago, IL 60637.

MED Gavin Haddin, ed. *Manhattan District History,* eight books, 1944 onward. See https://www.osti.gov/opennet/manhattan_district.jsp.

SCREG Papers of Subrahmanyan Chandrasekhar. The Special Collections Research Center, Regenstein Library, The University of Chicago, 1100 E. 57 Street, Chicago, IL 60637.

TWOEF "The World of Enrico Fermi." Archival notes for documentary of same name produced by Gerald Holton and the Canadian National Broadcasting Company. AIP Neils Bohr Center for the History of Physics, Ellipse Drive, College Park, MD, 20740.

BOOKS

Aczel, Amir D. *Uranium Wars: The Scientific Rivalry that Created the Nuclear Age.* New York: Palgrave Macmillan, 2009.

Alvarez, Luis. *Alvarez: Adventures of a Physicist.* New York: Basic Books, 1987.

Amaldi, Ugo. "Slow Neutrons at Via Panisperna: The Discovery, the Production of Isotopes, and the Birth of Nuclear Medicine." In Bernardini et al., *Proceedings of the International Conference "Enrico Fermi and the Universe of Physics,"* 145–168.

Badash, Lawrence, John O. Hirschfelder, and Herbert P. Broida, eds. *Reminiscences of Los Alamos 1943–1945: Studies in the History of Modern Science 5.* Boston: D. Reidel, 1980.

Badoglio, Pietro. *The War in Abyssinia.* London: Methuen, 1937.

Baggot, Jim. *The First War of Physics: The Secret History of the Atom Bomb, 1939–1949.* New York: Pegasus Books, 2010.

Barker, A. J. *The Rape of Ethiopia.* New York: Ballantine, 1971.

Bassani, G. Franco, ed. *Fermi, Maestro e Didatta: Celebrazione del Centenario della Nascita di Enrico Fermi.* Varenna, Italy: Societa di Fisica Italiana, 2001.

Battimelli, Giovanni. *L'eredita di Fermi: Storia fotografica del 1927 al 1959 dagli archive di Edoardo Amaldi.* Rome: Editori Reuniti, 2003.

———. "Funds and Failures: the Political Economy of Fermi's Group." In Bernardini et al., *Proceedings of the International Conference "Enrico Fermi and the Universe of Physics,"* 169–184.

Bernardini, Carlo. "Enrico Fermi e il trattato di O. D. Chwolson." In Bassani, *Fermi, Maestro e Didatta*, 15–19.

Bernardini, Carlo, and Luisa Bonolis, eds. *Enrico Fermi: His Work and Legacy.* Bologna: Societa Italiana di Fisica, 2004.

Bernardini, Carlo, L. Bonolis, G. Ghisu, D. Savelli, and L. Falera. *Proceedings of the International Conference "Enrico Fermi and the Universe of Physics." Rome, September 29–October 2, 2001.* Rome: Accademia Nazionale dei Lincei and Istituto Nazionale di Fisica Nucleare, 2003.

Bernstein, Jeremy. *Hitler's Uranium Club: The Secret Recordings at Farm Hall.* Woodbury, NY: American Institute of Physics, 1996.

———. *A Palette of Particles.* Cambridge, MA: Belknap Press of Harvard University, 2013.

Bird, Kai, and Martin J. Sherwin. *American Prometheus: The Triumphs and Tragedy of J. Robert Oppenheimer.* New York: Vintage, 2006.

Blumberg, Stanley A., and Gwinn Owens. *Energy and Conflict: The Life and Times of Edward Teller.* New York: G. P. Putnam and Sons, 1976.

Bohr, Niels. "Atomic Stability and Conservation Laws." In *Convegno di Fisica Nucleare,* 119–130.

Bonolis, Luisa. "Enrico Fermi's Scientific Work." In Bernardini and Bonolis, *Enrico Fermi,* 315–393.

Boorse, Henry A., Lloyd Motz, and Jefferson Hane Weaver. *The Atomic Scientists: A Biographical History.* New York: Wiley, 1989.

Born, Max. *Experiment and Theory in Physics.* New York: Dover, 1956.

———. *My Life & My Views.* New York: Charles Scribner's Sons, 1968.

Bosworth, R. J. B. *Mussolini's Italy: Life Under the Fascist Dictatorship, 1915–1945.* New York: Penguin Books, 2006.

Brode, Bernice. *Tales of Los Alamos: Life on the Mesa 1943–1945.* Los Alamos, NM: Los Alamos Historical Society, 1997.

Brown, Laurie M., Max Dresden, and Lillian Hoddeson, eds. *Pions to Quarks: Particle Physics in the 1950s. Based on a Fermilab Symposium.* Cambridge: Cambridge University Press, 1989.

Brown, Laurie M., and Lillian Hoddeson, eds. *The Birth of Particle Physics. Based on a Fermilab Symposium.* Cambridge: Cambridge University Press, 1983.

Bruzzanti, Giuseppe. *Enrico Fermi: The Obedient Genius.* Translated by Ugo Bruzzo. New York: Springer (Birkhauser), 2016.

Cabibbo, Nicola. "Fermi's Tentativo and Weak Interactions." In Bernardini et al., *Proceedings of the International Conference "Enrico Fermi and the Universe of Physics,"* 305–316.

———. "Weak Interactions." In Bernardini and Bonolis, *Enrico Fermi,* 138–150.

Canali, Mauro. *Le spie del regime.* Bologna: Societa editrice il Mulino, 2004.

Cannistraro, Philip V., and Brian Sullivan. *Il Duce's Other Woman.* New York: William Morrow, 1993.

Carr, Alan B. *The Forgotten Physicist: Robert F. Bacher, 1905–2004.* Los Alamos, NM: Los Alamos Historical Society, 2008.

Casalbuoni, Roberto, Giovanni Frosali, and Giuseppe Pelosi. *Enrico Fermi a Firenze: Le "Lezione di Meccanica Razionale" al bienno propedeutico agli studi di Ingegneria: 1924–1926.* Florence: Firenze University Press, 2014.

Chamberlin, Owen. "A Brief Reminiscence of Enrico Fermi." In Cronin, *Fermi Remembered,* 186–187.

Chew, Geoffrey. "Personal Recollections." In Cronin, *Fermi Remembered,* 187–189.

Close, Frank. *Half-Life: The Divided Life of Bruno Pontecorvo, Physicist or Spy.* New York: Basic Books, 2015.

———. *Neutrino.* New York: Oxford University Press, 2010.

Colangelo, Giorgio, and Massino Temporelli. *La banda di via Panisperna: Fermi, Majorana e I fisici che hanno cambiato la storia.* Milan: Ulrico Hoepli, 2014.

Compton, Arthur Holly. *Atomic Quest: A Personal Narrative.* New York: Oxford University Press, 1956.

Conant, Jennet. *Tuxedo Park: A Wall Street Tycoon and the Secret Palace of Science that Changed the Course of World War II.* New York: Simon & Schuster, 2002.

Convegno di Fisica Nucleare Ottobre 1931-IX. Rome: Reale Accademia d'Italia/Fondazione Alessandro Volta, 1932.

Crease, Robert P., and Charles C. Mann. *The Second Creation: Makers of the Revolution in 20th-Century Physics.* New York: Macmillan, 1986.

Cronin, James W., ed. *Fermi Remembered.* Chicago: University of Chicago Press, 2004.

Cropper, William H. *The Great Physicists: The Life and Times of Leading Physicists from Galileo to Hawking.* Oxford: Oxford University Press, 2001.

D'Abro, A. *The Rise of the New Physics: Its Mathematical and Physical Theories.* 2 vols. New York: Dover, 1951.

Dardo, Mauro. *Nobel Laureates and Twentieth Century Physics.* New York: Cambridge University Press, 2004.

Dash, Joan. *A Life of One's Own: Three Gifted Women and the Men They Married.* New York: Harper & Row, 1973.

Davis, Nuel Pharr. *Lawrence and Oppenheimer.* New York: Simon & Schuster, 1968.

De Latil, Pierre. *Enrico Fermi: The Man and His Theories. A Profile in Science.* New York: Paul S. Eriksson, 1966.

Del Gamba, Valeria. *Il raggazzo di via Panisperna: L'avventurosa vita del fisico Franco Rasetti.* Turin: Bollati Boringhieri, 2007.

Dyson, Freeman. *From Eros to Gaia.* New York: Pantheon, 1992.

Eiduson, Bernice T., and Linda Beckman, eds. *Science as a Career Choice.* New York: Russell Sage, 1973.

Enz, Charles P. *No Time to Be Brief: A Scientific Biography of Wolfgang Pauli.* New York: Oxford University Press, 2002.

Farmelo, Graham. *Churchill's Bomb: How the United States Overtook Britain in the First Nuclear Arms Race.* New York: Basic Books, 2013.

———. *The Strangest Man: The Hidden Life of Paul Dirac, Mystic of the Atom.* New York: Basic Books, 2009.

Fermi, Enrico. *Collected Papers. Volume I: Italy 1921–1938. Volume II: United States 1939–1954.* Chicago and Rome: University of Chicago Press and Accademia dei Lincei, 1962, 1965.

———. *Elementary Particles.* London: Geoffrey Cumberlege Oxford University Press, 1951.

———. *Notes on Quantum Mechanics.* Chicago: University of Chicago Press, 1961.

———. *Thermodynamics.* New York: Dover, 1956.

Fermi, Laura. *Atoms in the Family: My Life with Enrico Fermi, Architect of the Atomic Age.* Chicago: University of Chicago Press, 1954.

———. "The Fermis' Path to Los Alamos." In Badash, Hirschfelder, and Broida, *Reminiscences,* 96.

———. *Mussolini.* Chicago: University of Chicago Press, 1961.

Fermi, Rachel, and Esther Samra. *Picturing the Bomb: Photographs from the Secret World of the Manhattan Project.* New York: Harry N. Abrams, 1995.

Fernandez, Bernard, and Georges Ripka. *Unravelling the Mystery of the Atomic Nucleus: A Sixty Year Journey, 1896–1956.* Berlin: Springer, 2012.

Feynman, Richard. *"Surely You're Joking, Mr. Feynman!" Adventures of a Curious Character.* New York: Norton, 1985.

Forbes, Nancy, and Basil Mahon. *Faraday, Maxwell, and the Electromagnetic Field: How Two Men Revolutionized Physics.* Amherst, NY: Prometheus Books, 2014.

Ford, Kenneth W. *Building the H Bomb: A Personal History.* Singapore: World Scientific Publishing, 2015.

Franklin, Allan. *Are There Really Neutrinos? An Evidential History.* Cambridge, MA: Perseus Books, 2001.

———. *What Makes A Good Experiment? Reasons and Roles in Science.* Pittsburgh, PA: University of Pittsburgh Press, 2016.

Friedman, Robert Marc. *The Politics of Excellence: Behind the Nobel Prize in Science.* New York: Henry Holt, 2001.

Frisch, Otto. *What Little I Remember.* Cambridge: Cambridge University Press, 1979.

Galison, Peter. *How Experiments End.* Chicago: University of Chicago Press, 1987.

———. *Image & Logic: A Material Culture of Microphysics.* Chicago: University of Chicago Press, 1977.

Gamow, George. *Thirty Years that Shook Physics: The Story of Quantum Theory.* New York: Dover, 1985.

Garwin, Richard. "Working with Fermi at Chicago and Postwar Los Alamos." In Cronin, *Fermi Remembered,* 143–148.

Gell-Mann, Murray. "No Shortage of Memories." In Cronin, *Fermi Remembered,* 149–152.

Gleick, James. *Genius: The Life and Science of Richard Feynman.* New York: Pantheon, 1992.

Glick, Thomas F., ed. *The Comparative Reception of Relativity.* Dordrecht, Holland: D. Reidel Publishing, 1987.

Goodchild, Peter. *J. Robert Oppenheimer: Shatterer of Worlds.* Boston: Houghton Mifflin, 1981.

Greenberg, Daniel S. *The Politics of Pure Science.* 2nd ed. Chicago: University of Chicago Press, 1999.

Gregor, A. James. *Giovanni Gentile: Philosopher of Fascism.* New Brunswick, NJ: Transaction Publishers, 2001.

Groves, Leslie R. *Now It Can Be Told.* New York: Harper, 1962.

Guerra, Francesco, and Nadia Robotti. *Enrico Fermi e Il Quaderno Ritrovato: 20 marzo 1934—La vera storia della scoperta della Radioattività indotta da neutron.* Bologna, IT: Società Italiana di Fisica, 2015.

Hahn, Otto. *My Life. An Autobiography by the Great German Physicist.* London: Macdonald, 1970.

Heisenberg, Werner. *Physics and Beyond: Encounters and Conversations.* New York: Harper Torchbook, 1972.

Herken, Gregg. *Brotherhood of the Bomb: The Tangled Lives and Loyalties of Robert Oppenheimer, Ernest Lawrence, and Edward Teller.* New York: Holt/John McRae, 2002.

Hewlett, Richard G., and Oscar E. Anderson Jr. *A History of the United States Atomic Energy Commission, Volume I: The New World, 1939 / 1946.* University Park: Pennsylvania State University Press, 1962.

Hibbert, Christopher. *Il Duce: The Life of Benito Mussolini.* Boston: Little, Brown, 1962.

Hoddeson, Lilian, Paul W. Henriksen, Roger A. Meade, and Catherine Westfall. *Critical Assembly: A Technical History of Los Alamos During the Oppenheimer Years, 1943–1945.* New York: Cambridge University Press, 1993.

Holl, Jack M., with the assistance of Richard G. Hewlett and Ruth R. Harris. *Argonne National Laboratory 1946–96.* Urbana: University of Illinois Press, 1997.

Holton, Gerald. *The Scientific Imagination.* 2nd ed. Cambridge, MA: Harvard University Press, 1998.

———. *Victory and Vexation in Science: Einstein, Bohr, and Others.* Cambridge, MA: Harvard University Press, 2005.

Jammer, Max. *The Conceptual Development of Quantum Mechanics.* New York: McGraw-Hill, 1966.

Johnson, George. *Strange Beauty: Murray Gell-Mann and the Revolution in Twentieth Century Physics.* New York: Alfred A. Knopf, 1999.

Jones, Vincent C. *Manhattan: The Army and the Atomic Bomb, Special Studies, United States Army in World War II.* Washington, DC: Center of Military History, US Army, 1985.

Kaiser, David. *No End Save Victory: How FDR Led the Nation into War.* New York: Basic Books, 2014.

Kragh, Helge. *Quantum Generations: A History of Physics in the Twentieth Century.* Princeton, NJ: Princeton University Press, 1999.

———. *Dirac: A Scientific Biography.* Cambridge: Cambridge University Press, 1990.

Kumar, Manjit. *Quantum: Einstein, Bohr, and the Great Debate About the Nature of Reality.* New York: W. W. Norton, 2008.

Lanouette, William, with Bela Silard. *Genius in the Shadows: A Biography of Leo Szilard, the Man Behind the Bomb.* New York: Charles Scribner's Sons, 1992.

Lee, T. D. "Reminiscences of Chicago Days." In Cronin, *Fermi Remembered,* 197–201.

Libby, Leona Marshall. *The Uranium People.* New York: Crane, Russak/Charles Scribner's Sons, 1979.

Lightman, Alan. *Dance for Two: Selected Essays.* New York: Pantheon, 1996.

Lilienthal, David E. *The Journals of David E. Lilienthal. Volume One: The TVA Years 1939–1945. Volume Two: The Atomic Energy Years 1945–1950.* New York: Harper & Row, 1964.

Lyttelton, Adrian. *The Seizure of Power: Fascism in Italy 1919–1929.* New York: Charles Scribner's Sons, 1973.

MacPherson, Malcolm C. *Time Bomb: Fermi, Heisenberg, and the Race for the Atomic Bomb.* New York: Dutton, 1986.

Magueijo, João. *A Brilliant Darkness: The Extraordinary Life and Mysterious Disappearance of Ettore Majorana, the Troubled Genius of the Nuclear Age.* New York: Basic Books, 2009.

Maltese, Giulio. *Enrico Fermi in America: Una biographia scientifica: 1938–1954.* Bologna: Zanichelli, 2003.

McMillan, Elsie Blumer. *The Atom and Eve.* New York: Vantage Press, 1995.

Mehra, Jagdish. *The Solvay Conferences on Physics*. Dordrecht, Holland: D. Reidel, 1975.

Meier, Carl Alfred, ed. *Atom and Archetype: The Pauli/Jung Letters, 1932–1958*. Translated by David Roscoe. Princeton, NJ: Princeton University Press, 2016.

Miller, Arthur I. *137: Jung, Pauli, and the Pursuit of a Scientific Obsession*. New York: W. W. Norton, 2009.

Monk, Ray. *Robert Oppenheimer: Inside the Center*. New York: Random House, 2013.

Moore, Walter John. *Schrodinger: Life and Thought*. Cambridge: Cambridge University Press, 1989.

Moss, Norman. *Men Who Play God: The Story of the H-Bomb and How the World Came to Live with It*. New York: Harper & Row, 1968.

Nelson, Craig. *The Age of Radiance: The Epic Rise and Dramatic Fall of the Atomic Era*. New York: Charles Scribner's Sons, 2014.

Orear, Jay. *Enrico Fermi: The Master Scientist*. Ithaca, NY: Internet-First University Press (an affiliate of Cornell University), 2004.

———. "My First Meetings with Fermi." In Cronin, *Fermi Remembered*, 201–203.

Painter, Borden W. *Mussolini's Rome: Rebuilding the Eternal City*. New York: Palgrave Macmillan, 2005.

Pais, Abraham. *Inward Bound: Of Matter and Forces in the Physical World*. New York: Oxford University Press, 1986.

———. *Niels Bohr's Times, in Physics, Philosophy, and Polity*. Oxford: Clarendon Press, 1991.

Parisi, Giorgio. "Fermi's Statistics." In Bernardini and Bonolis, *Enrico Fermi*, 67–74.

Peierls, Rudolf E. *Atomic Histories: A Walk through the Beginnings of the Atomic Age with One of Its Pioneers*. Woodbury, NY: American Institute of Physics, 1997.

Pickering, Andrew. *Constructing Quarks: A Sociological History of Particle Physics*. Chicago: University of Chicago Press, 1984.

Polenberg, Richard, ed. *In the Matter of J. Robert Oppenheimer: The Security Clearance Hearings*. Ithaca, NY: Cornell University Press, 2002.

Pontecorvo, Bruno. *Enrico Fermi: Ricordi d'allievi e amici*. Pordenone: Edizione Studio Tesi, 1993.

———. *Fermi e la fisica moderna*. Naples: La Citta del Sole, 2004.

Powers, Thomas. *Heisenberg's War: The Secret History of the German Bomb*. New York: Penguin, 1994.

Rabi, I. I., Robert Serber, Victor F. Weisskopf, Abraham Pais, and Glenn T. Seaborg. *Oppenheimer: The Story of One of the Most Remarkable Personalities of the 20th Century*. New York: Charles Scribner's Sons, 1969.

Raboy, Marc. *Marconi: The Man Who Networked the World*. New York: Oxford University Press, 2016.

Rhodes, Richard. *Dark Sun: The Making of the Hydrogen Bomb*. New York: Simon & Schuster, 1995.

———. *The Making of the Atomic Bomb.* New York: Touchstone/Simon & Schuster, 1986.

Rigden, John S. *Rabi: Scientist and Citizen.* New York: Basic Books, 1987.

Rosenfeld, Arthur. "Reminiscences of Fermi." In Cronin, *Fermi Remembered,* 203–205.

Rutherford, Ernest, James Chadwick, and C. D. Ellis. *Radiations from Radioactive Substances.* Cambridge: Cambridge University Press, 1930.

Schluter, Robert. "Three Reminiscences of Enrico Fermi." In Cronin, *Fermi Remembered,* 205–207.

Schram, Albert. *Railways and the Formation of the Italian State in the Nineteenth Century.* New York: Cambridge University Press, 1997.

Schweber, Silvan S. *Nuclear Forces: The Making of the Physicist Hans Bethe.* Cambridge, MA: Harvard University Press, 2012.

———. *QED and the Men Who Made It: Dyson, Feynman, Schwinger, and Tomonaga.* Princeton, NJ: Princeton University Press, 1994.

Sciascia, Leonardo. *La scomprosa di Majorana.* Milan: Adelphi, 1997.

Seaborg, Glenn T. *The Plutonium Story: The Journals of Professor Glenn T. Seaborg, 1939–1946.* Columbus, OH: Battelle Press, 1994.

Seaborg, Glenn T., with Eric Seaborg. *Adventures in the Atomic Age: From Watts to Washington.* New York: Farrar, Straus, & Giroux, 2001.

Sebastiani, Fabio, and Francesco Cordella. "Fermi toward Quantum Statistics (1923–1925)." In Bernardini et al., *Proceedings of the International Conference "Enrico Fermi and the Universe of Physics,"* 71–96.

Segrè, Claudio G. *Atoms, Bombs & Eskimo Kisses: A Memoir of Father and Son.* New York: Viking, 1995.

Segrè, Emilio. *Enrico Fermi, Physicist.* Chicago: University of Chicago Press, 1970.

———. *A Mind Always in Motion: The Autobiography of Emilio Segre.* Berkeley: University of California Press, 1993.

———. "Nuclear Physics in Rome." In Stuewer, *Nuclear Physics in Retrospect,* 35–62.

Segrè, Gino, and Bettina Hoerlin. *The Pope of Physics: Enrico Fermi and the Birth of the Atomic Age.* New York: Henry Holt & Co., 2016.

Serber, Robert. *The Los Alamos Primer: The First Lectures on How to Build an Atomic Bomb.* With Introduction by Richard Rhodes. Berkeley: University of California Press, 1992.

Shepley, James, and Clay Blair, Jr. *The Hydrogen Bomb: The Men, the Menace, the Mechanism.* Philadelphia: David McKay, 1954.

Sherwin, Martin J. *A World Destroyed: Hiroshima and Its Legacies.* 3rd ed. Stanford, CA: Stanford University Press, 2004.

Shils, Edward, ed. *Remembering the University of Chicago: Teachers, Scientists, and Scholars.* Chicago: University of Chicago Press, 1991.

Shurkin, Joel N. *True Genius: The Life and Work of Richard Garwin.* Amherst, NY: Prometheus Books, 2017.

Sigal, Leon V. *Fighting to a Finish: The Politics of War Termination in the United States and Japan, 1945.* Ithaca, NY: Cornell University Press, 1988.

Sime, Ruth Lewin. *Lise Meitner: A Life in Physics.* Berkeley: University of California Press, 1997.

Smith, Alice Kimball. *A Peril and a Hope.* Chicago: University of Chicago Press, 1965.

Smith, Alice Kimball, and Charles Weiner, eds. *Robert Oppenheimer: Letters and Recollections.* Stanford, CA: Stanford University Press, 1995.

Smith, Dennis Mack. *Mussolini.* New York: Alfred A. Knopf, 1962.

Smyth, Henry D. *Atomic Energy for Military Purposes.* Princeton, NJ: Princeton University Press, 1946.

Snow, C. P. *The Physicists: A Generation that Changed the World.* Boston: Little, Brown, 1981.

———. *The Two Cultures and the Scientific Revolution.* New York: Cambridge University Press, 1961.

Steeper, Nancy Cook. *Gatekeeper to Los Alamos: Dorothy Scarritt McKibbin, a Biography of a Great Lady of Santa Fe.* Los Alamos, NM: Los Alamos Historical Society, 2003.

Steinberger, Jack. *Learning about Particles: 50 Privileged Years.* Berlin: Springer, 2005.

Stille, Alexander. *Benevolence and Betrayal: Five Italian Jewish Families under Fascism.* New York: Summit Books, 1991.

Stimson, Henry L., and McGeorge Bundy. *On Active Service in War and Peace.* New York: Octagon Books, 1971. Reprint, New York: Harper & Row, 1948.

Strauss, Lewis L. *Men and Decisions.* Garden City, NY: Doubleday, 1962.

Stuewer, Roger H., ed. *Nuclear Physics in Retrospect: Proceedings of a Symposium on the 1930s.* Minneapolis: University of Minnesota Press, 1979.

Telegdi, V. L. "Enrico Fermi: 1901–1954." In Shils, *Remembering the University of Chicago,* 110–129.

———. "Reminiscences of Enrico Fermi." In Cronin, *Fermi Remembered,* 171–173.

Teller, Edward. *The Legacy of Hiroshima.* New York: Doubleday, 1962.

Ulam, Stanislaw M. *Adventures of a Mathematician.* New York: Charles Scribner's Sons, 1976.

United States Atomic Energy Commission. *In the Matter of J. Robert Oppenheimer: Transcript of Hearing before Personnel Security Board, Washington, D.C., April 12, 1954 through May 6, 1954.* Washington, DC: US Government Printing Office, 1954.

Van der Waerden, Bartel Leendert, ed. *Sources of Quantum Mechanics.* New York: Dover, 1968.

Veltman, Martinus. *Facts and Mysteries in Elementary Particle Physics.* River Edge, NJ: World Scientific, 2003.

Vergara Caffarelli, Roberto. *Enrico Fermi: Immagini e documenti.* Pisa: La Limonaia, 2002.

———. "Enrico Fermi al Liceo Umberto I di Roma e all'Università di Pisa." In Bassani, *Fermi, Maestro e Diddata,* 8–15.

Wali, Kameshwar C. *Chandra: A Biography of S. Chandrasekhar*. Chicago: University of Chicago Press, 1991.

Wattenberg, Al. "Fermi as My Chauffeur (Fermi at Argonne National Laboratory and Chicago, 1946–1948)." In Cronin, *Fermi Remembered,* 173–180.

Weart, Spencer R., and Gertrude Weiss Szilard. *Leo Szilard: His Version of the Facts. Selected Recollections and Correspondence. Volume II.* Cambridge, MA: MIT Press, 1978.

Weinberg, Steven A. *Dreams of a Final Theory: The Search for the Fundamental Laws of Nature.* New York: Pantheon, 1992.

Weiner, Charles, assisted by Elspeth Hart. *Exploring the History of Nuclear Physics. AIP Conference Proceedings No. 7.* New York: American Institute of Physics, 1972.

Weisskopf, Victor. *The Joy of Insight: Passions of a Physicist.* New York: Basic Books, 1991.

Wheeler, John Archibald, with Kenneth Ford. *Geons, Black Holes, and Quantum Foam: A Life in Physics.* New York: W. W. Norton, 1998.

Wigner, Eugene. *Symmetries and Reflections.* Woodbridge, CT: Oxbow Press, 1979.

Wigner, Eugene P., with Andrew Szanton. *The Recollections of Eugene P. Wigner as Told to Andrew Szanton.* New York: Plenum Press, 1992.

Wilczek, Frank. "Fermi and the Elucidation of Matter." In Cronin, *Fermi Remembered,* 34–51.

Wilson, Jane S., and Charlotte Serber, eds. *Standing By and Making Do: Women of Wartime Los Alamos.* 2nd ed. Los Alamos, NM: Los Alamos Historical Society, 2008.

Wright, Courtney. "Fermi in Action." In Cronin, *Fermi Remembered,* 180–182.

Wyden, Peter. *Day One: Before Hiroshima and After.* New York: Simon & Schuster, 1984.

Yang, Chen Ning. "Reminiscences of Enrico Fermi." In Cronin, *Fermi Remembered,* 241–244.

Yodh, Guarang. "This Account Is Not According to the Mahabharata!" In Cronin, *Fermi Remembered,* 245–253.

Zamagni, Vera. *The Economic History of Italy 1860–1990: Recovery After Decline.* Oxford: Clarendon Press, 1993.

Zuckerman, Harriet. *Scientific Elites: Nobel Laureates in the United States.* New York: Free Press, 1977.

ARTICLES

Agnew, Harold. "Scientific World Pays Homage to Fermi." *Santa Fe New Mexican,* January 6, 1955, 8.

Allardice, Corbin, and Edward R. Trapnell. "The First Pile." *International Atomic Energy Agency Bulletin* Special Number 4–0 (December 1962): 41–47. https://www.iaea.org/sites/default/files/publications/magazines/bulletin/bull4–0/04005004147su.pdf.

Allison, Samuel K. "Enrico Fermi, 1901–1954: A Biographical Memoir." In *Biographical Memoirs,* 123–155. Washington, DC: National Academy of Sciences, 1957. http://www.nasonline.org/publications/biographical-memoirs/memoir-pdfs/fermi-enrico.pdf.

———. "Initiation of the Chain Reaction: The Search for Pure Materials." *International Atomic Energy Agency Bulletin* Special Number 4–0 (December 1962): 12–14.

———. "Scientific World Pays Homage to Fermi." *Santa Fe New Mexican,* January 6, 1955, 8.

———. "A Tribute to Enrico Fermi." *Physics Today* 8, no. 1 (January 1955): 9–10.

Anderson, Herbert L. "Meson Experiments with Enrico Fermi." *Reviews of Modern Physics* 27, no. 3 (July 1955): 269–272.

———."A Tribute to Enrico Fermi." *Physics Today* 8, no. 1 (January 1955): 12–13.

Anderson, Herbert L., and Samuel K. Allison. "From Professor Fermi's Notebooks." *Reviews of Modern Physics* 27, no. 3 (July 1955): 273–277.

Belloni, L. "On Fermi's Route to the Fermi-Dirac Statistics." *European Journal of Physics* 15 (1994): 102–109.

Bernstein, Barton J. "The Atomic Bombings Reconsidered." *Foreign Affairs* 74, no. 1 (January–February 1995): 132–152.

———. "Crossing the Rubicon: A Missed Opportunity to Stop the H-Bomb?" *International Security* 14, no. 2 (Fall 1989): 132–160.

———. "Four Physicists and the Bomb: The Early Years, 1945–1950." *Historical Studies in the Physical and Biological Sciences* 18, no. 2 (1988): 231–263.

Bethe, Hans A. "Hans Bethe Reviews Fermi's Work." *Santa Fe New Mexican,* January 6, 1955, 2, 8.

———. "Memorial Symposium Held in Honor of Enrico Fermi at the Washington Meeting of the American Physical Society, April 29, 1955." *Reviews of Modern Physics* 25, no. 3 (July 1955): 249ff.

———. "Oppenheimer: Where He Was There Was Always Life and Excitement." *Science* 155, no. 3766 (March 3, 1967): 1080–1084.

Bretscher, Egon, and John D. Cockcroft. "Enrico Fermi 1901–1954." *Biographical Memoirs of the Fellows of the Royal Society* 1 (November 1955): 68–78.

Brown, Laurie M., and Helmut Rechenberg. "Field Theories of Nuclear Forces in the 1930s: The Fermi-Field Theory." *Historical Studies in the Physical and Biological Sciences* 25, no. 1 (1994): 1–24.

Cambrioso, Alberto. "The Dominance of Nuclear Physics in Italian Science Policy." *Minerva* 23, no. 4 (December 1985): 464–484.

Cassidy, David C. "Cosmic Ray Showers, High Energy Physics, and Quantum Field Theories: Programmatic Interactions in the 1930s." *Historical Studies in the Physical Sciences* 12, no. 1 (1981): 1–39.

Cesareo, Roberto. "Dalla Radioattività Artificiale alla Fissione Nucleare: 1934–1939." *Physis: Rivista Internazionale di Storia della Scienza* XXXVII, fasc. 1 (2000): 209–230.

Chandrasekhar, Subramanyan. "The Pursuit of Science." *Minerva* 22, nos. 3/4 (September 1984): 410–420.

Clark, George W. "Bruno Benedetto Rossi." *Proceedings of the American Philosophical Society* 144, no. 3 (September2000): 329–341.

Cockcroft, John D. "The Early Days of the Canadian and British Atomic Energy Projects." *International Atomic Energy Agency Bulletin* Special Number 4–0 (December 1962): 18–20.

Conversi, Marcello, Ettore Pancini, and Oreste Piccioni. "On the Decay Process of Positive and Negative Mesons." *Physical Review* 68 (1945): 232.

———. "On the Decay Process of Positive and Negative Mesons." *Physical Review* 71 (1947): 209–210.

Crawford, Elisabeth, Ruth Lewin Sime, and Mark Walker. "A Nobel Tale of Postwar Injustice." *Physics Today* 50, no. 9 (September 1997): 26–32.

Darrigol, Olivier. "The Origin of Quantized Matter Waves." *Historical Studies in the Physical and Biological Sciences* 16, no. 2 (1986): 197–253.

———. "The Quantum Electrodynamic Analogy in Early Nuclear or the Roots of Yukawa's Theory." *Revue d'histoire des sciences* 41 nos. 3/4 (July–December 1988): 227–297.

Debye, Peter. "The Scientific Work of Enrico Fermi." *Proceedings of the American Academy of Arts and Sciences, The Rumford Bicentennial* 82, no. 7 (December 1953): 290–293.

De Gregorio, Alberto G. "Neutron Physics in the 1930s." *Historical Studies in the Physical and Biological Sciences* 35, no. 2 (March 1995): 293–340.

De Hevesy, George. "The Reactor and the Production of Isotopes." *International Atomic Energy Agency Bulletin* Special Number 4–0 (December 1962): 37.

Dresden, Max. Letter in "Heisenberg, Goudsmit, and the German 'A-Bomb.'" *Physics Today* 44, no. 5 (May 1991): 93–94.

Dyson, Freeman. "A Meeting with Fermi." *Nature* 427, no. 6972 (January 22, 2004): 297.

Eklund, Sigvard. "Introduction." *International Atomic Energy Agency Bulletin* Special Number 4–0 (December 1962): 3–5.

Emelyanov, Vasily S. "Notes on the History of the First Atomic Reactor in the USSR." *International Atomic Energy Agency Bulletin* Special Number 4–0 (December 1962): 25–27.

Etzkowitz, Henry. "Individual Investigators and Their Research Groups." *Minerva* 30, no. 1 (March 1992): 28–50.

Farrell, James T. "Making (Common) Sense of the Bomb in the First Nuclear War." *American Studies* 36 no. 2 (Fall 1995): 5–41.

Farrell, Joseph. "The Ethics of Science: Leonardo Sciascia and the Majorana Case." *Modern Language Review* 102, no. 4 (October 2007): 1021–1034.

Fermi, Enrico. "The Development of the First Chain-Reacting Pile." *Proceedings of the American Philosophical Society* 90, no. 1 (January 1946): 20–24.

"Fermi Invention Rediscovered at LASL." *The Atom, Los Alamos Scientific Laboratory,* October 1966, 7–11.

Fermi, Laura. "Some Personal Reminiscences." *International Atomic Energy Agency Bulletin* Special Number 4–0 (December 1962): 38–40.

Festinger, Leon. "Cognitive Dissonance." *Scientific American* 207, no. 4 (1962): 93–107.

Fiege, Mark. "The Atomic Scientists, the Sense of Wonder, and the Bomb." *Environmental History* 12, no. 3 (July 2007): 578–613.

Frisch, Otto R. "Obituary: Prof. Enrico Fermi, For. Mem. R. S." *Nature* 175, no. 4444 (January 1, 1955): 18–19.

Galison, Peter, and Barton J. Bernstein. "In Any Light: Scientists and the Decision to Build the Superbomb, 1952–1954." *Historical Studies in the Physical and Biological Sciences* 19, no. 2 (1989): 267–347.

Gambassi, Andrea. "Enrico Fermi in Pisa." *Physics in Perspective* 5 (2003): 384–397.

Garwin, Richard L. "Fermi's Mistake?" *Nature* 355 (February 20, 1992): 668.

———. "Living with Nuclear Weapons: Sixty Years and Counting." *Proceedings of the American Philosophical Society* 52, no. 1 (March 2008): 69–72.

Glauber, Roy. "An Excursion with Enrico Fermi, 14 July 1954." *Physics Today* 56, no. 6 (June 2002): 44–46.

Goldberger, Marvin L. "A Leader in Physics." *Science* 169, no. 3948 (August 29, 1970): 847.

Goldschmidt, Bertrand. "France's Contribution to the Discovery of the Chain Reaction." *International Atomic Energy Agency Bulletin* Special Number 4–0 (December 1962): 21–24.

Goudsmit, Samuel A. "De ontdekking van de electronenrotatie." ("The Discovery of Electron Spin.") *Nederlands Tijdschrift voor Natuurkunde* 37 (1971): 386. http://www.lorentz.leidenuniv.nl/history/spin/goudsmit.html.

———. "The Michigan Symposium in Theoretical Physics." *Michigan Alumnus Quarterly Review,* Spring 1961, 178–182.

Gray, Robert H. "The Fermi Paradox Is Not Fermi's, and It Is Not a Paradox." Guest blog, *Scientific American,* January 29, 2016. https://blogs.scientificamerican.com/guest-blog/the-fermi-paradox-is-not-fermi-s-and-it-is-not-a-paradox/.

Gross, David J. "On the Calculation of the Fine-Structure Constant." *Physics Today* 42, no. 12 (December 1989): 9–11.

Guerra, Francesco, Matteo Leone, and Nadia Robotti. "When Energy Conservation Seems to Fail: The Prediction of the Neutrino." *Science and Education* 23, no. 6 (June 2014): 1339–1359.

Guerra, Francesco, and Nadia Robotti. "Enrico Fermi's Discovery of Neutron-Induced Artificial Radioactivity: The Influence of His Theory of Beta Decay." *Physics in Perspective* 11 (2009): 379–404.

———. "Enrico Fermi's Discovery of Neutron-Induced Artificial Radioactivity: Neutrons and Neutron Sources." *Physics in Perspective* 8 (2006): 255–281.

———. "Bruno Pontecorvo in Italy." In *Bruno Pontecorvo: Selected Scientific Works,* edited by S. M. Bilenky et al., 527–547. Bologna, Italy: Società Italiana di Fisica, 2013.

Hagar, Amit. "Introduction to the New Edition." In *The Science of Mechanics,* edited by Ernest Mach. New York: Barnes & Noble, 2009. Reprint, Chicago: Open Court Publishing, 1919.

Hahn, Otto. "Enrico Fermi and Uranium Fission." *International Atomic Energy Agency Bulletin* Special Number 4–0 (December 1962): 9–11.

Hanson, Norwood Russell. "Discovering the Positron (I)." *British Journal for the Philosophy of Science* 12, no. 47 (November 1961): 194–214.

———. "Discovering the Positron (II)." *British Journal for the Philosophy of Science* 12, no. 48 (February 1962): 299–313.

Heilbron, John L. "Quantum Historiography and the Archive for History of Quantum Physics." *History of Science* 7 (January 1, 1968): 90–111.

Hewlett, Richard G. "Beginnings of Development in Nuclear Energy." *Technology and Culture* 17, no. 3 (July 1976): 465–478.

Hoddeson, Lillian Hartmann. "The Entry of the Quantum Theory of Solids into the Bell Telephone Laboratories, 1925–1940." *Minerva* 18, no. 3 (Autumn 1980): 422–447.

Holton, Gerald. "Striking Gold in Science: Fermi's Group and the Recapture of Italy's Place in Physics." *Minerva* 12, no. 2 (April 1974): 159–198.

Knight, Amy. "The Selling of the KGB." *Wilson Quarterly* 24, no. 1. (Winter 2000): 16–23.

Konopinski, Emil Jan. "Fermi's Theory of Beta-Decay." *Reviews of Modern Physics* 27, no. 3 (July 1955): 254–257.

Leone, Matteo, Alessandro Paoletti, and Nadia Robotti. "A Simultaneous Discovery: The Case of Johannes Stark and Antonino Lo Surdo." *Physics in Perspective* 6 (2004): 271–294.

Leone, Matteo, Nadia Robotti, and Carlo Alberto Segnini. "Fermi Archives at the Domus Galilaeana in Pisa." *Physis: Rivista Internazionale di Storia della Scienza* XXXVII, fasc. 2 (2000): 501–534.

Ley, Willy. "The Atom and Its Literature." *Military Affairs* 10, no. 2 (Summer 1946): 58–61.

McLay, David B. "Lise Meitner and Erwin Schroedinger: Biographies of Two Austrian Physicists of Nobel Stature." *Minerva* 37, no. 1 (March 1999): 75–94.

Meitner, Lise. "Right and Wrong Roads to the Discovery of Nuclear Energy." *International Atomic Energy Agency Bulletin* Special Number 4–0 (December 1962): 6–8.

Mermin, N. David. "Could Feynman Have Said This?" *Physics Today* 57, no. 5 (May 2004): 10–11.

Metropolis, Nicholas. "The Beginning of the Monte Carlo Method." *Los Alamos Science,* Special Issue (Spring 1987): 125–130.

Monaldi, Daniela. "Mesons in 1946," in *Atti del XXV Congresso Nazionale di Storia della Fisica e dell'Astronomia, Milano, 10–12 novembre 2005.* Milan: SISFA, 2008: C11.1–C11.6.

Noddack, Ida. "Uber das Element 93." *Angewandte Chemie* 47, no. 37 (1934): 653–655.

Norris, Margot. "Dividing the Indivisible: The Fissured Story of the Manhattan Project." *Cultural Critique,* no. 35 (Winter 1996–1997): 5–38.

Oppenheimer, Frank. "In Defense of the Titular Heroes," review of *Lawrence & Oppenheimer,* by Nuel Pharr Davis, *Physics Today* 22, no. 2 (February 1969): 77–80.

Oppenheimer, J. Robert. "Scientific World Pays Homage to Fermi." *Santa Fe New Mexican,* January 6, 1955, 8.

Pauli, Wolfgang. "Remarks on the History of the Exclusion Principle." *Science* 103, no. 2669 (February 22, 1946): 213–215.

Pearson, J. Michael. "On the Belated Discovery of Fission." *Physics Today* 68, no. 6 (June 2015): 40–45.

Reuters. "Atomic Scientist, Family Disappear: Expert from British Staff Said to Be in Moscow Following Arrival in Helsinki," *New York Times,* October 22, 1950, 34.

———. "British Atomic Scientist Believed Gone to Russia," *New York Times,* October 21, 1950, 3.

Schuking, Engelbert L. "Jordan, Pauli, Politics, Brecht, and a Variable Gravitational Constant." *Physics Today* 52, no. 10 (October 1999): 26–31.

Schweber, Silvan S. "The Empiricist Temper Regnant: Theoretical Physics in the United States 1920–1950." *Historical Studies in the Physical and Biological Sciences* 17, no. 1 (1986): 55–98.

———. "Enrico Fermi and Quantum Electrodynamics, 1929–1932." *Physics Today* 55 (June 2002): 31–36.

———. "Shelter Island, Pocono, and Oldstone: The Emergence of American Quantum Electrodynamics After World War II." *Osiris* 2nd series, 2 (1986): 265–302.

Seaborg, Glenn T. "The First Nuclear Reactor, the Production of Plutonium and Its Chemical Extraction." *International Atomic Energy Agency Bulletin* Special Number 4–0 (December 1962): 15–17.

Segrè, Emilio. "Fermi and Neutron Physics." *Reviews of Modern Physics* 27, no 3 (July 1955): 257–263.

———. "A Tribute to Enrico Fermi." *Physics Today* 8, no. 1 (January 1955): 10–12.

Seidel, Robert W. "A Home for Big Science: The Atomic Energy Commission's Laboratory System." *Historical Studies in the Physical and Biological Sciences* 16, no. 1 (1986): 135–175.

Seitz, Frederick. "Fermi Statistics." *Reviews of Modern Physics* 27, no. 3 (July 1955): 249–254.

Sigal, Leon V. "Bureaucratic Politics & Tactical Use of Committees: The Interim Committee & the Decision to Drop the Atomic Bomb." *Polity* 10, no. 3 (Spring 1978): 326–364.

Smith, Alice K. "Dramatis Personae." *Science* 162 (October 25, 1968): 445–447.

Smyth, Henry DeWolf. "Publication of the 'Smyth Report.'" *International Atomic Energy Agency Bulletin* Special Number 4–0 (December 1962): 28–30.

Sparberg, Esther B. "A Study of the Discovery of Fission." *American Journal of Physics* 32, no. 2 (1964): 2–8.

Squire, C. F., Ferdinand Brickwedde, Edward Teller, and Merl Tuve. "The Fifth Annual Washington Conference on Theoretical Physics." *Science,* February 24, 1939, 180–181.

Steiner, Arthur. "Scientists and Politicians: The Use of the Atomic Bomb Re-examined." *Minerva* 15, no. 2 (Summer 1977): 247–264.

———. "Scientists, Statesmen, and Politicians: The Competing Influences on American Atomic Energy Policy 1945–46." *Minerva* 12, no. 4 (October 1974): 469–509.

Telegdi, Valentine L. "Prejudice, Paradox, and Prediction: Conceptual Cycles in Physics." *Bulletin of the American Academy of Arts and Sciences* 29, no. 2 (November 1975): 25–40.

Teller, Edward. "Scientific World Pays Tribute to Fermi." *Santa Fe New Mexican*, January 6, 1955, 8.

———. "The Work of Many People." *Science* 121, no. 3139 (February 25, 1955): 267–275.

Ter Har, Dirk. "Foundations of Statistical Mechanics." *Reviews of Modern Physics* 27, no. 3 (July 1955): 289–338.

Thorpe, Charles, and Steven Shapin. "Who Was J. Robert Oppenheimer? Charisma and Complex Organization." *Social Studies of Science* 30 no. 4 (August 2000): 545–590.

Turchetti, Simone. "'For Slow Neutrons, Slow Pay': Enrico Fermi's Patent and the U.S. Atomic Energy Program, 1938–1953." *Isis* 97, no. 1 (March 2006): 1–27.

Wang, Jessica. "Edward Condon and the Cold War Politics of Loyalty." *Physics Today* 54, no. 12 (December 2001): 35–41.

Wattenberg, Albert. "The Fermi School in the United States." *European Journal of Physics* 9 (1988): 88–93.

———. "December 2, 1942: The Event and the People." *Bulletin of the Atomic Scientists* 38, no. 10 (December 1982): 22–32.

Weinberg, Alvin M. "Chapters from the Life of a Technological Fixer." *Minerva* 34, no. 4 (December 1993): 379–454.

Weinberg, Steven A. "The Search for Unity: Notes for a History of Quantum Field Theory." *Daedalus* 106, no. 4 (Fall 1977): 17–35.

Wellerstein, Alex. "Oppenheimer, Unredacted: Part I—Finding the Lost Transcripts." Restricted Data: The Nuclear Secrecy Blog, January 9, 2015. http://blog.nuclearsecrecy.com/2015/01/09/oppenheimer-unredacted-part-i/.

———. "Oppenheimer, Unredacted: Part II—Reading the Lost Transcripts." Restricted Data: The Nuclear Secrecy Blog, January 16, 2015. http://blog.nuclearsecrecy.com/2015/01/16/oppenheimer-unredacted-part-ii/.

Wheeler, John Archibald. "Fission Then and Now." *International Atomic Energy Agency Bulletin* Special Number 4–0 (December 1962): 33–36.

Wigner, Eugene P. "Thoughts on the 20th Anniversary of CP-1." *International Atomic Energy Agency Bulletin* Special Number 4–0 (December 1962): 31–32.

Wilson, Jane. "Lawrence and Oppenheimer." *Bulletin of the Atomic Scientists* (January 1969), 31–32.

Zinn, Walter. "Fermi and Atomic Energy." *Reviews of Modern Physics* 27, no. 3 (July 1955): 263–268.

INDEX

Photo: Susan Schwartz

David N. Schwartz holds a PhD in political science from MIT. He is the author of two books, including *NATO'S Nuclear Dilemmas*, and has worked at the State Department Bureau of Politico-Military Affairs, the Brookings Institution, and Goldman Sachs. He lives in New York with his wife, Susan. His father, Melvin Schwartz, shared the Nobel Prize in Physics in 1988. For more information visit www.davidnschwartz.com.